THE VOX DEI

COMMUNICATION AND SOCIETY
edited by George Gerbner and Marsha Siefert

IMAGE ETHICS
The Moral Rights of Subjects
in Photographs, Film, and Television
Edited by Larry Gross, John Stuart Katz,
and Jay Ruby

CENSORSHIP
The Knot That Binds Power and Knowledge
By Sue Curry Jansen

THE GLOBAL VILLAGE
Transformations in World Life and Media in
the 21st Century
By Marshall McLuhan and Bruce R. Powers

SPLIT SIGNALS
Television and Politics in the Soviet Union
By Ellen Mickiewicz

TARGET: PRIME TIME
Advocacy Groups and the
Struggle over Entertainment Television
By Kathryn C. Montgomery

TELEVISION AND AMERICA'S CHILDREN
A Crisis of Neglect
By Edward L. Palmer

PLAYING DOCTOR
Television, Storytelling, and Medical Power
By Joseph Turow

THE VOX DEI
Communication in the Middle Ages
By Sophia Menache

THE EXPORT OF MEANING
Cross-Cultural Readings of DALLAS
By Tamar Liebes and Eliha Katz

THE VOX DEI

Communication in the Middle Ages

SOPHIA MENACHE

New York Oxford
OXFORD UNIVERSITY PRESS
1990

Oxford University Press

Oxford New York Toronto
Delhi Bombay Calcutta Madras Karachi
Petaling Jaya Singapore Hong Kong Tokyo
Nairobi Dar es Salaam Cape Town
Melbourne Auckland

and associated companies in
Berlin Ibadan

Published by Oxford University Press, Inc.,
200 Madison Avenue, New York, New York 10016

Library of Congress Cataloging-in-Publication Data
Menache, Sophia.
The vox Dei : communication in the Middle Ages / Sophia Menache.
p. cm.
Includes bibliographical references.
ISBN 0-19-504916-0
1. Communication—Religious aspects—Christianity—History
of doctrines—Middle Ages, 600–1500. I. Title.
BV4319.M46 1990
261.5′2′0902—dc20 89-39662

9 8 7 6 5 4 3 2 1

Printed in the United States of America
on acid-free paper

In memory of my beloved brother
Nestor S. Moaded (Kike)
"disappeared" in Buenos Aires, 1976.

The belief in the divine source of human knowledge was perpetuated in the Church of the Madeleine at Vézelay in Burgundy (c.1120–1132), whose great sculpture on the West front presents the outpouring of the divine truth through Christ, and its communication to all men. The tympanum shows Christ in a mandorla while rays emerging from his hands descended on the heads of the Apostles, seated on either side. In the eight compartments around the central scene and on the lintel, there are strange groups of exotic people to whom the words of the Gospels are to be carried: Lepers, cripples, the blind, and the inhabitants of faraway lands, such as the heathen sacrificing a bull, pygmies so small that they mount horses with ladders, Scythians with monstrous ears...On the trumeau is St. John the Baptist, and on the jambs four Apostles, including St. Peter. Medallions with the signs of the Zodiac and the labors of the months crown the whole design.

ACKNOWLEDGMENTS

Six years ago I met Ms. Marsha Siefert from the Annenberg School of Communications, in the relaxing atmosphere of the University of Pennsylvania. She was then editing a short article I wrote for the Journal of Communication and faced me with the challenging proposal of writing a more detailed study on medieval communication. Her proposal eventually gave birth to the *Vox Dei.* From then until today, Ms. Siefert accompanied the various stages of this book, giving her encouragement and experienced advice. To her I express my deep gratitude. Research for this book was supported by the University of Haifa Research Fund. Furthermore, without the assistance of colleagues or friends who often directed me to the relevant sources, this work would never have been completed. Among those who shared their scholarly knowledge with me are Professor Charles Wood from Dartmouth College and Professor Michael Goodich, Dean of Humanities at Haifa University, who in the first stages of this book contributed invaluable ideas concerning methods and problems of research. My former teacher, Professor Aryeh Grabois from Haifa University, read the entire manuscript. His constructive criticism and scholarly advice improved the final version of this study. Professor Nurith Keidar, from Tel Aviv University, gave helpful assistance in selecting the plates that illustrate this book. I am also thankful to the American Numismatic Society for allowing me to reproduce some coins of its rich collection, and to the following institutions which generously authorized the reproduction of some plates that illustrate this book: The British Library (plate 24), the Bibliothèque municipale of Rouen (plate 2), the University of Chicago Press (plates 1, 18 and 26) and Weidenfeld and Nicholson (plate 9). The greatest appreciation must go to my husband, Rami, and my children, Ishai and Yael, who patiently suffered the ups and downs of a wife and mother whom they have had to share with the computer for the last six years.

Haifa, Israel
September 1989

S.M.

CONTENTS

THE VOX DEI

Introduction

In his letter to Charlemagne (c. AD 798), Alcuin, the king's adviser, condemned the popular maxim of *Vox populi, vox Dei*, the voice of the people is the voice of God, since, he argued, "the opinions of the populace are always close to insanity."[1] Such a prevalent disregard for the "voice of the people" reflects the authoritarian approach characteristic of ecclesiastical writers in the feudal period, when the "voice of God" was the unique source of legitimization in the sociopolitical sphere and in the field of communication as well. Consequently, only those who enjoyed special grace were allowed to voice the *Vox Dei*, namely, prelates, kings, or pseudoprophets on whom God Himself had conferred such a right. Both the people, their opinions, and their expectations, nevertheless, could not have been completely neglected. Alcuin further adduced that the people had to be led, *populus ducendus est*. The success of mass movements such as the Peace of God, the Gregorian Reform and the Crusades, indeed, ultimately depended on the Church's capability to achieve massive identification among large social strata with the goals it propagated. The pursuit of favorable public opinion, however, was not restricted to the ecclesiastical order. On the contrary, it became an essential feature in the process of state-building from the eleventh century onward. Particularly, during the lasting Investiture Contest between popes and monarchs, the competition between secular and religious authority carried through to their battle for public opinion. The medieval public was thus faced with different and, sometimes, opposing viewpoints, all presumably deriving from the grace of God. The institutional use of propaganda by the state and the Church thus assumed a massive scope. That is "perhaps unique among the great cultures."[2]

This book deals with the development of communication in Western European society between the eleventh and fourteenth centuries, as an integral part of the process of social, economic and political integration that medieval society was experiencing at the time. The geographical and chronological delimitations of this research reflect a unique historical process through which European society evolved from the universal concepts of early Christianity into an embryonic national consciousness. This process created a fruitful arena for the widespread use of manipulation and propaganda both by supporters and critics who exploited all the means of communication at their disposal: the written word, the oral message, and the pictorial representation. The socioeconomic changes characteristic of this period left their mark on the development of communication systems which, from an historical perspective, heralded the forms and methods of modern communication. This study aims to give an insight into this rich world that also included symbols, stereotypes, communication habits, and slogans, thus providing the essential features of communication which, though modified, are still in use today.

How does one investigate and analyze communication in medieval society? Approaches differ depending upon which aspect of medieval society is considered central. While some researchers emphasize extreme localism as a central trait, others point to the Christian Commonwealth, which reached its peak in the Crusades, as the most faithful expression of its universal essence.[3] Defining communication terms is equally difficult. For example, no satisfactory answer has yet been given as to whether public opinion is immanent of the historical process,[4] or, the "climate of opinion" in a given space and time could be evaluated independently of the factual level.[5]

Another problematic issue focuses on identifying the leading communicators in the Middle Ages. In medieval society, which like so-called traditional societies lacked professional communicators, people often turned to opinion leaders in order to learn how to evaluate, interpret, and respond to the information they received. Information, therefore, usually flowed along the lines of the social hierarchy, or according to the particular patterns of social relations in each community.[6] The characterization of the Middle Ages as the *Age de la foi*, on the other hand, epitomizes the crucial role some ascribe to the ecclesiastical establishment as the leading force in medieval communication.[7] Teaching and propaganda went then side by side, for in medieval society, to educate was actually to convert. As Haskins argues, "that the Church was the chief source of unity for medieval society is a commonplace which is not open to dispute."[8] Yet since no single power was in control of the means of communication, some scholars counter with the diversity of opinion among "bishops, scholars, radical reformers, heretics, statesmen, warriors, love poets and Jews."[9] Others conceptualize it as a two-way transfer of information. In order to foster supportive public opinion, the sociopolitical elites were receptive to the patterns of the lower social classes and were thus influenced by the so-called *culture populaire*.[10] Reciprocally, the cultural patterns of the upper classes tended to be popularized among the lower social strata.[11]

This study assumes that these differences are actually a result of emphasis on different stages in the evolution of medieval society, particularly from the eleventh to fourteenth centuries, a period conceptualized as the Central Middle Ages. The prevailing approaches to medieval communication assume, in one way or another, the stability and conservatism of medieval society as a whole. A more dynamic perspective of medieval society is still missing. One of the premises of this study is that a communication system always reflects the sociopolitical structures within which it operates, together with the prevailing assumptions of contemporaries concerning the nature of mankind, society, and state, and the interrelationship between them.[12] Research into communication in the Middle Ages thus includes the historical process and, at the same time, the development of media in medieval society. Another premise of this study is that a communication system evolves in societies that are at a fairly advanced stage of socioeconomic and cultural integration. The communication system becomes more important as the socioeconomic process begins to alienate the individual from the community. The isolation inherent in feudal society creates, therefore, an apparent paradox between feudalism, as a conservative order fostering autarchical tendencies, and the emergence of a communication system.

A more dynamic perspective of the historical development, however, solves this paradox quite convincingly. Beginning in the eleventh century, the static structures of the feudal system were gradually replaced by a new economic interaction between the countryside and the urban centers, the peasant and the merchant, the artisan and the lord.[13] Medieval society then combined widespread static rural conditions, a low level of education, and primitive modes of travel with a social structure that still required a certain amount of communication between widely separated units of the same type. The composite nature of medieval society was further influenced by the gradual evolution from feudal structures toward a monetarian-urban economy, a development fostered by the process of state-building. In the political sphere as well, Western society combined authoritarian and populistic tendencies, its leaders backed by the grace of God but still looking for consent. Although authoritarianism and populism could be regarded as antonyms, they actually complemented each other in a changing society whose leaders were in actual practice the promoters of change. This process of change encouraged new attitudes in the field of ideas as well. The former claim of Alcuin that "the opinions of the populace were always close to insanity" was actually questioned by the Roman dictum that *Quod omnes tangit ab omnibus approbari debet*, (what touches all should be approved by all those concerned), a claim that bestowed an aura of juridical consent on public matters. The development of a communication system in the Central Middle Ages thus appears as one facet of this process of change. The socioeconomic evolution in turn encouraged the cultivation of public opinion, particularly given the new sources of authority developing on the political sphere. A communication system therefore emerged in response to the need for integration as well as to the needs of competing voices of authority to place their claims for supremacy before the public.

Using concepts from the field of communication within the framework of medieval society allows for new insights in both fields. For the purposes of this study, communication can be summarily defined as a basic social function that may be intrapersonal, but that ultimately involves interaction between two or more persons.[14] It may develop through both verbal and nonverbal symbols,[15] and usually aims at influencing others' beliefs and/or behavior. Intention is characteristic of both the communicator and his audience, the former by giving the message and the latter by receiving it.[16] Communication involves at least two basic concepts: the discovery of ideas (categorization and conceptualization) and their transmission through the use of symbols. It allows human beings to engage in the processing of information necessary for the maintenance of their physical and spiritual conditions, while forming their own *Umwelt*.[17]

This study deals, to a large extent, not just with communication but with political communication. By political communication we mean the deliberate passing of a *political message* by a *sender* to a *receiver*, with the *intention* of making the receiver behave in a way that otherwise he might not do. By its very nature, political communication systematically aims to reach large audiences and to influence them through controlled information sent out over a long period. Repetition and simplification are also characteristics of this process, though they are not always used. The methodical regularity of the *sender—message—receiver* pattern suggests a permanent feature in political communication that seldom exists in real life and in which the sender has most of the advantages. He has the initiative, and he decides *when* to say something as well as *what* to say. He chooses his audience and his *method of communicating* and, usually, he will also take into consideration the *response* he expects to achieve.[18] These premises led to four major questions that have been elaborated in each part of this study:

1) *Who* is communicating?
2) Is there an *institution* or an organized group under whose auspices the communicator is operating?
3) What *channels of communication* were actually used?
4) Which sort of *audience* did the communicators have in mind?

In the framework of political societies, one can further point to three systems of interaction which, though separate, are closely related, namely:

1) *The governmental decision making process*, through which policy is formulated and into which existing public opinion is integrated by the decision makers.
2) *The opinion submitting process* that occurs whenever opinions are conveyed to or impressed upon decision makers by individual members or segments of the public.
3) *The opinion making process* whereby ideas are formed and circulated.[19]

In order to make the communication process clear in its historical context, each section of this study opens with a short introduction to the main historical processes it deals with. To allow for some original research and avoid generalizations, each section provides two case studies in which a historical process was analyzed from the perspective of communication.

The first chapter introduces the term *media* in the framework of medieval society and presents the main methodological problems in the field of communication. Questions such as the length of time needed for information to reach one place to another, the time frame within which people placed what they thought was happening, the leading communicators and the means of communication at their disposal were discussed.

Section I deals with the ecclesiastical message and the media used by the different components of the clergy to achieve support and obtain the full cooperation of their contemporaries. The definition of the Church as sacerdotal-sacramental implies its monopoly on the way to salvation and in the world to come. As Vicars of God on earth, the popes and the ecclesiastical hierarchy as a whole enjoyed the full use of divine prerogatives as reflected in the granting of indulgences. In this wide field, the fear of death became a cult, used by the clergy to reinforce its status. Both because of the unprecedented use of propaganda and the massive response they achieved, the Crusades provided another case study in the field of ecclesiastical communication.

Section II deals with the medieval monarchy. The biblical tradition endowed medieval kings with the grace of God and, as anointed kings, monarchs became not only the representatives of God on earth but also enjoyed therapeutic powers. In the Central Middle Ages, the emergence of embryonic national entities in France and England brought about the widespread cult of monarchy, followed by the emergence of national symbols. Yet both the kings and the symbols they used had to overcome times of crisis while adjusting themselves to the changing circumstances. Of the many crises of the fourteenth century, the endless confrontation between England and France known as the Hundred Years War provides a suitable case study for the development of political communication. The use of the Crusade theme and the emergence of stereotypes to justify the war by both conflicting sides has been analyzed in detail.

Section III deals with heresy. In the normative-hierarchic essence of medieval society, people who challenged the status quo were, at best, relegated to the margins of society and, at worst, condemned to death. Yet from the perspective of communication, the so-called heretical movements represented one of the most vigorous movements of the time, which adapted themselves to the changing circumstances while developing their own channels. The affair of the Templars and the Franciscan Struggle suggest the propaganda mechanisms developing at the time and the struggle for favorable public opinion both by the ecclesiastical establishment and those it rejected from the safe wings of Catholic orthodoxy.

As the progression of case studies comprising this study will show, the emergence of communication systems and the use of propaganda and manipulation in the Central Middle Ages were concomitant with the decline of the feudal regime. The development of towns and the transition into a monetary economy gradually divorced medieval society from the traditional frameworks. The consolidation of urban populations turned these medieval towns into the most fruitful arena for legitimizing a new order while encouraging the use of propaganda and the manipulation of popular opinion in its modern sense, though by more primitive means. New concepts of the individual, from Ockham's claim for absolute freedom to the Pseudo-Apostles' demand to allow each man to live according to his own understanding, also contributed to the disintegration of traditional society and were adopted by the developing communication system.

Despite the seemingly large gap between the Central Middle Ages and modern society, a better understanding of medieval communication opens new perspectives regarding the development of communication, propaganda, and manipulation in contemporary society. The weight of communication and the crucial importance it acquires in our own days appears as a direct result of the ever-growing atomization together with the sophistication of the mass media. Yet the medieval heritage provides additional facets which have not hitherto been analyzed in full. This book unites the perspectives of the medievalist and the communication researcher to bring about a better understanding of the historical process and precedent for communication in our own time.

1

The Media in the Middle Ages

The use of the term *media* with regard to medieval society is problematic. Media are related not only to the technique of communication but also to the socioeconomic framework which brings about their development. Contemporary media, for example, reflect appeal to a society that is oriented around them, so communicators can achieve some degree of identification with the characters or messages they broadcast by television, press, or radio. This identification compensates for the alienation that might otherwise be experienced by the individual in society. The reach and breadth of contemporary media has also significantly reduced the number of communication channels that must be utilized to reach large audiences. The chances of communication media with similar characteristics developing in the Middle Ages, however, were almost nil. In contrast to the audience anonymity inherent in the modern mass media, medieval communication was characterized by the immediate contact between the communicator and his audience. Moreover, medieval society as a whole was based on the solidarity and homogeneity of the socioeconomic group (corporation), be it in the framework of rural communities, guilds, assemblies, universities, ecclesiastical orders, and so on within the different social strata. This corporate structure cemented the individual to his niche and defined him first as a group member rather than as an individual. One may go further and connect the appearance of media, in their modern sense, with the decline of the corporate structure inherent in medieval society. Once medieval man divorced himself from the familiar confines of the corporate framework, he became more exposed to propaganda, thus encouraging the emergence of a communication system. The basic differences between modern and medieval societies during the Middle Ages brought about diverging roles of the media beyond the technical level. In medieval society, whose members were often limited to the narrow confines of the village or the manor, a mere encounter with a stranger on the road turned into a communication event. According to Coulton, an ordinary man in tenth-century Normandy probably encountered between 100 and 200 people in his whole life, and his vocabulary contained only some 600 words.[1]

The corporate structure of medieval society, its behavioral norms, and the physical factors of everyday existence comprised the basic conditions on which medieval communication emerged. As Huizinga claimed, every event was given greater emphasis 500 years ago, since God, being an integral component of everyday life, bestowed a transcendental dimension on the historical process.[2] People too, played a very active role in the exaltation of actual events, through an extreme extroversion of their feelings. Public expressions of joy and grief were considered *bon ton*, since medieval society had not yet acknowledged the concept of privacy. The village square, the church, the parade, the shepherds' tent, the tavern, the lighted kitchen—all these served as meeting places. The sociability of medieval people thus fostered the exchange of information, while the small distances between houses, the thin walls, which enabled neighbors to listen and to spy on each other, and the considerable time spent outside the house, whether around the well, in the mill, or during the cheese fair, gave rise to frequent and spontaneous meetings.

To conceptualize communication in the Middle Ages requires a generalization about the sociocultural climate. If European society were compared to a large forest in which there were isolated inhabited areas, the history of communication in the Middle Ages could be described as the gradual contact between those separate areas. And, as the separate areas became closer, they required more developed channels of communication. This process both encouraged greater awareness of the need for communication and, in addition, brought about the development of new channels for that communication. In the beginning, medieval communicators were forced to use several channels, each with only a relatively small degree of message diffusion, in order to reach large audiences. Over the long term, the development of several channels brought about the need for their coordination, from which emerged a communication system. This state of affairs justifies a different conceptualization of the term media in the Middle Ages, when everyday practices assumed the significance of communication channels. The use of the term media with regard to the Middle Ages, therefore, refers to the different means of communication elaborated at the time, without the socioeconomic implications they have acquired in modern society. The essential liaison between the different media means that they cannot be categorized according to institutional patterns alone. This chapter deals with the social environment, the main communicators, and the media they used, while their communication channels will be analyzed in further detail according to institutional criteria in the three sections of this study.

Communication in medieval society was conditioned by the isolation of one group from another within the hierarchy of the feudal system. The almost complete lack of communication among large sectors of the population, however, did not include the political elite whose perspective was beyond the local sphere. Benzinger approaches the feudal pyramid as a *Kommunikationsmodell*, in which the amount of information assimilated by the different social strata was conditioned by the social status and the political functions they fulfilled.[3] While peasants or craftsmen had to content themselves with scanty information, the sociopolitical elite dealt with a considerable range of reports. Moreover, the faster the information was received, the more accurately it could be translated into political practice. Medieval sources faithfully reflect the political elite's awareness of the advantages of receiving more information more quickly. When John of Salisbury, for instance, warned Pope Adrian IV of the prevailing criticism of the Church in England ("*loquitur populus*"), the pope manifested his concern on the matter and asked to be properly informed of further developments.[4] The growing importance ascribed to public opinion further encouraged the development of communication channels, according to the means available at the time. Closely surrounded by counselors who were able to provide information, medieval rulers also made use of the services of "well-informed circles" such as merchants, minstrels, ecclesiastics or wandering preachers whose mobility turned them into sources of information that no ruler could ignore.[5] Scherer regards them as "wandering reporters," whose salary was conditioned by their education or their ability to transmit or supply information.[6] Edward II, for instance, paid 20 shillings to Laurence of Ireland, messenger of the house of the Bardi, for reporting the election of Pope John XXII at Lyons in 1316, a mere ten days after it actually happened. Only one month later, the king received the formal announcement from the Avignon legate, who was rewarded with 100 pounds; a considerable sum in clear contrast to his promptness.[7]

Medieval ruling classes, therefore, had the strongest imperative to develop a communication system. Their purpose was both to manipulate large masses of people, which emerged as a new social category from the eleventh century onward, and to receive and transmit information as an integral part of their rule. Any attempt to compare the use of communication by the political elite of the Middle Ages to that of our days, however, would be anachronistic. In modern society, political communication systems mirror the process of integration in both the sociopolitical and economic spheres. In the Middle Ages, on the other hand, whether conducted by the Church, the monarchy, or the aristocracy, political communication resulted from the particularism and autarky characteristic of the times. In other words, the essential localism inherent in medieval structures encouraged the development of communication channels at the higher levels. This state of affairs is faithfully corroborated by the role ascribed to Latin as the international language of medieval communication. Until the twelfth century, the use of Latin expressed the separatism inherent in medieval society, which needed an external language, unknown to large sectors of the population, for communication. The use of Latin in the Central Middle Ages could be compared, *mutatis mutandis*, to the use of English in India and the African countries in the last century, where it acted, in fact, as an external medium not always concomitant with the needs of much of the local population.

The time required for the transmission of information provides an additional feature unique to medieval communication. A few examples will substantiate the reasons and impact of transmission time. The notification of Frederick Barbarossa's death in Asia Minor (June 10, 1190) reached Germany only four months later, while the news about the imprisonment of Richard Coeur de Lion near Vienna (December 11, 1192) arrived in England after a few weeks. In times of danger, when the rapid transmission of accurate news was vital to safety, horsemen often succeeded in covering long distances at a remarkably high speed. During the Wars of the Roses, the news of the murder of James I of Scotland (Perth, February 21, 1437) reached London in time for Cardinal Beaufort to compose a suitable letter to the pope on February 28. The messenger must have covered the 440 miles between Perth and London at an average speed of over 40 miles a day.[8] The long distances and the slowness of transmission, therefore, did not prevent the flow of communication between distant places and important news was actually transmitted from one continent to another. Yet delays in transmission often brought about significant distortions in the original news. In the words of Tacitus, *Quae ex longinquo in maius audiebantur*, people often tend to maximize events that happened in distant places.[9]

The relative slowness of transmission, indeed, led to and even encouraged the further distortion of news. The lack of tested channels of communication, the mist which covered most information, the fears of the unknown, and the time necessary to transmit news, contributed to the spread and credibility of rumors.[10] The news about the Mongol conquest of the Holy Land in 1300, for instance, reached Europe a few months later, together with rumors about the Mongols' supposed readiness to entrust the land into Christian hands following their expected conversion. Unlike the massive publicity enjoyed by this "non event," however, only a few chronicles later referred to the Mameluks' offensive and the resulting Mongol defeat in 1301, which actually put an end to the Christians' hopes.[11] The significant gap between rumor and reality thus reflects not only real impediments such as long distances and primitive means of transmission but also the willingness of medieval people to rely on whatever information they received, with or without substantiation. Both Boccaccio and Chaucer were aware of the spread of rumors in medieval society and depicted all kind of bizarre situations engendered in actual practice.[12] Medieval society thus was an extremely fertile ground for the transmission and reception of rumors. Rumors were regarded by medieval rulers as an integral part of their information system and, consequently, as a component of the services to be regularly rendered by their officers. When Bartholomew de Burghersh was sent to the papal court in Avignon in 1328, he sent a messenger from Dover to Stamford to inform the king of certain rumors that he considered important to the royal interest.[13] In a letter sent from Valencia in June 1308, Jayme II of Aragon asked his procurator in the papal court, Johannes Burgundi, to keep him informed about any rumors circulating there.[14]

The reception of different amounts of information and the different uses they were assigned by the various social strata justify the distinction between two large sociocultural categories, which the School of the *Annales* has popularized through the nomenclature of *culture savante* and *culture populaire*.[15] The gap between these two is not based only on the differing amounts of information that each recorded, but also on their differing attitudes toward this information and their choice of communication channels. The prevailing attitude of lower social strata was characterized by a suspicion of the messages, particularly written documents, diffused by established institutions. They favored immediate human contact, probably due to archaic German traditions that questioned the acceptability of the written word. The cultural elite, on the other hand, was more influenced by the Roman dependence upon precedents and the advantages inherent in written documents. The use of written documents and the reliability ascribed to them thus distinguished *culture savante* and *culture populaire*. However, the vague categories of popular and elite cultures are otherwise inaccurate, because most media were common to both cultures and were therefore adapted according to their diverging needs.

The *written message* thus appears as the most concrete media of the cultural elite. By the mid-twelfth century, records of all sorts were becoming more plentiful and the information they provided more precise. This was but the cultural aspect of the socioeconomic development of the times, now commonly called a demographic, economic, or industrial *revolution*. Innovations in agriculture and warfare, and in devices for the efficient use of water and air power, helped to create the conditions of economic progress and the remonetization of markets and exchange. Coinage appeared in quantity, and markets surfaced in nascent commercial centers. The number of commercial transactions increased, and this fostered the widespread use of cursive script which allowed merchants to write more quickly. Prices were determined more and more by supply and demand, and men gradually distinguished between inherited status and contractual obligations. As Stock pointed out, "Money or commodities with a monetary value, emerged as the chief force for objectifying economic concerns, just as, in the cultural sphere, the written text helped to isolate what man thought about from his process of thinking."[16] Whether in the form of a book, a pamphlet, or a letter, the written text gradually became an integral part of everyday existence among the cultural elite, either for registering contracts or for giving thoughts a concrete and more ordered expression. Monks and priests lost their former monopoly over the transcription of books and their writing, and written texts actually spread widely outside the ecclesiastical order.

Yet books remained an expensive product that only the upper classes could afford (see plate 1). The *Livre d'heures* of the Duke of Berry (c. 1416) exemplifies the considerable amounts of money and time invested in books, amounts that might be compared to the colossal film productions of today. As to the most popular books, in spite of the religious essence of medieval society as a whole, there was no significant correlation between the reading lists of the clergy and the laity. The particular interest of the clergy in matters of dogma found no echo outside the ecclesiastical order, while the laity were more interested in anecdotal sections of Holy Scriptures such as Kings and Acts.[17] Beside the essential distinction between clergy and laity, the socioeconomic factor also played an important role with regard to the degree of literacy. Cities became the centers of literacy par excellence while, in the framework of

PLATE 1. Front cover of the Lindau Gospels. Codex Aureus of Lindau. Reims (?), c. 870. Reliefs in bossed gold showing Christ on the cross, sun and moon, St. John and St. Mary, flying angels. Precious stones, mounted on arcades and lions' feet, pearls, filigree (320 × 255 mm.).

large cities, the interchange with Jews and Moslems widened the sociocultural horizons of townsmen and facilitated their further acquaintance with the classical heritage.[18] In rural settlements, on the other hand, the few books, if any, of the local priest were often the only source of knowledge and, in most cases, were of an exclusively religious nature.

Research into written communication requires further inquiry into the degree of literacy in medieval society as a whole, a field in which much research has still to be done. Harper claims that until about the eleventh century, illiteracy was more the rule than the exception. Not only much of the lower social strata but most kings and princes were unable to read or write.[19] Her conclusions have been corroborated by Toussaert's research on eleventh-century Flanders, where, he claimed, "illiteracy governed everywhere and the great lords took pride in their ignorance."[20] In a thirteenth-century French village such as Montaillou, there were only four people who could read and write, out of a population of 250. The education of the young was monopolized by the extended family (*domus*), while teaching became the prerogative of age and social class. The old taught the young, the lord his peasants, the priest his congregation. Yet the criterion of age prevailed over social status, and the existence of a priest in the family could not undermine the ascendancy of its older members, nor is it significant evidence as to the cultural influence of the young in the extended family.[21] The Fourth Lateran Council (1215) encouraged the establishment of additional elementary schools but the main source of education still remained within the family which postulated a predominantly religious message. From the twelfth century onward, however, the establishment of universities in Paris, Toulouse, Montpellier, Orléans, Oxford, Cambridge, Padova, Bologna, Salamanca and Coimbre, among others, brought about an educational revival in the framework of the nascent urban centers.[22] Between the years 1300 and 1500, more than fifty new universities were established: eleven in France, nine in Spain, eight in Italy, three in England and the rest in Eastern Europe, after the establishment of the University of Prague. Around 1400, there were some 800 students in the German Empire, and their number reached about 4,000 toward 1520. The development of intellectual centers adjacent to the cathedrals also attracted students from all parts of Europe, while the relatively high mobility of students and lecturers further contributed to the transmission of knowledge between distant locations. At this time, Archbishop Raymond of Toledo encouraged translations from Arabic to Latin, while Archbishops Theobald and Thomas of Canterbury strengthened the links between England and the Continent. Beside the university frameworks, in Flanders and Italy emerged schools of a new kind that supplied the merchant class with the instrumental knowledge they needed in mathematics, accounting, and geography. Giovanni Villani gives a detailed report of the educational revival in fourteenth-century Florence: the number of boys and girls who learned reading and writing reached the considerable total of 8,000–10,000; some 1,000–1,200 young people studied mathematics in one of the six municipal schools; about 550–600 specialized in grammar and rhetoric in four different schools; some sixty aspired to become judges, another 600, lawyers, while about sixty studied medicine and surgery.[23] Between the years 1300 and 1400, eighty new schools of this kind were established all over Europe, and in France alone the total number of students increased to about 500.

The improvement in the level of education led to the widespread use of *correspondence*, which among the ruling class acquired the weight of information exchange, due to the nature of the medieval government system.[24] Yet, in the Middle Ages, letters acquired peculiar characteristics of their own as they originated as oral messages, when distance made speech impossible. According to Ambrose, "the epistolary genre was devised in order that someone may speak to us when we are absent."[25] In their most part, medieval letters were quasi-public literary documents, written to be collected and publicized in the future, and intended to be read by more than one person. They were, therefore, designed to be correct and elegant rather than original and spontaneous, and often followed the form and content of model letters in formularies. According to the *artes dictaminis* of the later Middle Ages, a letter should have five parts arranged in logical sequence: the *salutation* was followed by the *exordium*, which consisted of some commonplace generality, a proverb, or a scriptural quotation. Then came the *narratio*, namely, the statement of the particular purpose of the letter, the *petitio*, deduced from the *exordium* and *narratio*, and finally the phrases of conclusion. Medieval correspondence could further be evaluated according to two major categories, namely, letters used by ecclesiastical or secular institutions, and those written by private persons. The papal court was a leading user of correspondence, whether by means of *encyclicals* or *bulls*. An encyclical was a circular letter sent to all the churches of a given area. In early times the word might be used to denote a letter sent out by any bishop, but the term was gradually restricted to such letters sent by the pope. Bulls were written mandates of the popes sealed in earlier times with the pope's signet ring, but from the sixth century either seal-boxes of lead or signets stamped in wax were used. Focusing on the most important issues of dogma and ecclesiastical discipline, bulls were widely diffused in the Middle Ages, especially among the clergy and the political elite, which were the major targets of papal correspondence. Popes also encouraged the spread of their bulls within the framework of the nascent university centers. A copy of Innocent III's bull *Per venerabilem* (1202), for instance, was sent to the Faculty of Law in Bologne in order to encourage the students to spread the papal message.

Although letter writing was known by the political elite in the seventh century, its generalization was concomitant to the process of state-building in the Central Middle Ages. The *letters patent* were addressed in a general way and always opened with the royal name and titles as follows: "*Edward, par la grace de dieu, roi d'Engleterre, seigneur d'Irlande, et duc d'Aquitain, a touz ceux qui ces presentes lettres verront ou orront, saluz et connoissance de verite.*"[26] The introductory sentence, whether in the vernacular or in its Latin version, *notum sit omnibus fidelibus presentem paginam inspecturis,* reflects the public nature of the letter and the hopes of its author to reach large audiences. Yet, one can reasonably assume that these words were not intended for the general public but rather for a small, interested group, who knew Latin or could afford the services of translators.[27] *Letters close* were usually addressed to a prince of superior or equal rank; they also opened with the royal name and titles as exemplified by a letter addressed by Edward III to Philip VI: "*Excellentissimo principi, domino Philippo, dei gracia regi Fran-*

corum illustri, consanguineo suo carissimo, Edwardus eiusdem gracia, rex Anglie, dominus Hibernie, et dux Aquitanie, salutem felicibus semper successibus abundare."[28] Letters of this kind were used to request or dispense political decrees, privileges and commissions, legal mandates, dispensations, and contracts. Some letters were of general interest such as that of Louis VII concerning the status of the Jews in France (1144) or, one dated from Soissons in 1155, which bestowed royal protection on the churches and villages in the domain. The very existence of such letters mirrors the weakness of the French monarchy at the time, when kings were actually unable to enforce their laws throughout the whole kingdom and were therefore forced to develop alternative communication channels to advance their interests. Letter writing was taught in medieval schools, and eventually, in the early Renaissance, a body of practical manuals and theoretical literature evolved, the *ars dictaminis*.

Another category of written media were the *Heavenly Letters,* believed to have come in a miraculous way from heaven, thus becoming the most direct expression of the *Vox Dei*. Letters of this kind enjoyed wide circulation, being diffused by mystical personages such as Peter the Hermit, or among the Flagellants. Such a "Heavenly Letter" arrived at the Parliament of Carlisle in 1307 signed by the anonymous Christian knight Peter, son of Cassiodorus, who mourned the sufferings of the English Church, "Daughter of Jerusalem," "Virgin of Zion," seeking in vain for its salvation in the papal curia.[29] Although the impact of this letter on the parliament cannot be properly evaluated, it represents the prevailing criticism of the papal policy in England and illustrates another channel of communication which fitted into the religious spirit of the times. The use of Heavenly Letters was usually restricted to controversial subjects that left no room for open criticism. They provided their authors with the privileges of anonymity while conferring on their contents an enviable halo of the *Vox Dei*.

On the other hand, the spread of trade on an international scale encouraged the use of letters in commercial transactions as well, which very often acquired the weight of formal documents. Besides, both Boccaccio's *Decameron* and Chaucer's *Canterbury Tales*, to quote but two examples, illustrate the widespread use of letters for private needs. Correspondence between lovers, for instance, seems to have been a fairly common practice toward the end of the Middle Ages.[30] This state of affairs had significant consequences as to the Church attitudes toward female literacy. Although the Church favored the education of women in the basics of morality and religion, some writers such as the knight of La Tour Landry wanted women to be able to read but thought it unnecessary for them to write. Philippe de Navarre, furthermore, was completely against the idea of education for women of all classes, since once women could read, he argued, they would be able to receive letters from their lovers.[31] The use of correspondence, nevertheless, was quite widespread in medieval society and though it mostly involved men, we also have examples of female correspondence, such as the letters written by Heloise to Abelard. Giraldus Cambrensis (c.1146–1223), archdeacon of Brecon, made frequent use of letters in his endless conflict with his nephew and expressed confidence in their efficiency, despite the poor condition of England's roads at the time.[32] Leclerq has published two stirring testimonies on the intimate correspondence between two monks in thirteenth-century England. The anonymous writer advises his friend as to the ways to reach his peace of mind while releasing himself of psychosomatic sufferings.[33] Still, one has to bear in mind Leclerq's reservations as to the authenticity of such letters because of the didactic value they were ascribed. Many correspondents, indeed, elaborated their letters as scholarly treatises whose resemblance with the original can hardly be discerned.[34] The correspondence both of Bernard de Clairvaux and Abelard, for instance, raises questions with regard to the degree of spontaneity or elaboration of their original, which have not been preserved.[35]

The use of letters and the increase in literacy, however, did not essentially change the prevalence of *oral delivery* in medieval society. Oral delivery as a means of communicating and storing facts was well-suited to a society that was still regionalized, highly particularized, and more conscious of inherited status than of achievement through pragmatic social roles. Medieval oral culture was therefore essentially conservative; it suited small, isolated communities with a strong network of kinship and group solidarity. Oral delivery was often stereotyped through national or psychological categories and provides evidence of existing communication habits. Gerhoh von Reichersberg, for example, disapproved of the practices widespread among the Romans "who speak loudly and, often, without much sense."[36] Others criticized in detail all kind of habits common to medieval man. Hugh of St. Victor compared the human body to a *respublica* whose members have specific functions to accomplish in due measure and decency. Those who move their fingers while speaking, open their mouths when listening, make thousands of gestures, or move their arms "as monsters," were strongly disapproved of, as they undermined the inner harmony of the body's imaginary state.[37] One celebrated medical treatise of the early twelfth century, the *Secret des Secrets*, tries to identify the characteristics of human personality according to the tone of voice: a low voice is pleasant, eloquent, and characteristic of gentlemen, but too much sweetness hints at malevolence or even stupidity and ignorance. Those who speak too quickly are suspected of treachery and perfidy.[38]

Oral delivery was constantly adapted for changing needs across social strata. The *conjuratio* or oral oath requested from townsmen, in which they stated mutual obligations and rights, indicates how oral discourse was integrated into the communication framework of the nascent urban centers. Medieval princes also made frequent use of messengers carrying oral messages that were not always corroborated by written documents, particularly with important subjects in which absolute secrecy was required. The Hunnish king, for instance, sent a message to the Eastern emperor via an envoy who memorized and mechanically repeated the words. Messengers carrying oral messages were frequently used and, though the organizational level of transmission was low, simple message exchange was of a fairly high functional efficiency. Working on an irregular basis and without a salary, young children, servants, and vagrants served this purpose after they had memorized the facts, thus facilitating the communication between commoners as well.[39]

The use of oral communication in medieval society should not be evaluated, therefore, as a function of *culture populaire* vis-à-vis *culture savante* but, rather, of the communication habits and the tendency of medieval man to share his intellectual experiences in the corporate framework. Besides, the high cost of books and the difficulties in obtaining them also fostered a "sociabilization of culture." Crosby argues that "in the Middle Ages the masses of the people read by means of the ear rather than the eye, by hearing others read or recite rather than by reading by themselves."[40] Reading in company or, to be more precise, listening to reading, was customary in both the family and the academic forum. The illiterate asked for oral delivery of written books, while, conversely, oral traditions were written down.[41] Medieval authors wrote with oral delivery in mind, adding auditive effects such as direct address, the constant repetition of words, phrases, and situations, the frequent use of swearing, or benedictions at the beginning or at the end of the piece.[42] In this regard, visual and auditory memory, and the oratory talent of the communicator, acquired cardinal importance for the success of a book.

From the eleventh century onward, an increasing number of social groups or fraternities was devoted to literature, poetry, and liturgical theatre. According to Gerhoh Reichensberg, poetry was of crucial importance as "the whole land rejoices in the praises of Christ, in songs in the vernacular as well, especially in German, which is a particularly good language for singing songs in."[43] The Church had fostered the study of music as an integral part of the *quadrivium*, one of the basic spheres of knowledge integrated in the seven *Artes liberales* (see plate 2). In medieval society, music chanted by a harmonious chorus was regarded as pleasant to the ears of God and, as such, it became an integral part of the ritual. Liturgical theater too, was performed in the church building. The oldest known text is that of the *Regularis concordia ad fidem indocti vulgi ac neophytorum corroborandam*, which represented the passion of Christ and explained the symbolism of the divine office.[44] Written in the eleventh century for both clergy and laity, it was performed by the monks of Fleury-sur-Loire during Easter Week, Good Friday, and Easter Sunday, timing which secured *a priori* maximal reception. In the universities as well, it became customary to read new treatises in public in order to spread or receive new ideas. Both Pierre le Chantre and Thomas Aquinas emphasized the importance of public lectures and disputations.[45] A considerable number of such events had been recorded: in 1215, Buoncampagni da Signa recited his *Rhetorica antiqua* at Bologna where it was forthwith "approved and crowned with laurel." The same work was read in 1226 at Padua before the doctors and students, the papal legate, and the bishop and chancellor of Milan. Master Lawrence of Aquileia read his *Ars dictamen* at the University of Paris and received the approval of masters and scholars. Similar reactions followed the recital of Rolandinus of Padua's *History of the Trevisan Mark* at the University of Padua in 1262, when the public "applauded, approved and

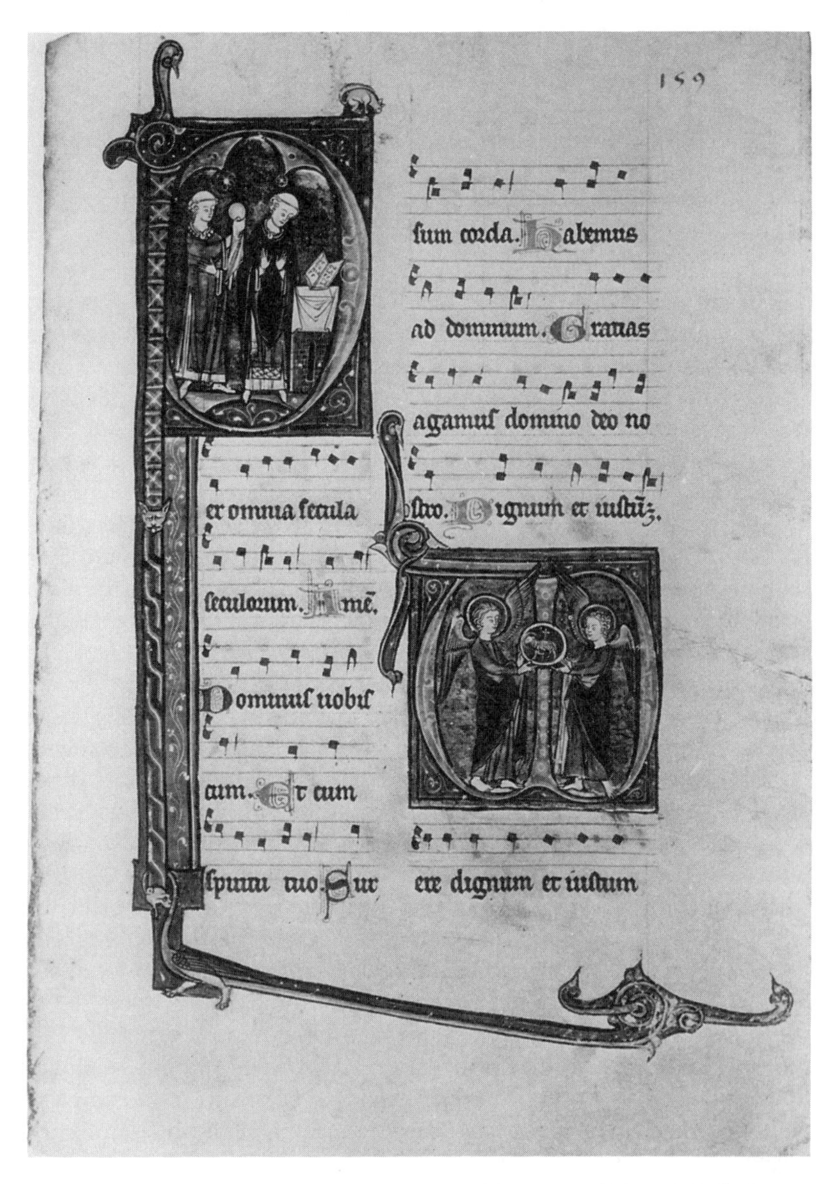

PLATE 2. Missal for Rouen Cathedral, composed between 1235 and 1245. *Per omnia* and *Vere dignum,* initials (23 × 16.2 cm).

solem ... ɔr chronicle by their magistral authority."[46] ... *Topographia Hiberniae* at Oxford for three ... than public recitations were private readir ... ıow he read aloud from his *Meliador* to the C(... friends to read his work in his presence to allc ... ·licizing it.[49]

Oral delivery thus became an essential trait inherent in medieval communication without becoming the sole prerogative of any socioeconomic class. At the height of this rich tradition of oral delivery was minstrelsy, the performances of a professional storyteller, variously named *troubadour, jongleur*, or *minstrel*, who was neither cleric nor aristocrat. His work, mainly, the recreation of known tales, is distinguishable from other contemporary genres by the brisk pace, the simple and vivid characterization, and the use of archaic, formulaic phraseology and of familiar motifs. Among the most notable epics were the *Chansons de geste*, the *Nibelungenlied*, and *Kudrun*, which embodied legend, folklore and fairy tale. In palaces as well as in churches, during journeys, riding on horse or walking, in front of peasants, knights, or kings, the minstrels played a key role in medieval communication. Their search for patronage and their duty to follow their benefactors on journeys facilitated the diffusion of ideas and information over Europe. They contributed to the spread of French and Provençal poetry in Eastern and Central Europe as well, thus encouraging the preponderancy of French as the courtly language of a considerable part of Christendom. The Archpoet, a German-Latin poet of unknown name, for instance, accompanied Rainald of Dassel, arch-chancellor of Frederick Barbarossa on his journeys. His rich experience on the roads of medieval Europe brought about his famous confession that "*meum est propositum in taberna mori*," my aim is to die in a tavern, a goal that expressed quite plainly his attitude to real life. Although the minstrels' performances were enjoyed by the sociopolitical elite and commoners alike, only the most wealthy classes were actually capable of hiring the services of minstrels on a regular basis. Edward II, for instance, received four minstrels in his chamber at Westminster, and they became an integral component in the court of Edward III as well.[50] The number of minstrels was naturally adapted to the importance of the occasion. On the wedding of Princess Margaret, daughter of Edward I, the presence of 426 minstrels was reported. Their repertoire was also adapted to the occasion and covered a wide spectrum that included the *Lives of the Saints*, well-known epics or moral stories and fairy tales, which still retained a considerable pagan character.[51]

Despite the importance ascribed to minstrelsy in the field of communication, *preaching* was the most powerful form of communication of the times. The cardinal importance of preaching was further reiterated by the Church's legislation on the matter, which specified the timing as well as the skills required. Until the appearance of the friars at the beginning of the thirteenth century, preaching had been mainly the privilege of the secular clergy. The second Council of Limoges (1031) established that every priest should preach at his parish on Sundays and holidays.[52] The Council of Albi (1254) enforced this dictum while introducing the duty of children aged seven years or more to come to the church, in order to be educated in the principles of the Christian faith and learn the *Pater noster* and the *Ave Maria*.[53] The continuous legislation on the subject, however, hints at the inexpedience of former regulations. As for England, the Peckham Constitutions (1281) encouraged the priest to preach at least four times a year, during Easter, Whit Sunday, Christmas, and for the processions of Palm Sunday and Rogation Sunday.[54] On the other hand, the importance ascribed to preaching by the friars in general and the Dominicans in particular, brought about an average of 250 sermons a year in the Dominican churches, which enjoyed full attendance.[55]

In the first chapter of his *Summa de Arte Praedicatoria*, Alan of Lille defined preaching as "the teaching of religion and customs at the service of the faithful, based upon logics and rooted in an authoritative source."[56] According to St. Augustine, "It is the duty of the interpreter and teacher of Holy Scripture, the defender of the true faith and the opponent of error, both to teach what is right and to refute what is wrong, and in the performance of this task to reconcile the hostile, to rouse the careless, and to tell the ignorant both what is occurring at present and what is probable in the future."[57] The main goal of preaching, therefore, was not self persuasion and/or persuasion of others on the basis of assumptions and opinions, but, rather, the manifestation of the *Vox Dei* and its uncompromising teaching. Augustine's book *On Christian Doctrine* (AD 426) contributed the theoretical basis of preaching, which was further enriched by the sermons of Gregory the Great, Haimon d'Auxerre, and Bernard de Clairvaux, among others. During the eleventh century and up to the end of the Renaissance, hundreds of manuals appeared that taught the *Artes praedicandi*, namely, the methods of preaching, the different audiences with the problems they caused, and the structure of sermons. The final product of preaching, the sermon, has been considered by the renowned English scholar Gerald R. Owst as one of the medieval precursors of humanism.[58]

Although the preachers themselves were among the intellectual elite, preaching cannot be evaluated simply as a manifestation of *culture savante*. In order to attract large audiences and influence their way of life, preachers increasingly had to adjust their vocabulary to everyday reality and speak a language understandable to the ordinary parishioner. Thus in the twelfth century, Latin, which was retained for sermons preached to the clergy,[59] gave way to vernacular sermons preached to the laity. St. Augustine (354–430) is the first known example of a priest allowed to preach because of his knowledge of the vernacular, since his bishop ignored the local language.[60] The preacher usually wrote his sermon after his appearance in public, sometimes a few weeks later and, if he wanted his sermons to be preserved, he rewrote them in Latin notwithstanding the original language of their delivery in public. The friars' success, however, increased the tendency to preach in the vernacular, as this proved to be a most useful means to attract large audiences. Maurice de Sully, bishop of Paris (1120–1196), for instance, dictated the models for sixty-four sermons in French, while Samson, abbot of Bury St. Edmunds (1182–1212) preached in French and in the local Suffolk dialect as well as in Latin.

Preaching in the vernacular required not only translation into a language comprehensible to most of the audience, but also the clarification of all those terms, events, and personalities which were known to the cultural elite but not to the common people. The *Sermoni Subalpini*, for instance, reports in the Latin original: "We read in the Pentateuch of Moses that after the death of Moses, Joshua sent two men to spy out the Promised Land, since he wanted to know about its citizens and the city of Jericho, which was a big and rich city over the Jordan." In the vernacular, this message was elaborated to explain concepts such as the Pentateuch, Joshua, and the Promised Land. The final product therefore runs as follows:

> *Seignor frare, nos legem en un deil cinc libres que Moyses escris que apres la soa mort si fo fait un so hom, quia avea nom Iosue, dus e regeor del povol dei Jue. E quest Iosue si tramis doi homes de l'ost que il anesen espier la cita de Iericho, qui era munt rica e de grant renomenaa. Or cil aneren e paserun flum Iordan, mas si veneren tart a la cita.*[61]

In the vernacular version, the preacher clarified the meaning and avoided deviating from the subject. Sometimes the audience's limited attention led preachers to avoid the use of fancy expressions and rhetorical ornaments that might hinder immediate understanding. The emphasis on direct communication was reflected in terms such as "lessen," "we beg you," "you have to," "you are expected," "avoid," and so on, which hint at the immediate response that the communicator expected from his public.

Communication research over the last few decades has shown that the difficulty and reluctance in grasping abstract, generalized ideas, and the lasting adherence to well-known norms are not necessarily a peculiar trait of the primitive mind.[62] They are rather behavioral trends typical of Western civilization.[63] Medieval audiences were not always sympathetic or attentive, so the preachers had to understand them in order to make their sermons more effective. Alain of Lille (d. 1203), an active preacher in Southern France, differentiated between the various social and professional categories in his audience—soldiers, notaries, intellectuals, priests, nobles, women in general, and widows and maidens in particular—to whom his sermons were accordingly addressed.[64] Humbert de Romans was more sensitive to moods and education, life experience, innocence, kindness, or cruelty among his public, besides the prevailing criteria of age, sex, and social status.[65] Such an awareness of the heterogeneous character of their public was a direct result of the many problems the preacher faced in practice to gain and hold audience attention. Caesarius of Heisterbach (1180–1240), for instance, tells of a Cistercian abbot whose audience was drowsing peacefully during his sermon. He stopped suddenly and started saying, "Once upon a time there was a king named Arthur. . . ." This sentence immediately caused a general awakening and tense attention. The abbot then complained: "When I talk of God you drowse, but you all wake up to hear fairy tales!"[66] Criticism of the Church also proved to be a very effective way of attracting the attention of medieval audiences. Bernardino of Siena testifies that when a preacher broaches this subject, his hearers forget all the rest; everybody instantly becomes attentive and cheerful.[67] In medieval England as well, medieval sources report with disapproval the sleepy audiences and the whispers of women who formed the major part of the preachers' public. It was quite customary to interrupt the sermon with applause, whistling, or other expressions of agreement or contempt. Sometimes, the preacher had to face inquiries that questioned his basic knowledge on the subject matter of his sermon. Bromyard (d. 1390), a Dominican preacher and chancellor of Cambridge, described his own experience in a rather pathetic manner:

> The men are delighted when the preacher harangues against the womenfolk, and vice versa. Husbands are pleased when their wives' pomposities are denounced in the sermon, how perchance they may spend the half of their wealth upon their own adornment. Wives rejoice to hear the preachers attacking their husbands, who spend their earnings in the ale-house. Those who know that they are guilty of some crime try to get the detractors denounced in the pulpit, because they think that men will talk of their deeds. And so what is preached against the vices of others gives pleasure, but what is said against their own, displeases. Thus when the preacher attacks all vices, everyone is displeased.[68]

The Church's long experience with the various difficulties of preaching and its continuous efforts to reach a large audience brought about a voluminous conciliar legislation on the subject. Already the Council of Mayence (813) stated that it was the preacher's duty to preach in a simple language that would be understandable to the common folk.[69] This rule was sustained throughout the Middle Ages. The need to arouse attention sometimes led to over-emphasis on the entertaining aspects of the sermon at the expense of content, logic, or accuracy. The content of the Holy Scriptures was therefore adapted to medieval reality, while Christ, the Apostles, and the saints acquired a tint of feudalism suitable to the times. Besides, the principle of compensation and punishment in the world-to-come became the leitmotiv of medieval preaching, confining its rhetorical essence to a marginal plane and turning preaching into "sacred eloquence" or, rather, "divine communication."

In spite of the many difficulties that plagued Christian communicators in everyday practice, personages such as Bernard de Clairvaux and Jacques de Vitry inflamed their audiences and brought about their immediate response shown in the massive recruitment for the Crusade. According to Jacques de Vitry, "if any one, out of hatred or anger, deprives the people of preaching, he is like the foolish and malicious man who, to spite his wife, mutilated himself, and so harmed himself rather than others."[70] The massive audiences that came to listen to thirteenth-century preachers such as the Franciscans Antony of Padua and Hugues de Digne, or archbishop Bourges, Philippe Berruyer, forced them to preach outside the church buildings. When a popular preacher such as Haymo of Faversham (d. 1244) called on the faithful to confess their sins, the clergy of St. Denys needed three whole days to listen all the confessions. A similar phenomenon occurred in Milan 200 years later, when the Milanese clergy was not able to deal with the great number of confessions resulting from the preaching of St. Bernardino of Siena. Technological improvements and the higher level of education at the end of the Middle Ages, therefore, did not undermine the high reputation of preachers, although they lost their former monopoly in the field of communication. The Franciscan Richard, for instance, preached in 1429 in Paris for ten consecutive days from 5 A.M. until 11 P.M. Antoine Fradin, Olivier Maillard, Vincent Ferrer, and Savonarola, to mention but the most outstanding examples, enjoyed enormous influence over their audiences, the effect of their rhetoric might nowadays be regarded as mass hypnosis.

Beside the rhetorical skills of preachers, their success resulted also from the structure characteristic of medieval sermons, established in St. Victor during the twelfth-century. Every sermon opened with a statement on the subject, followed by three parts; namely, the translation or clarification of the evangelical sentence to be discussed; the interpretation of the sentence in the allegorical or the tropological sense, and, finally, the clarification and emphasis of the moral message. Gilbert de Nogent's interpretation of Jerusalem illustrates the pattern of medieval sermons: *literally*, it is the city of that name; *allegorically*, it represents the Holy Church; *tropologically*, it represents the faithful soul of those who aspire to the vision of eternal peace; and *anagogically*, it denotes the life of the pious in heaven who see God revealed in Zion.[71] Being aware of the heterogeneous nature of medieval audiences, Pope Gregory the Great justified this method on the grounds that "the word of God both exercises the understanding of the wise by its deeper mysteries, and by its superficial lessons nurses the simple-minded. It presents openly that wherewith the little ones may be fed; it keeps in secret that whereby men of loftier range may be rapt in admiration."[72]

The *exempla*, moralizing anecdotes inherited from the classical world, were also an important means utilized by medieval preachers while addressing the laity and the clergy alike. These anecdotes actually suggest the vulgarization process promoted by the Church in its endless fight against the remains of paganism, still alive in the popular culture. Contemporary sources summarize through six criteria the main patterns followed by the *exempla*: *Authenticity*, had to be testified by the authority of one of the Church Fathers or a faithful witness. *Credibility* meant they had to be true to ensure their acceptance. *Conciseness* specified maintaining attention throughout the full sermon. *Pleasure* meant they were expected to teach the right path in a pleasant way. *Memoranda* indicated their influence was meant to last and improve the way of life of the faithful. *Allegory* was the appeal to the imagination and the sentiments of the audience through the use of metaphors. As claimed by

Etienne de Bourbon, the *exempla* should instruct the common people in a simple and easy way, while leaving their mark on their memory to the long term.[73] Through allegories, folk tales or *exempla*, the sermons thus became one of the main means at the disposal of medieval communicators in the religious and in the sociopolitical fields.

Although the Church was undoubtedly the main promoter and beneficiary of preaching and tried to restrict its use to the ranks of the clergy, it did not succeed. Preaching became also an important communication channel for nonconformist groups such as the so-called heresies and peasants' movements, usually led by some wandering preacher who conferred upon himself the aura of the biblical prophets. These pseudoprophets voiced the resentment of many, thus creating a new literary style that can be categorized in the genre of protest.[74] The wide publicity they achieved is exemplified by the moralizing cliché invoked by John Ball at Blackheath (1381), which became a sort of hymn to later generations:

> "When Adam delved and Eve span,
> Who was then the gentleman?"

The increasing awareness as to the lack of class differentiation in the act of Genesis encouraged a slow but consistent process of change through which the silent illiterates of the past claimed their right to express the *Vox Dei* according to their own understanding.

The growing importance of large social sectors in the social and political arena brought about a further development in the visual media whose message reached both the intellectuals and the illiterate: *art*, either by means of painting, drawing, or sculpture, provided another useful channel to transmit the *Vox Dei*.[75] Medieval art displayed before the believer the mysteries of Holy History, Church doctrine, the lives of the saints, or the main values and vices, as suitably categorized by the Holy Mother the Church.[76] According to Duby, medieval art played the role of sacrifice in the dialogue between life and death, mediating between the powerful and harmonious world of the Creator and the mysteries surrounding the helpless human being.[77] Gregory the Great (540–604) was among the first communicators who encouraged the use of art for educational goals. In his letter to Serenus, the pope claimed that painting was meant to educate the ignorant, who could not read by themselves, by teaching them Holy History. Sculpture or painting, moreover, might excite the believers and tighten their links with the Holy Church while strengthening the principles of the true faith.[78] In spite of his objection to displaying statues in the chapels of the Cistercian Order, Bernard de Clairvaux too, supported the decoration of churches with sacred scenes "to facilitate the labor of bishops, who, due to their office, have to teach the principles of Christian doctrine to everybody, educated as well as illiterate. Thus, the bishop should stimulate the imagination of the common-folk by material objects, since the spiritual way is rather difficult for them." The words of St. Bernard suggest the Church's ambivalent approach to the artistic media, considered as a "second best" for transferring the *Vox Dei*. These prevailing reservations against the frequent use of images reflect the fears, quite justifiable on the other hand, regarding the tendency to fetishism still alive behind the worship of images.

The use of artistic media, however, was both too attractive and too effective for it to be completely avoided. On the contrary, around the year 1000, the gates of European cathedrals gradually became "open books," telling the believer the main chapters of Holy History. At the entrance to Hildesheim Cathedral in Saxony, two bronze gates were installed in 1015, representing the crucial events of the Old and New Testament, such as the banishment from the Garden of Eden and the resurrection and passion of Christ (see plate 3). The linterns of the Church of the Holy Sepulchre also reflected Christocentric elements interpreted within the world of the Crusaders in the first half of the twelfth century: the banishing of the merchants from the temple by Christ (Matt. 21:12–14) thus foretold the purification of the Holy City by the Crusaders and the release of Jerusalem from the oppression of the nonbelievers.[79] The two statues placed at the south gate of Strasbourg Cathedral in the thirteenth century illustrate the didactic message implied in medieval art. The woman on the right represents the *ecclesia triumphans* watching with a winner's pride the defeated woman on the left who represents the *synagoga*; her covered eyes symbolize the Jews' refusal to acknowledge the Christian truth (see plate 4). Many episodes of Holy History were also transmitted to the faithful through the frescos displayed in church buildings from the eleventh century onward. They usually represented scenes from the New Testament, namely, pictures of Christ's childhood together with the lives of local saints, and events connected with the secular world, such as the Crusades or the conversion of Constantine. The monumental works of art in the late Middle Ages further consolidated these trends and also presented new elements. Around 1300, the gates of Rouen Cathedral portrayed the emotional aspects of the last moments of Christ, such as the sorrow of Mary and the Apostles.[80] The use of the artistic media, however, was not limited to the Church, and sometimes it served its opponents as well. The symbol of the Lamb of God Triumphant in thirteenth-century Languedoc, appearing in the municipal seals of Béziers, Carcassonne, Narbonne, Rieux, and Toulouse, hints at the identification of townsmen with the sacrifice of the Lamb of God who had suffered, and been crucified, but was still expected to come in a second Resurrection. The political message implied thereby became quite significant in light of the Languedoc failure to resist the Northern domination. It suggests the prevailing hopes of townsmen for an ultimate victory in the political and in the religious sphere as well.[81]

Numbers, too, had a special significance in the field of medieval communication. The number *twelve* symbolized the universal Church. Its inner composition (*three* x *four*) was also meaningful: *three* stands for the essence of Holy Trinity and symbolizes spiritual things, while *four* is the number of the basic elements (water, fire, earth, wind), thus representing material things. Similarly, the number *seven* represented the double nature of mankind and its

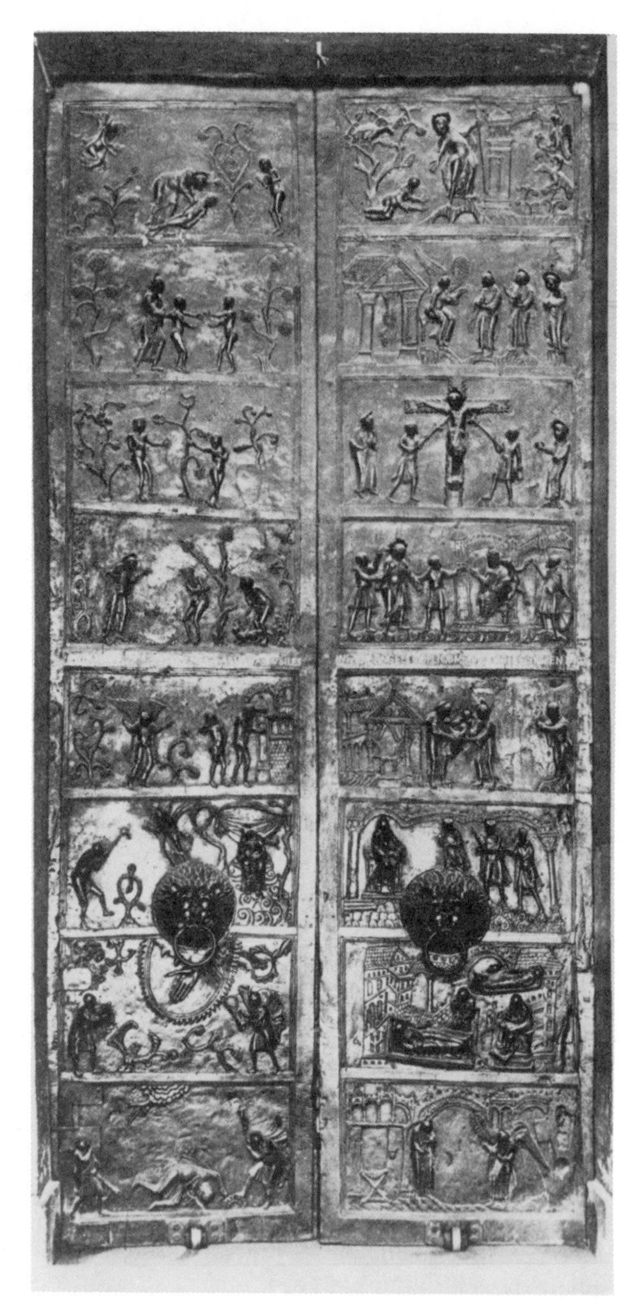

PLATE 3. Bronze doors of Bishop Bernward. Hildesheim Cathedral, 1015. The doors' size is impressive, 16.5 feet. The left wing shows eight scenes from Genesis, the right has the same number taken from the New Testament. Some of the scenes adjacent on the two wings are related in a significant way: original sin is next to crucifixion. Eve suckling Cain is facing the Virgin and Child in the Adoration scene, thus contrasting motherhood arising from sin with that from immaculate conception.

PLATE 4. Column-figures of the *Ecclesia* and *Synagoga* from the south transept portal of Strasbourg Cathedral, carved about 1230. On the extreme left of the entrances was the Church with chalice and pennanted cross; on the right was the synagogue, with her eyes bandaged, the lance broken, and the Tables of the Law slipping from her hand.

dualistic attitude toward the universe. Although this symbolism was not commonly known, its expression in one way or another touched all social strata, thus paving the way for the emergence of a cultural denominator common to all Christendom. The particularist tendencies inherent in the feudal background were therefore suitably balanced by the universal essence of the *Vox Dei*.

The imperative to voice the *Vox Dei* was further emphasized by the sound of *church bells,* which might well be compared to the role of the telephone, radio, newspapers, clock, calendars, and telegrams in our days. Both in the nascent urban centers and in the countryside, they measured the time in daily practices and served as a channel to transmit the most important messages. From the end of the fourteenth century, the "Angelus" or "Gabriel Bell" was rung at dawn in London and reminded all Christians to say one *Pater noster* and five *Ave Marias*. It also announced the first Mass, which was performed at four or five o'clock in the morning. This served to indicate the beginning of a new working day as well, when cows and pigs were let into the streets to be driven outside the walls by the common herdsmen. Bells thus shaped the opening and closing of markets, and indicated the times at which outside competitors might begin to buy or sell. Market bells rang in the big malt markets of London at nine o'clock in the morning and probably gave the signal for foreigners two hours later. A church clock or bell conveniently situated often saved the expense of setting up a "market bell." At York, for instance, "the clock at the Chapell on House bryg" striking the hour of ten gave the foreign fish merchants the opportunity to offer their merchandise.[82] Similarly, bells also served to announce victories or defeats in the battlefield, deaths and celebrations, processions, and sudden crises. The Common Bell was rung to summon all freemen of the town to the city's Council. At Bristol, in 1310, a special bell in Trinity Church was used for this purpose, but in many towns a bell was hung in the Guild Hall from the time that the building had been erected. In the countryside as well, the church bells measured the time for peasants who, according to the testimony of Jean de Garlande, "can not estimate the time but with the help of bells"[83] (see plate 5). Bells thus became a most important source of information, while rural priests became both the clock and the calendar of their parishioners as they announced the proximity of holidays and important events.

In the face of celebrations or crises, the church bells thus enforced the solidarity group of medieval men, which was given a corporal expression via the *processions*. Processions were an integral part, if not the most important feature, of medieval culture. In the symbolical plan, they represented the militant Church headed by Moses on his way to Christ. The growing participation of monks fostered changes in the architectural structure of monasteries, whose chapels were now designed to allow the entry of processions into the nave. The profusion of processions during the Middle Ages also turned them into a suitable forum for exchanging information and, at the same time, for fostering massive identification with the Church or with the secular ruler who had promoted them and his policy. In medieval Flanders, for example, processions were held every Sunday after Mass, and were established by the secular or religious authorities for a variety of reasons, namely, following a scarcity of rains, a general amnesty, or peace agreements.[84] Between May and July 1412, daily processions were held in Paris to ensure divine support for the immedi-

PLATE 5. Church of St. Maria della Rocca at Ascoli, 1330.

ate defeat of the Armagnacs. Although these pageants alone could hardly improve the anarchical situation in the city, they nevertheless contributed to the consolidation of an esprit de corps among the suffering townsmen. Processions could thus serve the goals of interested political parties at times of crisis and many were, indeed, designed to this end. Still, they remained a genuine expression of social anxiety. They acquired a singular importance in light of the many plagues and long periods of famine that were part and parcel of each generation. According to Delumeau, phenomena of this kind brought about the "*dissolution de l'homme moyen*," and left no room for the ordinary man, helpless in face of the unknown.[85] Boccaccio vividly recorded the antisocial sequences of the Black Death (1346–1348) in daily life, as "one citizen avoided another, hardly any neighbor troubled about others, relatives never or hardly ever visited each other. Moreover, such terror was struck into the hearts of men and women by this calamity, that brother abandoned brother, and the uncle his nephew, and the sister her brother, and very often the wife her husband. What is even worse and nearly incredible is that fathers and mothers refused to see and tend their children as if they had not been theirs." In the midst of the plague, when people found it difficult to explain the catastrophe which resulted "through the influence of the heavenly bodies or because of God's just anger,"[86] processions facilitated the socialization of fears, bestowing on the individual a feeling of group solidarity.

The expedience of processions is further supported by their high frequency following the waves of famine that struck Western Europe between 1315 and 1317. In Paris, for instance, people of all classes joined the processions to St. Denys to ask for divine grace, walking barefoot, as individuals or grouped in guilds or religious bodies. One contemporary chronicler reported that "We saw a large number of both sexes, not only from nearby places but from places as much as five leagues away, barefoot, and many even, excepting the women, in a completely nude condition, with their priests, coming together in procession at the Church of the Holy Martyr, and they devotedly carried bodies of the saints and other relics to adore." During July 1315, these processions lasted about fifteen days while similar pageants took place at Chartres and Rouen.[87] In England, too, a great number of processions were performed following prolonged floods and the loss of crops. The archbishop of Canterbury ordered the clergy of London to participate barefoot every Friday in a procession to the Church of the Holy Trinity, carrying the *Corpus Domini* and other relics.[88] The growth of processions indicates the relative geographical mobility of medieval man who could thereby become acquainted with new places and customs. The religious essence of daily practices gave this mobility a religious significance without limiting its further implications in social life. The mere participation in a procession thus mirrors a new dynamic of social life: it reflects the increasing needs for a suitable arena to exchange information across wider distances.

A *pilgrimage* offered another important means of mobility against the static nature inherent in the feudal regime. Etymologically, the term pilgrimage (*peregrinatio* in classical Latin) means the wandering of people across the fields, *per agere*, indicating vagabondage. Adopted by early Christianity, the term acquired a positive sense, to mean the journey of those faithful who left their families and the ordered life in order to search for salvation in union with God. The search for indulgence, for the special favor of a prestigious saint, for the social esteem that glorified returning pilgrims, and the pursuit of adventure, turned pilgrimages into another outstanding feature in medieval society.[89] Basic Christian premises and the widespread belief in miracles encouraged the emergence of communication channels that fitted the psychological atmosphere of the times. The miracles ascribed to the saints were commonly reported by pilgrims, some of whom returned joyously to the shrine to report on wonderful events that had happened to them. Others, who painfully traveled to the shrine, there experienced a miracle, sometimes in the very sight of the scribe who was appointed to record them. Another less common source of information was the letter from a distant prelate, usually copied into the record of the saint's miracles by the local scribe. Pilgrimages offered, therefore, not only the chances of knowing new places and thus fulfilling the pursuit of adventure, but also of enjoying special favors, as they were expected to bring about an immediate divine response.

Being the birthplace of Christ Who lived and suffered crucifixion there for the sins of mankind, the Holy Land was regarded as the pinnacle of pilgrimage. As early as the fourth century, St. Jerome wrote to Paulinus from the Holy Land that "people are flocking here from all the world. The whole of mankind fills the city while the great multitude of both sexes leaves no escape from them."[90] In the Middle Ages, however, the Holy Land turned into a distant and perilous destination, and new alternatives were needed to satisfy the prevailing expectations. From the beginning of the twelfth century onward, new centers of pilgrimage spread all over Europe, at the graves of St. Thomas in Canterbury, of St. James in Compostela, and of St. Peter and St. Paul in Rome. The Crusader movement, on the other hand, facilitated pilgrimage to the Holy Sepulchre and other shrines in the Holy Land. The increasing number of pilgrims overseas gave rise to a regular transport service from Venice. "It is the rule," reports a fourteenth-century traveler, "that the Venetians send every year five galleys to the Holy Land. They all reach Beirut, which is the port for Damascus in Syria. Thence two of them bring the pilgrims to Jaffa, which is the port for Jerusalem."[91]

Apart from the limitations of time and money, a pilgrimage was available to every free person and, indeed, many spared neither money nor effort to go on one. Yet the wealthy classes sometimes opted for more comfortable alternatives and paid a third person who traveled on their behalf. The resulting agreement brought about the emergence of a new socioeconomic group, the *palmers*, who traveled to places as far away as Rome or Jerusalem, in the name of other pious and still wealthy Christians, who paid their expenses and provided them with a generous salary for their journey. The pilgrimage season in England extended from Christmas, Easter, Whitsun until Michaelmas Day (September 29) or, alternatively, from the Easter Week, Whitsun, and lasted until after the harvest. Although the Church had been the initiator and main promoter of pilgrimage, it tried to limit the length of time and the number of clerics and monks going on pilgrimages who, according to a twelfth-century English bishop, had become too many.[92] In a Hereford source of about 1200, the right of ecclesiastics to go on pilgrimage was carefully regulated. Provided that a canon had resided at the cathedral for a year, he was allowed a three-week pilgrimage in England and, in addition, once in his life he might cross the sea on pilgrimage. He was given seven weeks for a visit to St. Denys in Paris, sixteen weeks for Rome or Compostella and, if he went to Jerusalem, the chapter allowed one year's leave.[93]

The importance ascribed to pilgrimage in medieval life justifies further inquiry as to the maintenance of *roads*, their physical condition, and the degree of security they provided. Some medieval roads had remained from Roman times or dated from an even earlier period, and a few of them still serve as transportation routes today.[94] The bad condition of a considerable number of roads, however, made it quite difficult to establish any timetable with accuracy. In 1319, Edward II invited the scholars of the King's Hall in Cambridge to spend Christmas at York with him. Scholars who were strong enough made the journey by road on horseback in five days. Since the youngest aged seven or little more were hardly capable of making such a journey, they were taken by boat from Cambridge to Spalding. From Spalding they rode on horseback to Boston, their luggage following them in carts. At Boston they took a boat again and went the rest of the way by water. They finally reached York three days late for the Christmas feast for which they had been invited (see map 1). This miscalculation may suggest that such a journey was an exception rather than the rule. In spite or because of the poor maintenance of the roads, the slowness of travel became a major factor in social life, since roads became in actual practice a suitable meeting place. A twelfth-century traveler reported that he encountered 200 people on horse and on foot along a fifteen-mile stretch of Sussex road.[95] Most of the *exempla* written by Jean Goby in the fourteenth century were actually based on stories told while walking on the road.[96]

The fear of robbers and the many feudal tolls on bridges and central squares sometimes led to a preference for the *maritime routes* as they allowed a more accurate planning of schedules. The increasing use of maritime routes was further fostered by the spread of commerce on an international scale, which required more reliable means of travel. This state of affairs is faithfully reflected in the strict schedules established in commercial transactions of the thirteenth and fourteenth centuries, which left little room for unexpected delays.[97] The capability of the Italian city-states such as Venice, Pisa, and Genoa to fulfill these requirements ensured in actual practice their supremacy over the western Mediterranean. From the thirteenth century onward, Portuguese and Catalonians also settled on the Western coasts of the Mediterranean and navigated to the Atlantic Ocean, Ceuta becoming one of their key harbors.[98] Improvements in maritime transport and the increasing use of the sea as the most suitable and rapid way, however, did not reduce the basic reluctance of ordinary men to travel by sea; the sea was considered the "number one danger," probably due to the lack of maritime experience among most German peoples, the Normans excepted. From the perspective of many, the sea embodied the unexpected and the frightening, as it had engendered catastrophic events such as the Black Death, and had facilitated bloody invasions

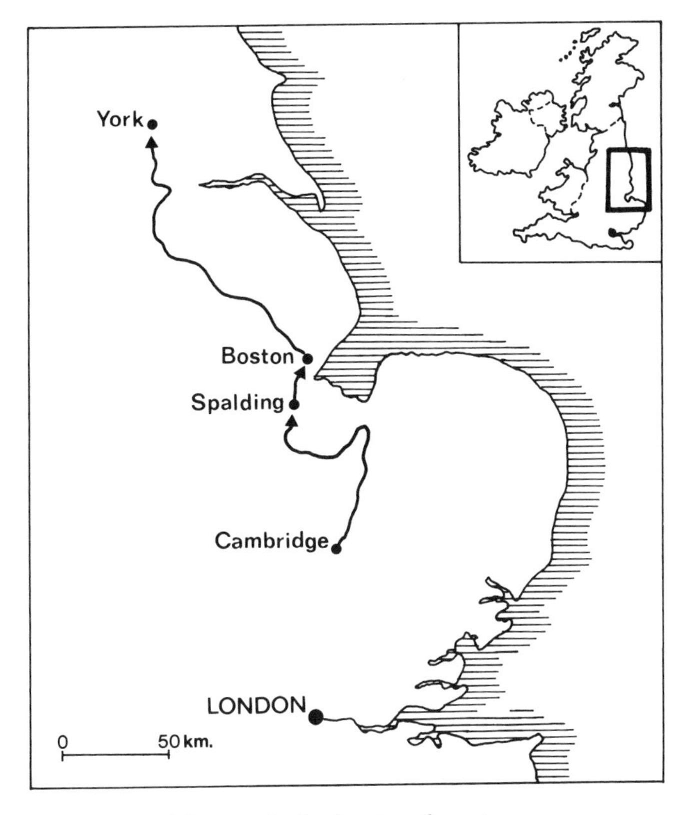

A journey in the fourteenth century.

such as those of the Normans. Historical memory thus turned the liquid mass into the "Temple of the devil," which, according to St. John, was destined to disappear at the end of the times (Apoc. 1:20). The reports of those who had been forced to navigate added to these irrational fears: Jacques de Vitry, for instance, related in detail the hardships on his journey to Acre in 1216.[99] Storms in mid-sea were also experienced by those who returned with St. Louis from the Sixth Crusade in 1254 and were pathetically reported by Joinville in his *History of St. Louis.*[100] Descriptions of this kind deepened the lack of security and the mystery through which the ordinary man related to the watery depths. Beyond the sphere of irrational fears, the economic and political development encouraged the widespread use of maritime routes, which, by the Central Middle Ages, had become an important, if not the most important, means of transport at the disposal of merchants, pilgrims, missionaries, or mere adventurers.

The thirst for faraway countries, indeed, brought many to follow in the steps of Marco Polo to the Far East, and encouraged a new consciousness as to the advantages implied in the instruction about foreign customs and places. One popular fourteenth-century author advised, "Learn the situation and know the regimes of the world, mainly, of this region." He further adduced the example of Achilles who, notwithstanding his courage, was replaced by Ulysses because of his better knowledge of the country and the language."[101] The geographical knowledge of ordinary people, however, remained quite poor and, in most cases, was limited to the intellectual elite and the merchant class. Atlases written at the beginning of the twelfth century reflect the uncertainty of geographical knowledge and the wide room for misconceptions.[102] Nonetheless, the increase in travel further fostered the learning of *foreign languages*, thus improving the mutual knowledge and understanding between people from different countries. Although Latin remained the language of the cultural elite, it had ceased to be a dynamic language since the fall of the Western Roman Empire in 476, while most people outside Italy were ignorant of Latin. In the realm of Mediterranean trade, Italian, Catalan, and Greek became the most important languages, while in the Empire German prevailed among the Italian merchants. From the twelfth century onward, French acquired a preponderant role in the fields of diplomacy and social life and was therefore the language chosen by Marco Polo to write his experiences. Danish nobles who could afford it sent their sons to Paris and, already from the eighth century, rudimentary dictionaries were at the disposal of foreign students who came to the city. Such manuals usually enclosed some practical advice as how to ask for food or for information on the road. The Old Norse *Speculum regum* encouraged its readers: "if you wish to become perfect in your knowledge, study all languages, and more than any others Latin and French, for they are the widest known, but do not neglect your native tongue."[103] Such advice, however, did not solve the many problems in actual practice and Hugo of Trimberg, for example, reported that travelers who were in the verge of hunger in distant places were forced to study such difficult languages as Czech, Italian, or Hungarian.[104] Thence the admiration aroused by those personages who knew several languages, such as Charlemagne and Otto the Great.[105] A twelfth-century chronicler stated that Henry II spoke all the languages between the Atlantic and the Jordan; this exaggeration indicates, nevertheless, the king's substantial linguistic knowledge.[106] Frederick II answered Greek and Arabic scholars in their own languages and was acquainted with Slavonic idioms as well.[107] The awareness as to the political utility inherent in wide linguistic knowledge is substantiated by Edward III's requirement from all his nobles to achieve at least an elementary knowledge of French at the beginning of the Hundred Years War.[108]

The learning of foreign languages was further fostered by the increasing missionary efforts in Africa and the Far East, which, together with the large number of pilgrimages, encouraged the Church to promote the study of foreign languages in the ecclesiastical order as well. The *Liber Calixtinus*, the official codification of the saint's cult from about 1139, offered a description of most important pilgrimage roads between France and western Spain. It also enclosed a short Latin-Basque vocabulary of fifteen words for the use of pilgrims to Santiago de Compostella; it provided the translation of elementary words such as bread, wine, meat, fish, wheat, and water. Although the Church basically opposed the use of languages other than Latin for the ritual, already in the ninth century it allowed the translation of the liturgy into Slavonic idioms in order to facilitate the mission in the northeastern areas of Europe. By the fourteenth century, similar efforts in Moslem countries brought about the widespread study of Arabic, and five linguistic schools were conducted at the time by the Dominicans. Pope Clement V supported, at the Council of Vienne, the founding of academic chairs in Arabic, Hebrew, or Chaldean, at the Universities of Paris, Oxford, Salamanca, and Bologna in order to lead "wanderers into the path of truth."[109]

Looking for new places, studying new languages, developing all the media at their disposal, medieval men thus challenged the isolation inherent in the structure of feudal society. Furthermore, using an endless number of alternative channels, they paved the way for a communication system which, in its first stages, improved the links between the isolated inhabited areas all over Europe while gradually bringing about new contacts with remote countries and cultures. The development of a communication system thus appears as one aspect of the socioeconomic process that followed the gradual decline of the corporate framework. When medieval men abandoned the narrow confines of the village or the manor, they actually became more communication-oriented and, as such, helped to integrate existing means of communication. The move to the nascent urban centers further deepened the exposure of medieval man to a variety of messages that had been unavailable in the familiar bonds of the corporate framework. This resulted in a process of continuous change whose richness and variety found full expression in the Church, the most faithful representative and the acknowledged communicator of the *Vox Dei*.

Europe about 1250.

I

The Church

2

The Ecclesiastical Message

In the introductory remarks to his study on *Western Society and the Church in the Middle Ages*, Southern argues,

> The history of the Western Church in the Middle Ages is the history of the most elaborate and thoroughly integrated system of religious thought and practice the world has ever known. It is also the history of European society during eight hundred years of sometimes rapid change, when the outlines of our institutions and habits of thought were drawn. . . . Church and society were one, and neither could be changed without the other undergoing a similar transformation. This is the clue to a large part of European history whether secular or ecclesiastical.[1]

Yet the unquestionable interplay between the Church and secular society in the Middle Ages did not guarantee immediately that the laity would identify with the goals of the Church. Duby, for instance, dates the mass impact of the evangelical message to the late Middle Ages.[2] The identification of Western society with the goals spread by the Catholic Church cannot therefore be evaluated as a trend inherent in medieval society per se but, rather, as the outcome of a long process of almost 1,000 years, which reached its zenith toward the end of the period. Whether as an outcome of deeply rooted pagan traditions, a different understanding of the Christian dogma, or conflicting political interests, there were, indeed, many contenders to the emergence of a monolithic Christian society.

Pope Gregory VII well understood the heterogeneous nature of medieval society, when he came to define the aims of the Reform named after him. The final goal of the Gregorian Reform focused on the consolidation of Christendom, which, as such, was subjected to papal authority and received from the ecclesiastical order its means of salvation. The emergence of such a monolithic Christian society, however, was preconditioned by a drastic reform of the ecclesiastical order itself, while suppressing the divergences of opinion that were part and parcel of prelates, clerics, and monks as well.[3] The heterogeneous essence of medieval society in general and of the Church in particular, undermined the chances of a complete identification between the Church and secular society, without questioning the reciprocal influence between them. On the contrary, clergy and laity together constituted the *Societas Christiana* and, as such, were integral members of the mystic body of Christ. The crucial role adjudicated to the priesthood and the laity in the emergence of the papal monarchy further created a common denominator between them both as the target of the papal message. Yet the changing interests of the papacy sometimes brought about an unstable interplay between them. Although the monopoly of the sacraments had elevated priesthood above ordinary men, the Reformer Popes did not hesitate to encourage the believers to challenge those priests who defied papal authority, thus overturning the otherwise passive role of the laity in the ecclesiastical order.[4]

This chapter deals with the main premises of the ecclesiastical message that legitimated the status of the Church in the Middle Ages and provided its coherence and power. Both clergy and laity constituted the audience for messages, orders, information, and plans that originated in the papal court. If *repetition* and *constancy* are key words in political communication, there is no doubt that the Church's success in spreading its message was significantly influenced by its allegiance to the same basic principles, repeatedly transmitted over a wide range of communication channels. The basically same premises, indeed, accompanied the medieval Church in the long process of Christianization of the German peoples and the emergence of the papal monarchy later on. The arguments developed by Pope Gelasius I (492–496) on the clergy's supremacy were further developed by Innocent III (1198–1216) and given canonical sanction in the bull *Unam Sanctam* of Pope Boniface VIII (1294–1303), who systematically summarized the basic beliefs of the medieval Church with the pope at its head. One thousand years of historical development, however, left their mark on the Church's ability to accomplish its claims in actual practice; the defensive position of the papacy in face of the Eastern emperor in the fifth century was rather different from the aggressive stand of Innocent III or Boniface VIII before kings and emperors alike. Yet despite the changing historical circumstances, the Church message invariably returned to the Holy Scriptures as the most reliable reflexion of the *Vox Dei*. One can further argue that its normative essence was one of the main traits of the Church message, one endowed with a halo of eternity as against the ephemeral nature of the historical process.

The two main axioms of the ecclesiastical message focused on the Church's privileged status: it was perceived as the most faithful reflection of the will of God, while the pope at its head was also depicted as the living personification of the same will. Both axioms found a faithful expression in the bull *Unam Sanctam*, published by Boniface VIII on November 18, 1302, at the peak of his struggle against the king of France, Philip the Fair.[5] The selection of the *Unam Sanctam* as a case study of the ecclesiastical message is largely justified in light of both the large diffusion it achieved and, still, the controversial nature of the reactions it fostered in contemporary society:

We are compelled by the faith to believe and to hold—and we do firmly believe and sincerely confess—that there is One Holy Catholic and Apostolic Church, outside of which there is neither salvation nor remission of sins; her Spouse proclaiming it in the Canticles: "My dove, my undefiled is but one; she is the only one of her mother, she is the choice one of her that bare her" (Song 6:9), which represents one mystic body, whose head is Christ, but of Christ, God. In this Church there is one Lord, one faith and one baptism (Eph. 4:5). At the time of the flood there was one ark of Noah, indeed, symbolizing the one Church; this was completed in one cubit and had one, namely, Noah as helmsman and captain, outside of which all things upon earth, we read, were destroyed (Gen. 7:21). This Church, moreover, we venerate as the only one. . . . She is that seamless garment of the Lord which was not cut but which fell by lot (John 19: 23–24). Of this one and only Church, therefore, there is one body and one head—not two heads like a monster—, namely, Christ and the vicar of Christ, St. Peter, and the successor of Peter. For the Lord Himself said to Peter, "Feed my sheep" (John 21:15). "My sheep," He said, in general, not designating these or those sheep; from which it is plain that He committed to him all His sheep. . . .

We are told by the word of the Gospel that in this His fold there are two swords, the spiritual and the temporal. For when the Apostles said "Behold here are two swords"—when, namely, the Apostles were speaking in the Church—the Lord did not reply that this is too much, but "It is enough" (Luke 22:38). Surely he who denies that the temporal sword is in the power of Peter misunderstands the word of the Lord when He says: "Put up again thy sword into his place" (Matt. 26: 52). Both swords, therefore, are in the power of the Church, the spiritual and the material; but the latter is to be wielded for the Church, the former by her; the former by the priest, the latter by kings and knights, but at the will and by the permission of the priest. The one sword, moreover, ought to be under the other, the temporal authority to be subjected to the spiritual. . . .

But that the spiritual exceeds any earthly power in dignity and nobility we ought the more openly to confess that more spiritual things excel temporal ones. . . . For, the truth bearing witness, the spiritual power has to establish the earthly power, and to judge it if it be not good. Thus concerning the Church and her power is the prophecy of Jeremiah fulfilled: "See, I have this day set thee over the nations and over the kingdoms" (Jer. 1:10), etc. If, therefore, the earthly power err, it shall be judged by the spiritual power; and if a lesser spiritual power err, it shall be judged by the greater. But if the supreme power err, it can be judged by God alone, not by man. For the testimony of the Apostle is: "Do ye not know that the saints shall judge the world? and if the world shall be judged by you, are ye unworthy to judge the smallest matters?" (1 Cor. 6:2).

For this authority, although given to man and exercised through man, is not human but rather divine, being given by divine lips to Peter and founded upon a rock for him and his successors through Christ Himself whom he confessed; the Lord Himself saying to Peter: "Whatsoever thou shalt bind. . . . " (Matt. 16:19). Whoever, therefore, resists this power thus ordained of God, resists the ordinance of God. He believes, like the Manicheans, that there are two beginnings. This we consider false and hereti-

> cal since, by the testimony of Moses, not in the beginnings, but "In the beginning" (Gen. 1:1) God created the heaven and the earth. Furthermore, we declare, state, define and pronounce that it is altogether necessary for salvation for every human creature to be subject to the Roman pontiff. The Lateran, 14 November, in our 8th year. As a perpetual memorial of this matter."[6]

The opening paragraph of the *Unam Sanctam* sums up the whole ideology of the Catholic Church as it evolves from the Nicene Council (AD 325) up to this day: *there is but one, holy, Catholic and Apostolic* Church outside of that there is neither salvation nor remission of sins. The papal argumentation was based on the principle of unity that lies at the heart of the divine plan and serves as a central pivot to the human race. Relying on the principle of unity, Boniface reverted to the theory underlying early Christianity according to which there is but one God, one shepherd, one ark, one captain. Following the Decretalists, who denied any dual approach,[7] Pope Boniface condemned as *Manichaeism*,[8] any attempt to harm the unity of the Church and its leader, the Vicar of St. Peter, namely, the pope. Aristotelian philosophy bestowed in the thirteenth century a cosmic dimension on the papal theory. According to James of Viterbo (d.1308) the principle of unity, essential to the cosmos, justifies its application in the moral order and the existence of a supreme authority that ensures the common good. The Aristotelian concepts of "cosmic order" and "common good" had yet to be universally accepted when he drew the conclusions appropriate to Christian society: "There is but one Jesus, one Church, one husband, one wife, one shepherd, one herd, one king, one kingdom, one prince, one principality, one leader, one republic."[9] The Franciscan Francis of Meyronnes (d.1327) pointed out the advantages of a single leadership with regard to the achievement of the universal Christian peace.[10] John of Paris (d.1306), the sworn opponent of Pope Boniface as well, claimed that the papal authority was directly ordained by Christ and represented the unity of the Church.[11]

The monolithic approach to reality bestowed holiness on the Church: being created by God, it monopolizes the means of salvation, namely, the sacraments. Christ's words to Peter: "That thou art Peter, and upon this rock I will build my church; and the gates of hell shall not prevail against it" (Matt. 16:18) turned into the act of founding the Church while Christ, through Peter, bestowed on it infallibility, as "the gates of hell shall not prevail against it." These postulates reflect the essential traits of the Church's message throughout the Middle Ages. The original sin of Adam and Eve had uprooted the original identification between man and God as manifested in Genesis: "So God created man in his own image, in the image of God created he him" (Gen. 1:27). Although the sacrifice of Christ and His passion had paved the way for salvation, it still required the self-contribution of the faithful, which was implemented through the sacraments. The sacraments constitute an integral part in the path to salvation and could not be ignored after the original sin. The original sin had therefore submitted the whole human race to the mercy of the sacraments, the Church, and the pope at its head.[12] Although the sacraments resulted from sin, as Adam and Eve had no need of them in the very act of Genesis, they also represent the means to overcome sin through which the union of God and man is perpetuated. The complete dependence of humanity on the Church was thus assured, as the path to salvation became conditional on its mediation through the sacraments. This resulted in the individual's feelings of impotence and helplessness on which the Church established its own status as mediator between man and God. The seven sacraments—namely, baptism, confirmation, Eucharist, penance, extreme unction, orders, and matrimony—had already been acknowledged by Petrus Lombard in the twelfth century,[13] were further corroborated by St. Thomas Aquinas, and formally sanctioned at the Councils of Florence (1439) and Trent (1545–63).[14] By means of the seven sacraments, the Church followed people from birth (baptism) until death (extreme unction) through all the stages of life (confirmation, Eucharist, penance, matrimony) and, to those most qualified, it also offered the chances to share in the holy service while becoming members of the clergy (orders). Against the vulnerability of the human race and its powerlessness to face the mysterious powers of the universe, the Church thus provided a perfect ordered world, one which welcomed all faithful into the "Ark of Noah." Against the multifarious and anarchical nature of everyday existence, moreover, the Church offered an uniform perspective that deleted all nuances and upheld the unique truth. In this perfect, ordered Christian society, there is but *one* Lord, *one* baptism, *one* faith, *one* fold, and *one* Shepherd. The Church thus became *Catholic*—the term applied to the faith of the whole Church, that is, the doctrine believed "everywhere, always, and by all," which is also the only orthodox faith, as distinct from heretical and schismatic. Belief in the *Apostolicity* of the Church refers to the clergy's inheritance, by a continuous succession, of the Apostles' ministry whose functions they fulfilled in actual practice—therefore, the continuity of doctrine and the further communion of the faithful with the See of Peter.

The wide consensus regarding the unity of the Church, concretely manifested by the pope at its head, paved the way for the emergence of the papal theocracy. Before the status of the pope had been clearly defined, his supporters proved the consistency between the papacy and the plans of providence as faithfully expressed by the words of Christ to Peter, the first pope: "and I will give unto thee the keys of the kingdom of heaven: and whatsoever thou shalt bind on earth shall be bound in heaven: and whatsoever thou shalt loose on earth shall be loosed in heaven" (Matt. 16:19) and "Feed my sheep" (John 21:15). These words substantiate the plenitude of power conferred by Christ on Peter and through him to all his legitimate successors in the Holy See after him. Christ refrained from limiting the extent of power conferred on Peter but chose to express Himself in absolute terms: "whatsoever," in the Latin version *quodcumque*, thus implying the wholeness of Peter's power, which was not restrained to spiritual things but incorporated all facets of human life. Christ announced the immediate effect of Peter's decisions on heaven, thus bestowing His vicar with absolute power on the world to come as well. He did not specify what sheep Peter had to feed but referred to "*my sheep*," thus implying that the whole flock of God, namely, the whole human race, had been delivered to Peter's guidance. Nevertheless, the connection between Peter and the Roman bishops who claimed his inheritance had still to be proved. The tradition of the "*quo vadis Domine?*" provided a meaningful tool at this regard as it indicated the presence of Peter in Rome and Christ's wish to establish there the See of His vicar. Peter's martyrdom in Rome, moreover, bestowed Christian legitimacy on the pagan city. Since Peter was the first bishop of Rome, all the bishops after him inherited his status without being accredited with the many qualities that brought about his election as prince of the Apostles. The popes remained, therefore, the *indignus heres*, "unworthy heirs," of Peter, who nevertheless received all his prerogatives in actual practice.

The main conclusions to be drawn from the papal plenitude of power were focused on the Church as the perfect model of the desired universal order. The Second Council of Lyons (1274) dealt with the pope's authority over all the churches in the world, a manifestation of the Petrine heritage, underlying the essence of the papal plenitude of power.[15] As the heir of St. Peter, leader of the Apostles, the pope was the source of authority for all churches all over the world, thus giving the Holy See supreme authority in the ecclesiastical order, for "the Roman Church appoints all the prelates of every status and condition."[16] This is a unilateral relationship, creating an almighty papal authority. The people, who were the source of imperial authority, could dismiss the Roman emperor; not so in the case of the pope. All the churches of the world combined could not dismiss the pope, for it was not they who had vested their authority in him but, rather, had received their authority from him. The pope, therefore, was the representative of all the churches in the world and the source of their authority.[17] Petrus Aureoli regards this dictum as exempting the pope from any liability towards the Holy Scriptures and Conciliar decrees,[18] an extreme conclusion that could release the papacy from the limitations laid down in the Decretum and accepted by former Canonists.[19] The principle of papal plenitude of power thus became supreme, relegating the Holy Scriptures, the Christian tradition, and the Conciliar decrees to a marginal role.[20] Even the few attempts to contain the papal plenitude of power did not undermine the main predications of the papal theocracy. Gilles of Rome (1243–1316) assumed that the pope, being the Vicar of God, should imitate God in accordance with necessity and justice, for "wherever there is a holy intention there is the spirit of God and it implies freedom."[21] Yet the limits delineated by "necessity" and "justice" remained too ambiguous to restrain the ubiquity of the papal monarchy whose claims spread beyond the Church. The prevailing belief that the divine goals had to be imitated and achieved on earth extended the monolithic approach to universal dimensions. As a mirror of the celestial, cosmic plan, the papal monarchy was deemed to include the entire human race. In the name of unity, the dividing lines between the spiritual and the material, the priest and the monarch, the Church and the State, already blurred, almost disappeared. St. Peter and his heirs thus gained complete authority over the two swords.

Gilles of Rome stressed the existence of a single source of power, radiating like a mystic body over all authority while brandishing the two swords, the spiritual and the temporal.[22] The images of Melchizedek (Gen. 14:18–20) and Job (Job 29:25) were perceived as indisputable proof of the misconception inherent in a dualistic outlook. Furthermore, Deuteronomy clearly determines the right of priests to pass judgment "between blood and blood, between plea and plea, and between stroke and stroke" (Deut. 17:8–10), thus indicating the authority of the priest over both the spiritual and the temporal spheres.[23] The superiority of the ecclesiastical authority had not been undermined in the New Testament. Although Christian priests still kept their authority over temporality, they had, nevertheless, relinquished its actual implementation. This resulted in the division of labor between priest and king, *rex et sacerdos*, which ensured the achievement of the praiseworthy goals of the Christian state as defined by the Church—namely, the defense of the Church and the faith. If these meritorious objectives were not convincing enough, Gilles of Rome produced a further, decisive argument: the subordination to the pope brings the subordinate closer to true freedom, for such subjugation does not mean enslavement, but freedom.[24] The concept of subordination as a basic stage on the path to freedom had already appeared in the philosophical foundations of early Christianity, and since the pontificate of Pope Gregory I, the papal curia had made great use of the title "*Servus servorum Dei*," the pope regarding himself as the servant of God's servants.[25] These pronouncements, however, related the Christian freedom with the surrender of all the faithful to God, the pope at their head. Gilles of Rome, on the other hand, regarded the Vicar of God as the object of surrender, thus delegating God, in fact, to a secondary plan. This reasoning paved the way for the concluding sentence of the *Unam Sanctam*, which declared the salvation of every human creature to be conditional on its submission to the Roman pontiff.

Despite the extremism inherent in the Bonifacian claim, it ultimately gave apostolic sanction to former arguments repeated over generations. Acknowledging the division of labor between priest and monarch, Pope Gelasius had already sustained the primacy of priesthood, due to its monopoly of the sacraments and the unquestioning "fact" that the clergy had to report before the Almighty the deeds of every human creature at doomsday.[26] The evangelical symbol of two swords (Luke 22:38) was well known to Gottschalk of Aachen (1076), the first Western philosopher known to have used this metaphor to represent the relationship between Church and state.[27] Moreover, the papal version of the two-sword symbol, including the demand for papal authority over the two swords, had appeared about 150 years before Boniface's time, in the writings of Bernard de Clairvaux,[28] Hugh of St. Victor,[29] John of Salisbury,[30] Alanus Anglicus,[31] and Pope Innocent IV.[32] Accordingly, the Franciscan scholar Bonaventure (1217–1274) had demanded complete obedience to the pope as the "universal prince of the world."[33] Roger Bacon (c. 1214–1292) upheld the subordination of the whole human race to the pope, the sole arbitrator between man and God.[34] This opinion was also shared by Arnold of Villanova and James of Viterbo in the fourteenth century.[35] Thence the relationship between the spiritual and the material was interpreted through the precedence of grace over nature, justifying a hierarchical attitude to actual reality based on the pseudo-Dionysian theory.[36] The resulting "ordered hierarchy" was legitimized by the subordination of the imperfect to the perfect, the dependence of the body on the soul, the universal order, and the Aristotelian recognition of spiritual goals.[37]

The political implications of this reasoning were soon developed in full. According to Gilles of Rome, the legitimization accorded by the pope to the state was, in fact, its sole sanction, which further raised Christian States from their former position of a "big band of robbers" to that of Christian Republics.[38] The anointing of princes by holy unction, legitimizing the secular government, actually subordinated the anointed monarch to the Church, as the ceremony had to be performed by a priest acting on the pope's behalf. The oath of allegiance of the emperor to the pope, therefore, indicates his dependence on the Church and also hints at the relations of dependence between priesthood and monarchy as a whole.[39] The words of God to the priest of Anathot: "See, I have this day set thee over the nations and over the kingdoms, to root out, and to pull down, and to destroy, and to throw down, to build and to plant" (Jer. 1:10) were thus used to justify the papal monarchy. Boniface and his supporters had forgotten, in the heat of the argument, the former reservations of Pope Innocent III, which eventually restricted the papal rule to matters of sin.[40]

The preceding reasoning fostered a broader definition of the pope's plenitude of power, while the Petrine heritage was enriched by the Roman concept of *plena potestas*, plenitude of power.[41] The words of Christ to Peter (Matt. 16:18–19) thus acquired a new significance, as they were now enforced by the weight of Roman law. The Roman principle of plenitude of power transferred to the pope, the representative, all the authority of God, the represented, both in the theoretical and the empirical sense. In this way, the will of God was redefined through Roman concepts, which tightened the ties between the pope and God: the former status of the Roman *procurator* actually strengthened the concept of the Christian vicar.[42] Thus, the principle of *Ecclesia vivit sub lege Romana*, the Church lives under the Roman law, included the pope as well, although the pope's immediate contact with God placed him above human laws. The Crusades further fostered this development while bringing the Vicar of Christ closer to the universal monarch. Papal authority was no longer limited to Christendom but spread to all mankind in equal measure.[43] As the initiator of the Crusades, the pope also drew from the Roman heritage an element of immunity that was vital to the consolidation of the papal monarchy. This process had been heralded by the *Papal Dictate* of Pope Gregory VII as early as 1075[44] and found its final crystallization in the message of the *Unam Sanctam*, for the pope "shall judge the world" though he himself is judged by God alone.

True, the continuity and perseverance of the Church message did not assure its universal reception, and one can point at the many contenders to the very orthodoxy of these claims.[45] Moreover, the process of state-building characteristic of the Central Middle Ages was undermining the monolithic approach of the Church and paved the way for a dualistic approach, which, while distinguishing more clearly between the religious and the political spheres, fostered the secularization process of Western society. Yet, the confrontation between Church and State was not an easy one, as laicization, in its modern sense, was completely anachronistic to the medieval climate of opinion. Beyond the new allegiance to political frameworks, medieval society as a whole remained a religious one, which, as such, looked to the Church for answers to all troublesome facets of life in this world and the world to come. Moreover, the logic and structure of the papal message facilitated its reception, further fostered by the widely accepted monolithic approach and its foundation on biblical sources. The monolithic approach allowed medieval people to overcome the mysterious nature of the universe whose meaning and course came always back to its unique source: God. The multifarious essence of nature and of the sociopolitical world was thus oversimplified, paving the way for a unique truth whose acknowledged communicator was the Church with the pope at its head. The *Unam Sanctam* left no room for doubts at this regard, and authoritatively laid down the monolithic approach through a constant repetition of words and ideas that were supported by a wide spectrum of images: there was *one* dove, *one* ark of Noah, completed in *one* cubit, whose *one* and only captain was Noah. Similarly, there is but *one* faith, *one* shepherd, *one* baptism. Furthermore, Genesis tells that "In the beginning. . . . ," thus suggesting the monolithic essence of the whole creation. The monolithic approach thus eliminates any possibility of nuances or social tolerance, but leaves room for a polarity that differentiates but between good and bad, true and false.

The impact of the monolithic approach was further strengthened by its being based on the Holy Scriptures, thus creating a complete identification between the *Vox Dei* and the ecclesiastical message on which it bestowed divine authority. The religious education of medieval men assured a priori the reception of biblical ideas and symbols, which resulted in a basic attentive attitude to the Church's message. Expressed in unequivocal and explicit terms, the ecclesiastical message presented both the rewards and the punishment that might be expected by the supporters or opponents of the Church, in this life and in the world to come as well. The identification of the Church with the ark of Noah suggests the weight inherent in the ecclesiastical message: as in the time of the flood, all life was preserved in the ark, and outside remained nothing except the kingdom of darkness and death, those who deliberately left the Church relinquished, therefore, their chance of salvation and fell into eternal death. Few communicators might express themselves in a similar, absolute way, with such complete confidence in the acceptability of their message. Furthermore, the gradual argumentation characteristic of the *Unam Sanctam* encouraged wide identification with its most radical postulates. The papal message opened with neutral statements about the unity inherent in the cosmos and the Church, the existence of two swords, the recognition of the division of labor between priest and monarch, and the preference for one single authority, issues that might enjoy a wide consensus in medieval society. Throughout his argumentation, the pope took care to emphasize the complete orthodoxy of his statements at every stage, as they reflected the Will of God from the time of Genesis. Only after a long and, to some degree, convincing predication did Boniface allow himself to reveal the main dictum of the papal monarchy, namely, the prevalence of papal authority and its divine origins that enforced the absolute subordination of the whole human race to the Roman pontiff. True, the pope's argumentation applied to a very narrow intellectual elite involved in the political conflict that the pope was sustaining at the time against the king of France. Nevertheless, the use of well-known symbols, such as the ark of Noah, Christ's garment, and the two swords, facilitated the reception of the papal message in a wide range of social strata. The Church's message thus spread beyond the narrow limits of the *culture savante* and also reached the *culture populaire*, through different levels of perception. In order to assure the reception of its message in the long term, however, the Church elaborated a broad spectrum of communication channels, which form the subject matter of the next chapter.

3

Ecclesiastical Communication Channels

The importance attributed by the Church to communication and the use of propaganda do not appear as traits characteristic to the Central Middle Ages, but rather, were imprinted at the very beginning of Christianity. Christ had instructed the Apostles to evangelize the gentiles (Matt. 28:19; Luke 24:47; Acts 10:34–48) while the missionary effort left its mark on the Church's policy throughout the Middle Ages. As early as in the seventh and eighth centuries, Catholic missionaries reached countries as far away as Ceylon, Malabar, and China. The Church's awareness of the crucial role of communication was epitomized by its use of *excommunication*, the exclusion of the sinners from the communion of the faithful, as the severest punishment for those who deliberately divorced themselves from the "Ark of Noah." Moreover, Christ's order to Peter, "Feed my sheep" (John 21:15), embraced the whole human race and bestowed universal dimensions on the Church's message. From the perspective of Church history, therefore, the uniqueness of the period between the eleventh and fourteenth centuries results from the *institutionalization of communication*, which grew out of the Reformer papacy's attempts to monopolize the communication channels to Christendom. The divergent needs and receptivity of clergy and laity on the one hand, and the geographical scattering of the Christian audience on the other, encouraged the development of a sophisticated communication system which is the subject matter of this chapter.

The *papal legates* provided one of the main means at the disposal of the medieval popes, evaluated in research as "the prolongued arms of the papacy." They were permanent delegates or messengers ad hoc, who represented the papal interests before secular and religious authorities or in well-defined geographical areas. Their appointment was performed in Consistory, the assembly of Cardinals summoned by the pope, and held in his presence. The nomination letter established the authority conferred on the legate, his destination and, when needed, his relationship with other legates. It did not define any time schedule, however, as the mission was conditioned by the performance of a specific task, and the pope's death did not justify its interruption. The increasing importance of legates in the Central Middle Ages brought about a clearer definition of the institution and the authority bestowed on it. Cardinals sent as legates were known as *legati a latere*; they were personal ambassadors of the pope, who usually had plenipotentiary jurisdictional powers and sometimes overrode episcopal and archiepiscopal jurisdictions, except in the major cases reserved for the pope alone. A papal legate not drawn from the ranks of the Cardinals was called *legatus missus* and had clearly defined powers, while papal nuncios were dispatched on special missions. Below these legates ranked the *legati nati,* who were of archiepiscopal status such as the archbishops of Canterbury, Reims, and Salzburg whose jurisdictional functions were circumscribed. They received permanent legatine status through a special papal privilege and enjoyed some liturgical prerogatives. Their main function was to act as a court of appeal.

Medieval Canonists acknowledged the ancient origins of the legatine institution, which they related to the Roman *legati*, messengers used for sending important dispatches or information.[1] Although there is no convincing proof of the uninterrupted sequence of the institution, it fits into the Church's tendency to adapt former Roman patterns to its needs. The different roles ascribed to legates, indeed, mirror the development of the Church and the papacy whose policy they propagandized while becoming an important source of information for the papal court. Between the fourth and eighth centuries legates were sent to the Eastern Empire as personal delegates of the pope, whom they represented in the Ecumenical Councils. In addition, they strengthened the Christian mission among the pagan tribes in Central and Northeastern Europe. St. Boniface (680–754), the "Apostle of Germany," for instance, laid the foundations of the Church and increased the apostolic influence both in Germany and France. The gradual conversion of the German peoples and the institutionalization of the papacy gradually transferred the focus of legatine activity into the field of diplomacy. Pope Innocent II (1130–1143) made wide use of legates to improve his relations with Christian rulers. Pope Alexander III (1159–1181) sent legates to defend his legitimacy against the antipopes supported by Emperor Frederick Barbarossa and to settle the Becket controversy in England. Innocent III (1198–1216) sent legates to ensure the inheritance of the young Emperor Frederick II, to settle the marriage arrangements of King Philip II of France, and to defend the papal position in his lasting conflict with King John of England. Legates also played a cardinal role in the endless Investiture Controversy and in the no less unconditional papal struggle against heresy, while they personified in actual practice the most faithful partisans of the papal monarchy. Another facet of legatine activity focused in enlisting the ecclesiastical hierarchy and ensuring its attendance to Ecumenical Councils. The legates informed the prelates of the main issues on the agenda and tried to secure universal participation. During the pontificate of Callistus II (1119–1124), the Cardinals Gregory and Pierleone were sent to France as *legati a latere* to ensure maximal attendance at the First Council of Lateran. Pope Innocent III, also, systematically sent legates throughout Christendom between 1213 and 1215 before the Fourth Council of Lateran. Such missions enabled the popes to receive accurate reports on the actual situation of churches and their more acute problems, thus becoming an important source of information at the disposal of the papal court prior to the council.

The legates' important contribution in the field of Church communication was further underlined by the lack of permanent embassies. Communication channels thus had to be both reliable and independent, as the considerable distances from the papal curia significantly reduced the chances of constant communication between the pope and his representatives. Yet, beside the importance ascribed to the legates in the field of diplomacy and the transmission of information, the question still remains regarding their ability to foster the reception of the papal message among different social strata. In other words, the degree of the legates' success or failure at communication had to be tested not only according to the expectations of the papacy but also according to the attitudes and reactions of contemporaries in actual practice. This need results from the weight ascribed in the Middle Ages to the personal contact between the communicator and his audience, one which turned the messenger's popularity into a crucial factor in how his message was received. The immediate contact between the communicator and his audience thus encouraged identification between the sender, his messenger, and the messages. The reception of the pope's message was therefore crucially influenced by the approaches to his legates, since most people were hardly capable of differentiating between the personality of the legates and the official function they were fulfilling on the pope's behalf. The resulting identification between the pope, his message, and his messengers actually harmed the ability of the legates to foster the wide acceptability of the papal message throughout Christendom. If a generalization has to be made, medieval sources as a whole suggest that the mere appointment of a legate aroused much hostility, foremost among the ecclesiastical order. Prelates and monks alike dreaded the possible harm the legates could bring to their authority and income, as the papal representatives stood above the whole local hierarchy which had to provide their expenses.[2]

This state of affairs was faithfully recorded by the synodal legislation that voiced the many complaints of the clergy in this regard. The Synod of Canterbury (1310), for instance, opened an investigation against William Testa, the papal legate in England, which was included in a long list of complaints presented to Pope Clement V at the Council of Vienne.[3] In their reservations against the papal legate, the prelates attributed to him the underlying cause of the many economic troubles affecting the Church of England at the time.[4] Contemporary chroniclers as well, put a great emphasis on the legates' greed, describing them as the "most clever money's dilapidators," forever committed to the exploitation of the Church of England, in open contradiction to the customs of the land.[5] Such criticism was not limited to English commentators, but spread on the Continent as well. An Italian chronicler blamed the legates for the decline of the whole Church and compared their behavior to that of dogs devouring their prey.[6] These reactions indicate the resentment of priests and monks who were expected to assist the legates in the implementation of their mission. Faced with the alternatives of assisting the legates at their own expense or fostering their own interests, most of the clergy opted for the second one. A negative image of the papal legates was therefore elaborated, one which made them more vulnerable to the criticism of the laity as well.

The prevailing resentment against the legates acquired greater significance among the laity whose attitudes suggest their degree of identification with the Church's message and, accordingly, their degree of support on the papal policy in actual practice. In Italy, particularly in the Papal State, conflicting interests between the laity and the legates were more acute and fostered utilitarian attitudes among the local population, which easily deteriorated into political opportunism. In Ferrara, for instance, the intervention of the papal legate, Arnaud de Pelagrua, had freed the city from the threat of the Venetian conquest in 1308. Yet a short time later, the legate himself and his partisans were described as "rapacious wolves," because of their tendency to strengthen the papal rule in the city.[7] The deteriorating relations between Arnaud and the local population finally led to the open revolt of 1317 and the complete refusal of the citizens to acknowledge the papal rule any longer.[8] Similar attitudes could be detected in Bologna and Modena, where the legates led a compulsory centralizing policy, which gave rise to reservations and contempt. In 1306, the papal legate Napoleon Orsini, threatened by the outcries of "death, death to the traitor," had to flee from Bologna.[9] Giovanni Orsini too, *legatus a latere* in Tuscany, was compelled to escape from Siena in 1326, followed by groups of armed men chasing him.[10]

The previously mentioned examples indicate the conflict of interests between the party spirit of the Italian city-states and the compulsory policy characteristic of fourteenth-century popes and their legates in Italy. The embryonic national awareness prevailing in Western Europe did not make the legates' labor easier there. In England, for instance, the papal claim for suzerainty sharpened the contradiction of interests between the legates and the whole population, which was voiced by Matthew Paris, one of the most influential chroniclers of the thirteenth century.[11] The prevailing criticism, moreover, did not remain in the field of literary protest but also left its mark in the political field, thus further harming the reception of the papal message. In 1312, Arnold, cardinal-priest of St. Sabina, and Arnold, bishop of Poitiers, were sent to England as *legati a latere* to intercede between King Edward II and his barons and to rescue the kingdom from the dangers of civil war. The legates settled in St. Albans and from there sent letters to the rebellious barons:

> Even though the barons had heard of letters delivered by foreigners, they received peacefully the messengers of the papal legates. Yet they refused to accept the letters, with the excuse that they were not learned men, only such that had been taught arms and fighting. They therefore did not even trouble to read the letters. The barons were then asked if they were willing to talk

> with the legates of the Holy See who had come to England to make peace. To this the barons replied that there were many bishops in the realm, both learned and honest, whose advice they were willing to accept, though not that of foreigners who knew nothing of the causes of their struggle.[12]

Without inquiring into the accuracy of the barons' claims, they reflect, nevertheless, the prevailing illiteracy among fourteenth-century aristocracy, which found it unnecessary to justify itself or to explain its xenophobic attitude. The defiance to the legates' intervention, furthermore, was not the result of occasional incidents, but rather of the basically differing interests that placed the pope and the barons on opposing sides of the political arena. As the political crisis in England worsened, the opposition to the legates reached its peak. In 1317, Pope John XXII appointed the Cardinals Gaucelin d'Eusa and Luca Fieschi as *legati a latere* in order to help Edward II to strengthen his shaky internal position.[13] Contemporary sources reported the arrival of the legates to London "in great splendor" and condemned their impudence in claiming the expenses of their mission, which resulted in complete failure.[14] The high annual income of the legates was also emphasized as it was obtained, according to the chroniclers, "not without much greed and uncontrolled ambition."[15] On their way to Durham for the consecration of Bishop Lewis of Beaumont, despite their ecclesiastical immunity, the papal legates were attacked and their possessions stolen.[16] Continuous conciliar legislation on the matter further indicates that incidents of this kind appeared to be more the rule than the exception, as the papal court actually failed to safeguard its representatives.[17]

The stubborn defiance the legates faced, both in theory and in practice, sometimes indicates the difficulties the Church encountered when large social strata began to express feelings of national awareness then often associated with xenophobia. Both in England and in the Italian city-states, the papal legates were regarded as intruders whose interests were not compatible with those of the whole population. Yet opposition to the legates was not limited to one particular period, but rather expressed the intolerance and xenophobia inherent in medieval society as a whole. In most extreme situations it brought about the martyrdom or the murder of the legate; this semantic differentiation suggests his death at the hands of the Infidel, thus achieving martyrdom—as was the case of Boniface in 754—or, in less heroic circumstances, as there was the case of Peter of Castelnau, probably murdered by heretics in southern France in 1208. The legates' arbitrariness, the economic burden they represented, and their attempts to change ancestral ways of life all encouraged a latent but still smouldering opposition against them which, at time of crises, easily flared up into open hostility.

The many difficulties that plagued the work of the legates indicate an essential weakness of medieval communication hardly unique to the Church. The personal essence of communication and the immediate contact with wide social strata exposed communicators to the criticism of their contemporaries, thus reducing the effect of their mission and the reception of their message.

On the other hand, in the particular case of the legates, this same wide criticism indicates their success in attracting attention, the most basic stage in the communication process, while they succeeded in placing themselves and their mission in the center of the political arena. This "promotional" success was due to the almost complete lack of alternative communication channels, which in most cases turned the arrival of legates into an extraordinary event that easily caught the attention of their contemporaries. The magnificent appearance of the legates' entourage, moreover, though arousing the anger of monastic chroniclers, transmitted to medieval men the powerful status of the papacy and strengthened the links with an otherwise distant pope.

The reluctance of large sectors of the clergy to collaborate with the legates encouraged the papal tendency to integrate the prelates into the Church communication system, in the framework of *Ecumenical Councils*. These were major legislative assemblies summoned by the pope, whose sanction ultimately bestowed juridical force and infallibility on the decisions. The agenda usually focused on issues of general interest, namely, the authoritative definition of the Articles of the Faith, the reform of the Church, the maintenance of peace in Christendom, and all matters the pope thought necessary. Of the twenty-one Ecumenical Councils recognized by the Church to date, seven had taken place between the eleventh and fourteenth centuries, a relatively high frequency that reflects the papal tendency to turn these assemblies into an integral part of their communication system at the time. Table 3.1 presents the place/name of the council, its date, the main subject-matter discussed and the estimated number of participants.

Callistus II was the first pope who summoned an Ecumenical Council in the West (1123) and established the patterns of participation, which included cardinals, archbishops, bishops, abbots, representatives of chapters, and general masters. Well-known theologians were also invited as counselors, and the presence of princes or their delegates was most welcomed. As pointed out by Pope Alexander III, an Ecumenical Council aimed to gather all those who

TABLE 3.1. Ecumenical Councils Between the Eleventh and Fourteenth Centuries

Place	*and Name*	*Date*	*Subject*	*Number of Participants*
1. Lateran	I	(1123)	Investiture Controversy	300–500
2. Lateran	II	(1139)	Arnold of Brescia	500–1,000
3. Lateran	III	(1179)	Papal Elections	300
4. Lateran	IV	(1215)	Albigenses	1,200
5. Lyons	I	(1245)	Frederick II	150
6. Lyons	II	(1274)	Reunion	several hundred
7. Vienne		(1311–1312)	Templars	150–200

might be affected by its legislation, a goal that turned the council into the most suitable audience to the papal message. In regard to the ecclesiastical order at least, it seems that a large attendance of prelates was indeed achieved. The records of the Third Council of Lateran indicate the participation of prelates from distant countries such as Spain, France, England, Ireland, Scotland, Germany, Denmark, Hungary, and the Holy Land, besides the customary predominance of Italian prelates.[18] Starting with the First Council of Lateran, most of the issues in the agenda were discussed in the framework of committees, an arrangement that reduced the plenary meetings to between three and six, and shortened the period of deliberations to about two weeks; I Lateran Council lasted from March 18 until April 6, 1123; II Lateran during the month of April, 1139; III Lateran between March 5 and 19 or 22, 1179; IV Lateran between November 11 and 30, 1215; I Lyons from June 26 until July 17, 1245; II Lyons from May 7 until July 17, 1274; while the Council of Vienne lasted from October 16, 1311 until May 6 of the following year. The tendency to shorten the deliberations aimed at preventing the many problems that might result from the massive absence of prelates. Pope Innocent III tried to reduce risks by laying down that two bishops should remain in each province, though they, too, had to be represented by suitable procurators. This procedure proved to be most effective, as it ensured the universal attendance of prelates while preventing a complete hiatus at the local level.

Ecumenical Councils thus provided the popes with an immediate link with the ecclesiastical hierarchy and a suitable opportunity to receive accurate information about the current state of churches throughout Christendom. Prior to the actual assembly, prelates were encouraged to record their own reports on all areas under their supervision, an arrangement that provided the pope with additional information. Toward the Second Council of Lyons, the *Collectio de Scandalis* was elaborated, containing a detailed list of the most outstanding irregularities affecting the Church at the time. Bishop Olmutz wrote a critical survey on contemporary priesthood while Humbert de Romans recorded the *Opus Tripartitum* with a faithful analysis of the Crusades and the factors which, in his opinion, had brought about the deteriorating situation in the Holy Land. The importance ascribed to these reports was reflected by Clement V's decision to institutionalize such proceedings toward the Council of Vienne; all prelates were thus requested to draw up suitable reports, which would be answered by the pope at the forthcoming assembly.

In addition to the information they provided to the papal court, the Ecumenical Councils also became a most suitable meeting place for prelates from all areas of Christendom: here they could voice their complaints or expectations and exchange ideas with other members of the clergy and with the secular rulers or their representatives. Yet the participation or representation of secular princes, though always encouraged and most welcomed by the Church, became significant only from the thirteenth century onward. At the Second Council of Lyons, for example, the Eastern emperor, and the kings of England, France, Sicily, and the Tartars were suitably represented. The increasing number of participants involved a larger number of assistants and retinues, whose size and status were in strict accordance with the political and economic position of the institution or personage they served. The Third Council of Lateran tried to restrain the size of clerical retinues (canon IV), a decision upheld by the Fourth Council of Lateran.[19] Continuous legislation on the matter, however, suggests the complete inefficacy of former edicts. Tens and sometimes hundreds of servants, secretaries, administrators, counselors, notaries, men-at-arms, and knights continued to accompany the prelates and the laity at the meeting place. Moreover, all those looking for a suitable opportunity to profit, such as wandering knights, minstrels, money-changers, merchants, and craftsmen, joined the prelates, thus turning the council into a first-rate social event, where it was possible to exchange the latest information, listen to rumors, or propagate new ones.[20]

During the council deliberations, the pope and his policy succeeded in focusing the interest of contemporaries as most prominent international events and festivals in our own days. From this perspective, Ecumenical Councils provided the papacy a suitable opportunity to emphasize its central role in the whole of Christendom. The universal influence of councils was further achieved by the spread of their legislation beyond the ranks of those who actually participated. Absent prelates and all those interested in learning of the council decisions could avail themselves of accurate reports, as each participant was required to make several copies of the acts and distribute them in a particular area.[21] Written reports became quite customary during the thirteenth century also in the diocesan and provincial frameworks, as the relative improvement of literacy among the clergy allowed the use of the written word. The increasing use of written reports heralds an important stage in the history of medieval communication. The more sophisticated needs of the ecclesiastical order could not be satisfactorily met by oral delivery alone, thus fostering changes in the Church communication system, which lost its former intimacy. The immediate contact between the sender and its audience was therefore gradually replaced by fostering the use of media, in the modern sense, between the papal court and the clergy.

With less participation than the Ecumenical Councils, but, from a communication perspective of more importance due to their greater frequency, were the *diocesan, provincial, or regional synods*, in which the clergy of a diocese, a province, or a whole region participated. The gathering of regional assemblies was a well-known practice from the Early Middle Ages, but it acquired greater significance following the efforts of the Reformer Papacy to strengthen links with Christendom as a whole, and especially with the ecclesiastical hierarchy. Synods were convoked by the local bishop or archbishop who presided over the meetings, unless a papal legate was present. The participation of a legate allowed the attendance of prelates from different dioceses or provinces, thus bringing about a regional synod. Throughout the twelfth and thirteenth centuries, the periodical gathering of regional synods was encouraged, providing the papacy and its legates with an additional communication channel in the local sphere. Regional synods, however, were not convened regularly but remained at the disposal of the legates, who promoted them as the need arose. Quite different was the status of diocesan and provincial synods, whose annual gatherings have been sanctioned by the Fourth Council of Lateran.[22] In England, for instance, diocesan synods usually took place in the autumn, but in certain dioceses there was an additional assembly in the spring. The bishop who presided over the meetings sometimes recorded the deliberations and circulated the records among the clergy of his diocese, as done by Eudes de Sully (Paris, 1200). The legislation of the Fourth Council of Lateran made the regular publication of synodial acts obligatory, a duty which brought about the collection of Richard le Poore (Salisbury, 1221–1227). As the practice developed, these collections became an important source of information for the local clergy; they provided accurate definitions of the Articles of the Faith, the cult, and the sacraments, in addition to clear instructions for the suitable composition of wills and performance of burials, excommunications, interdicts, and the like. Such collections provided the clergy with most accurate "reference manuals," as they covered a wide spectrum of issues that the average priest had to face in daily practice. Synods, whether diocesan, provincial, or regional, thus became an important forum for consolidating the group solidarity of the clergy in a particular area while providing a regular framework for deliberative assemblies. They also contributed to a better indoctrination of priests, whose minimal knowledge and sometimes almost complete illiteracy was clearly indicated by the agenda. Yet all advantages on the social and educational levels were conditioned by the extent of participation. Contemporary sources suggest that the attendance of the clergy was rather low, either because of the travel expenses or the dangers inherent in such journeys, or because of the impediments of age and physical conditions claimed by those summoned. From the thirteenth century onward, obstacles of this kind gave rise to the nomination of procurators, who participated in synods on behalf of one or several priests.[23]

Although the clergy constituted the main target, and its participation was accordingly mostly encouraged, attendance at synods was not limited to the priesthood but included a varying number of laymen who enlarged the audience of the Church's message. The *Peace Movement*, for instance, emerged in the framework of ecclesiastical synods and reflects the impressive success of this communication channel among different social strata.[24] As defined by the Synod of Charroux (989), the Peace Movement aimed to limit warfare in Christian society, namely, to restrain the more anarchic social elements while enlisting them into the Church's service, and protect those most exposed to the violence of the times, such as women, elderly people, children, traveling merchants, and priests.[25] The Peace Movement embodied two major and, to some extent, opposing goals: it encouraged the belligerent-feudal aristocracy to become a "defender of the Church" and, as such, to collaborate with the Church policy; and at the same time, it supported the mobilization of the peasantry in favor of the "defenseless poor" who, in many cases, were the victims of the aforesaid aristocracy. The Peace Movement thus appears as a challenge to the feudal system and to the static social structure inherent therein as it called upon the masses of the poor to take part in the peace process through integrating in the Peace Army. From a historical perspective, it represents one characteristic trend of Church communication in the Middle Ages: God was and remained the only source of legitimization for the sociopolitical process—in this particular case, the Peace Movement, which was accordingly promoted by His representatives. Still, the implementation of God's will was conditional upon the massive participation of contemporaries, achieved through the framework of regional synods. Medieval sources testify to the wide interest aroused by the Church's program and emphasize the "great concourse of people of both sexes from Poitiers, Limoges and the environs" who came to Charroux and integrated into the Peace Army.[26] The impressive involvement of wide sectors of medieval society in the Peace Movement presents another distinctive facet of the communication process at the time. Against the mostly passive or theoretical identification pursued by modern propaganda, the degree of reception of the Church's message was immediately translated into actual practice, thus narrowing the gap between the propaganda's success and the historical process. From the thirteenth century onward, however, further insistence on matters of ecclesiastical discipline gradually alienated the laity from synodial meetings and encouraged the search for additional communication channels.

All the communication channels discussed hitherto, namely, legates, councils, and synods, were used ad hoc, according to changing circumstances. Yet their effectiveness and influence in the long term were conditioned by the existence of well-established institutions at the local level, which guaranteed the reception of the Church's message and its diffusion throughout Christendom. In other words, the reception of the Church's message could not be based on circumstantial delegations or periodical meetings but depended rather on the existence of a regular administration. As a whole, the Church's organization into provinces, dioceses, and parishes reproduced former Roman patterns. This fact was acknowledged by Canon Law, which ascribed the foundation of the ecclesiastical provinces to the Apostles and to the formal sanction of Pope Clement I (d. c. AD 96). The establishment of dioceses was related to St. Peter himself or, at the latest, to Pope Dionysius (d. AD 268).[27] These claims, which were a matter of controversy in research, reflect, however, the long institutionalization process of the Church administration system, which prevented the emergence of clear geographical patterns. In contrast to the high density of ecclesiastical institutions in the Mediterranean area stood the distant and depopulated ecclesiastical provinces in Northern and Eastern Europe. While the archbishops of Brindisi or Trani, for example, ruled over provinces of a manorial size, the archbishops of Mayence or Gnesen were bestowed considerably large territories. By the middle of the thirteenth century, there were twenty-nine ecclesiastical provinces in Italy and twenty-six in Gaul, but only seven in Germany, two in Scotland, four in Spain, four in Ireland, two in England, two in Hungary, and only one in each of the Scandinavian countries. This distribution, however, might be modified, as the popes could rearrange the Church's organization following the evangelization process, the conquest of new territories, or changes in demographic growth.

The *episcopal* see or *diocese* constituted the primary cell of the Church organization and provided a crucial factor in ecclesiastical communication. Located in the very heart of medieval cities, cathedral buildings symbolized the vital energy of the nascent urban centers whose development the Church had significantly fostered. The spread of cathedrals all over Europe reflects this close interaction between the Church and the socioeconomic process, which had brought the French historian Duby to write on *Le Temps des Cathédrales* (Paris: Gallimard, 1976). Cathedrals transmitted through nonverbal means the message of the Church's power and became the material expression of the townsmen's pride (see plates 6,7,8). Royal propagandists were well aware of the attraction of cathedrals buildings on medieval society when they turned major cathedrals into a national symbol, which amalgamated the community of the realm under the wings of Christ, St. Mary, or a saint patron. Notre Dame of Paris, for instance, became the acknowledged emblem of the Kingdom of France and its capital, a situation which became characteristic of most European countries up to the present day.

The primordial authority in the diocese was conferred to the *bishop*. He was responsible for the education of the laity and the proper instruction of the clergy, and exercised the highest legal powers in civil as well as clerical feuds. In thirteenth-century England, for instance, an average diocese employed about 2,000 to 3,000 clergymen. The considerable number of clerics and the relatively primitive channels of communication raise questions as to the capability of medieval bishops serving as a communication channel with regard to the faraway papacy on the one hand, and to the diocesan clergy on the other. In addition to maintaining an unbroken flow of correspondence with the papal court and the missions of legates to the area, bishops were also required to come to Rome, *ad limina apostolorum*, to the thresholds of the Apostles at regular intervals. The custom to venerate the tombs of the Apostles and to report on the state of their dioceses to the pope originated in a decree of the Roman synod of 743, which enjoined such visits on all bishops ordained at Rome. Gregory VII extended this obligation to all archbishops and, from the thirteenth century, it was imposed on all bishops ordained by the pope himself or by his special representative. Communication between the pope and bishops or archbishops thus included a wide range of channels, which embodied both written messages and personal contacts.

A key factor in the flow of communication between the bishop and the diocesan clergy was provided by the annual visitations performed by bishops in all areas under their supervision.[28] The Household Roll of Bishop Richard Swinfield exemplifies such a visitation, carried out between April 9 and May 31, 1290 in the diocese of Hereford (see map 3). Although such visitations were fairly frequent in the Central Middle Ages, they lacked the regularity stipulated by Canon Law. The great distances to be covered, the bad condition of the roads, and the constant physical effort required from prelates who in most cases were of advanced age, brought about the widespread custom of consecrating a large number of churches and altars during one visitation, thus confirming the irregularity of this practice and the intervals between one visitation and the

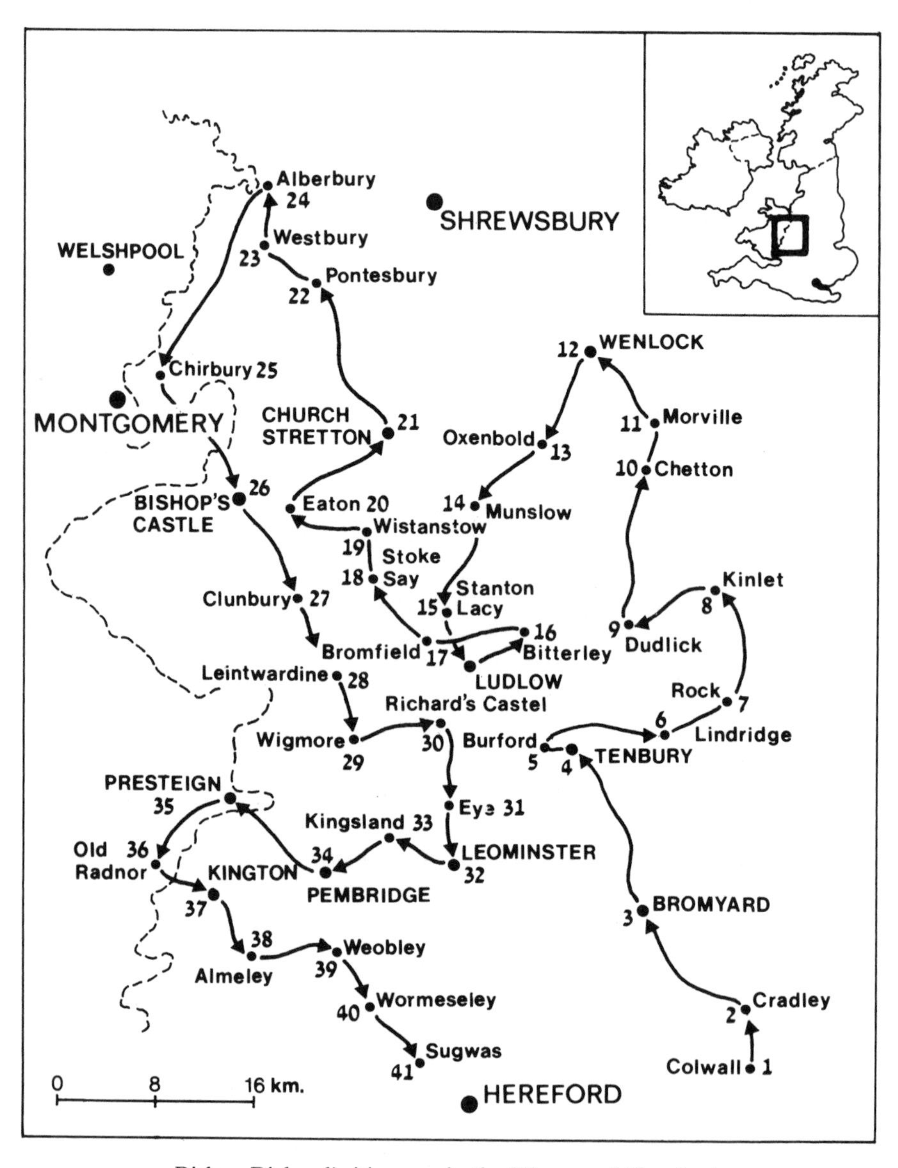

Bishop Richard's itinerary in the Diocese of Hereford.

other. Medieval sources testify to the importance ascribed to these visitations and their contribution in improving the fluent communication between the prelates and their flock. In a society still much influenced by the immediate contact between the communicator and his audience, the bishop's visitations provided the personal touch, greatly appreciated by medieval man. Besides, the sumptuous appearance of the bishop and his entourage confirmed the

PLATE 6. West façade, Angoulême Cathedral, before 1130. Extensive iconographic sculptural programs. The apocalyptic vision of the Second Coming is illustrated in reliefs placed all over the façade.

same nonverbal message of the cathedral building: the very presence of the bishop transmitted anew the powerful, eternal, and mythical status of the Church it represented. In contrast to the reluctance showed toward the papal legates, moreover, medieval sources reflect a more positive attitude toward the visitations of bishops; they were not regarded as intruders or foreign factors but, rather, as salutary means to prevent mischief. Matthew Paris once described Grosseteste as "the Bishop of Lincoln to whom quietness and repose are unknown."[29] This judgment was confirmed by other sources that reported in detail Grosseteste's visitations of "the monasteries, archdea-

PLATE 7. West façade, Notre Dame-la-Grande, Poitiers, second quarter of the twelfth century. The spandrels of the ground story have scenes based on a mystery play, including an early representation of the Tree of Jesse, arranged almost like a frieze.

conries, and deaneries of his diocese. In each he held a general chapter, preached a sermon and issued decrees. He also suspended many rectors of churches."[30] The visitation thus provided a suitable occasion for a more effective supervision of the diocesan clergy. As to the bishops themselves, they were also subjected to the visitation of their archbishops, which provided an additional means of communication and control. The letter of John Peckham, archbishop of Canterbury (1279–1292) to Roger Longespée, bishop of Coventry, faithfully illustrates the need for such a practice:

> Passing lately through your diocese we saw many things which we thought we had corrected during our visitation—incest, simony, misappropriation of

> churches, and children who had not yet been confirmed. These things need your attention, but you have been absent so long that you seem not to care. We therefore order you, on receipt of this letter, to take up residence in your diocese, so that—even if you are not competent to redress spiritual evils, you may at least minister to the temporal needs of the poor. If you cannot conduct confirmations yourself, you must provide some other bishop who knows the language, to go round the diocese and do what is necessary. Let us hear from you by the Feast of St. Thomas the Apostle that you have done this.[31]

Although the effectiveness of such visitations depended in fact upon the personal initiative and the commitment of prelates to their office, they still contributed a regular channel of communication at the local level, which was further complemented by the regular meeting of synods.

Subordinate to the bishop and in closer communication with the faithful stood the *priest* whose area was the *parish*. Canon Law defines the parish as a territory whose inhabitants were related by the bishop's order to a specific church and were committed to the spiritual leadership of the priest, permanently appointed to that church.[32] The parish thus represented the "society of the faithful" who actively participated in the religious ceremonies performed on Sundays and Christian holidays, as well as in the payment of tithes. According to Canon Law, poverty or inability to work entitled the parishioners to receive from one quarter to one third of the tithe income that was reserved for welfare.[33] Like the cathedral buildings in the urban centers, the church's belfry and its nearby graveyard became the symbol of homeland in the countryside. In large country areas such as those of medieval England, there were some 9,500 parishes with an average population of 300 people each. This impressive number of parishes, however, did not always fulfill all needs. In thirteenth-century Leicestershire, for instance, one priest alone was entrusted with the care of about 200 rural communities. The territory of an average rural parish is estimated at some 4,000 acres which, in addition to the village, included several scattered hamlets or berewicks, isolated farmsteads. Priests were involved in all aspects of daily life including the maintenance of law and order, sanitary services, welfare, and the collection of the tithe. They often represented the highest authority in the area and, in most cases, the only person who could read and write. In addition to the holy orders that had raised him above the community of the faithful, the priest's literacy emphasized the gap between him and his parishioners, as he usually represented the only person capable of developing written communication, thus extending the narrow limits of the parish to larger areas. Despite the fact that in urban parishes the cultural gap between the priest and his parishioners was considerably reduced, the prosperity of towns and the demographic growth encouraged the emergence of urban parishes, which grew further due to the development of some "quarter patriotism." The following figures indicate the significant growth of parishes in the most important towns: until the eleventh century, parishes were completely unknown in Cologne; in Frankfurt and in Montpellier respectively, there was but one, while in larger populated cities

PLATE 8. West facade of Reims Cathedral. Started in 1211, went on for about fifty years. Elaborate theological programs in stone, involving a large number of sculptures; it led gradually to the division of work, to specialization and not infrequently to something approaching mass production.

such as Ypres and Worms there were only four. Between the years 1080 and 1290, eighty new parishes were established in Paris, seventeen in Sens, and nineteen in Cologne, a number which did not change until 1803.[34] Twenty-six parishes were recorded in Metz by 1325 out of a population of 25,000 townsmen, and the same number was registered in Padua by 1378. By the thirteenth century there were about 400 or 500 urban parishes in England, a considerable number in light of the rather small size of towns in the kingdom. In London alone there were about 100 parishes, fifty in Norwich and about forty in Lincoln and York. In Exeter, Winchester, Bristol, Stanford, Ipswich, and Cambridge, there were about twenty. The average population of urban parishes was about 200 people.

The canonical duty of priests to establish themselves permanently in the parish encouraged the intimate and continuous communication between the priest and his flock, further fostered by the Church legislation, which forbade parishioners of moving from one priest to another. Only wandering scholars, sailors, travelers, and soldiers before battle were relieved of their basic commitment to their home parish and were allowed to confess to the nearest priest.[35] At the church, at home, or in the rather narrow confines of medieval villages and urban quarters, the priest and his flock were thus closely linked for life. This state of affairs was given canonical sanction by the constitution *Utriusque sexus* from the Fourth Council of Lateran, which established the duty of confession at least once a year.[36] The importance the Church communicators attributed to confession as an essential means on the path to salvation, was conveyed to the faithful by a wide spectrum of *exempla*, which emphasized the identification between confession and the will of God. The fourteenth-century Dominican, Jean Gobi, reported the story of two friars who met a lonely peasant in a desolate area of Ireland. He revealed to them that for the past thirty years he had served the devil, wearing his ring and doing whatever he had commanded. After explaining the severity of the sin and its worthlessness, the Dominicans convinced the peasant to confess, as confession both "cancels the power of Devil and moderates the divine punishment, while paving the way for good deeds." The peasant confessed in full repentance, an act which brought about the immediate disappearance of the devil's ring and transformed the penitent completely, so that even the devil was no longer able to recognize him. The penitent was therefore able to start a new life while devoting himself to charity and the salvation of others' lost souls.[37] Yet the question still remains whether a priest was allowed to tell a lie to encourage confession; Gerard du Pescher, for example, regarded any lie as a mortal sin.[38] By means of the canonical duty of confession, nevertheless, the communication between the faithful and the Church was bestowed a divine meaning, while providing the ecclesiastical order with a continuous source of information as to the prevailing climate of opinion in medieval society.

The administration of penance thus elevated the status of priests expected to personify the scale of values desirable to God. On the other hand, everyday practices encouraged much familiarity between the priest and his flock. In the almost complete blending of the sacred and the profane, any clear difference between the priest and his parishioners was not only undesirable but also anachronic. This interpolation was undoubtedly due to the crucial role played by the Church in all aspects of social life. Still, it reflects also the reluctance of medieval people to free themselves from the protective links with the representatives of God on earth, a state of affairs which deteriorated sometimes into a "legitimized sacrilege." Church buildings were often used for dancing, sports events, gambling, and commercial transactions as well, thus blurring even more the already confusing lines between the divine and the profane.[39] The Synod of Reims (1231) forbade dancing in churches and in graveyards under penalty of excommunication, thus indicating the widespread diffusion of such a practice.[40] The parish church, nevertheless, was and remained the Temple of God and, as such, devoted to prayer. According to Canon Law, all believers had to participate at least in the three Sunday services. The best attended was the Mass, held fairly early in the morning. People were also encouraged to be present at mattins, which preceded it, but many considered it sufficient to arrive only for Mass and, even then, were often late. For evensong, on Sunday afternoons, only a few attended. Furthermore, there were numerous absences from Mass, which resulted in empty churches, and medieval sources report the cases of believers who evaded Mass for three and even nine consecutive years.[41] At the other end of the scale, there were also those who could not live without attending Mass, turning it into a kind of addiction. There is a curious episode related to Henry III of England who, on his way to diplomatic negotiations with Louis IX of France, stopped to hear Mass at every single Church on his way. In order to expedite the opening of the negotiations, the no less pious St. Louis ordered the closing of all churches until the entourage of the king of England had passed thereby.[42]

Attending Mass, however, did not per se imply careful attention nor any deep involvement in the ceremony. For the few congregants who could read, help was provided in the form of primers and Mass books, but these were of no use to the vast majority of illiterate worshippers. Believers who came regularly to Mass often brought their babies and dogs with them, a somewhat curious habit that contributed very little to the solemnity of the ceremony. Scant attention was usually paid during the ritual, as prayers were pronounced in Latin, a language beyond the understanding of most of the faithful; the priest's words thus became a monotonous background against which the congregants could discuss more practical everyday matters.[43] Contemporary *exempla* depicted this situation in terms of retribution and punishment. "For those who willingly heed the words of God, God will listen to them, but God will not listen to those who pay no heed," claims the *Alphabetum narracionum*, following an edifying anecdote related by Jacques de Vitry. He further reports the story of a peasant who throughout his life was systematically deaf to the call of his priest. On his death, when a Mass for the salvation of his soul was being chanted, to the great amazement of all the congregation, the portrait of Christ on the cross shut His ears.[44] Both individual and collective punishments were brought into operation for the most serious cases in which the whole community was guilty of inattention or inconsideration toward the priest. In addition, the ringing of a bell came into use during the twelfth century. It was known as a *sanctus bell*, and was aimed at focusing attention during Mass, especially at the Sanctus and at the Elevation.[45]

Beside their leading role in the religious life of the parishioners, priests were also entrusted with the basic religious education of the faithful, particularly the children. Although they were expected to teach all the Articles of Faith, in most cases priests found it satisfactory to read a brief synopsis of the religious dogma, as established by the local synods. In the diocese of Carcassonne, priests taught the "*Credo*" and the "*Pater*" beside the main dogmas,[46] while in the diocese of Lincoln it was customary to teach also the "*Ave*" and to make the sign of the cross properly. At the Sunday gatherings, the priests explained the Creed, the Lord's Prayer, and the Ten Commandments and, sometimes, also the Seven Deadly Sins and the sacraments.[47] Richard le Poore, bishop of Salisbury, encouraged the clergy to instruct their most brilliant students more carefully in order to enable them to instruct others.[48] Regulations of this kind suggest the awareness of the prevailing low cultural level of priests and a realistic approach as to the ways of solving it. *Real-Politik*, however, was not an easy goal to achieve in the sociopolitical framework of the times, nor was the Church's approach immune to particular interests. On the one hand, the Church had a special interest to encouraging the basic religious education of the laity in order to strengthen orthodoxy and protect Christendom from the threat of heresy. On the other hand, the Church was also interested in maintaining the cultural gap between clergy and laity in order to ensure its own supremacy in the long term. These two contradictory goals forced the ecclesiastical order to search for a difficult compromise between improving the laity's desirable but still limited knowledge, and the clergy's prevailing illiteracy.[49] The many contradictions inherent in such a policy were reflected in the ambivalent attitude regarding the use of vernacular languages. From the thirteenth century onward, the Church allowed preaching in the vernacular in order to facilitate the reception of its message. Yet the same use of the vernacular on religious issues by the laity might cause them to be suspected of heresy.

The clergy's low level of educational attainment remained one of the main challenges facing the Church. Even if one neutralizes the satirical tone characteristic of medieval sources and their critical approach to the clergy, it seems that the large majority of priests had but a very rudimentary education, lacking the qualifications required by their office in both the religious and sociocultural spheres. Young priests sometimes received only a few lessons from a more experienced colleague, who instructed them in the performance of Mass and the proper conferring of sacraments. They usually also studied the list of the Seven Deadly Sins and their punishments, considered as a most essential part of their instruction. Yet despite continuous conciliar legislation, only a few priests actually completed their formal education. The fourteenth-century bishops of Grenoble complained that most of their priests did not keep the synodial constitutions nor any other books. The spread of universities in the late Middle Ages did not bring about a radical change in this unfortunate state of affairs. Until the sixteenth century, only about one-third of the clergy actually finished university studies, and the number of graduates ready to serve in the poorest parishes was much smaller.

One partial solution was found in encyclopedic manuals that aimed at instructing the clergy in matters related to the Christian dogma, as well as in practical issues of everyday life. The thirteenth-century Dominican, Raymond de Peñafort composed one of the first *confessionale*, which focused on the questions considered expedient to ask the faithful during confession. The *Bible des pauvres* and the *Speculum humanae salvationis* defined the main principles of Holy Scripture by means of illustrations, while the *Praecordiale devotorum* of Johannes Philippi de Basle included a long list of practical lessons regarding the proper implementation of the clergy's spiritual duties. Besides conventional advice related to the seven sacraments and the proper way to hear confessions and to question penitents, the *Oculis sacerdotis* (c. 1320–1328) of William of Pagula related to more delicate and intimate aspects of human behavior: "The priest ought to inquire of the penitent if he was accustomed to curse men or other creatures; for to be vehemently angry with God's creatures or with cattle and to curse them, even with the ill will to harm them, which country men often do, cursing men and innocent animals, is a great sin. . . . " There is also advice to be given to expectant mothers to avoid heavy work and, after birth, to suckle their own children, as prescribed both by the Scriptures and science. Parents are warned against letting young children get smothered or overlaid in bed, against rashly tying them to their cradles, or leaving them unattended by day or night. Problems of marriage and sexual morality are also dealt with.[50] The publication of these manuals reflects the Church's tendency to standardize practices within the ecclesiastical order, a goal which could not have been achieved by means of oral delivery or personal contacts alone. It also suggests the changes undergone by large sectors of the clergy who became more communication-oriented, and as such, more aware of the communication process in which they became active partners. The greater communication awareness within the ecclesiastical order encouraged a more frequent use of written messages in all facets of daily life. The effectiveness of these manuals, however, was actually conditioned by their degree of publicity among the clergy, which seems to have been rather low in light of the poor situation of most parishes both in the cultural and the economic spheres.

Alongside the attempts to improve the educational level of the average priest, the Church's increasing communication awareness left its mark on the canonization policy of the thirteenth and fourteenth centuries. By granting preference to all those committed to the salvation of their flock, *zelus animarum*, the Church actually declared a new scheme of priorities that turned bishops and priests into the most perfect model of religious life.[51] The individual holiness of former saints and martyrs was then relegated to a second plane by giving preference to the canonization of prelates who might be less holy but who were still totally committed to the guidance of the faithful.

By demoting the individual act of martyrdom, the divine aspects of communication were further emphasized, as the commitment to the faithful brought about both satisfaction in this world and the blessing of God in the world to come. At the same time, the Church encouraged a more active participation of the laity in the parish's practices, a process which suggests a new readiness to enlarge its ranks, thus favoring the reception of its message in Christian society.

Another manifestation of the Church's support for communication channels outside the ecclesiastical order was the development of *fraternities*. From the twelfth century onward, the fraternities represented the corporate world view of medieval society, one which reinforced the self-confidence of its members through the solidarity of the socioeconomic group. Fraternities crystallized on the basis of a common goal, often a religious one, and were further strengthened by the social status, profession or neighborhood common to all its members. Among the first known professional fraternities were those of the water (c.1168) and cloth merchants (1188) in Paris. As time passed and professional fraternities proliferated throughout Western Europe, they assumed religious activities that aimed at satisfying the devotional expectations of its members on a popular level. In fourteenth-century Hamburg alone, there were some 100 fraternities, with twenty-one in Wittenberg, while about eighteen were found in the small villages of Champagne.[52] In most cases the fraternity adopted a religious or moral goal that strengthened the links between its members and the patron saint whose tutelage they sought. Dominican fraternities, for instance, were devoted either to the salvation of the soul, "*pro salute animarum*," the fight against heresy, or the spreading of the Marian cult.[53] Yet fraternities under ecclesiastical tutelage could be considered a rather rare phenomenon in light of the widespread tendency to ensure the laity a considerable freedom of action. Most fraternities were established on a voluntary basis, a feature quite exceptional in the sociopolitical structure of the times.

The comradely framework was clearly depicted by the use of nicknames and the peculiar manners developed in each fraternity whose members were unconditionally committed to mutual assistance. The statutes of the fraternity of Saint Nicolas de Guérande (1350) established that "all the brothers had to love each other by faith and oath, to be faithful and show rejoicing and signs of cognizance wherever they meet, without thinking or doing any harm to each other in any way."[54] The duty of mutual assistance as well was carefully defined in the statutes of each fraternity and further conditioned by its income. The statutes of Saint Dominique de Prato established that its rectors had to visit sick brothers within the first three days of their illness "to bring them some consolation in the name of God and the aforesaid fraternity."[55] The rectors were also responsible for all funeral arrangements of deceased members. After the family informed the fraternity of the death, the rectors had to spread the news through the streets of the village or the town, by a beadle ringing one bell or, if possible, two. All brothers were requested to participate in the funeral procession, the expenses of which were paid by the fraternity in accordance with its income. Beadles carrying one or two bells were also used for calling the brethren to the annual meeting. Certain fraternities held monthly gatherings in which the brothers discussed current problems, such as assistance to the widows and orphans of deceased brethren, and the number of masses to be said for their souls.[56] The beadle running along the streets reminded the brothers of the approaching meeting, which usually took place on the day of the patron saint. As fraternities developed, the annual meeting became an impressive event in which all the brethren marched in colorful costumes, carrying their flags and symbols, thus increasing their feelings of solidarity and group pride. The annual parade turned the fraternity and its patron saint into the center of public attention, if only for one day.

In light of the absence of a proper welfare system, elderly people who joined the fraternities found there a kind of "social security insurance," which covered their own needs, those of their families, and their last journey on earth as well. This state of affairs often brought about an increasing demand for membership that exceeded the financial and administrative capabilities of most fraternities. Each fraternity established its scope of membership that varied from about twenty to a total of eighty to 140 members in larger fraternities. Candidates to brotherhood had to be "*bonnes gens*," namely, good Catholic men and women. Non-Christians, excommunicated people, and all those leading a dubious way of life were *ipso facto* excluded from the brotherhood. The statutes usually established also that the candidates should be in good health and capable of maintaining themselves, a realistic requirement in view of the prevailing tendency among the sick and the poor to turn the fraternity into a convenient escape for all ills. On his acceptance by the brotherhood, each candidate had to swear about his unconditional commitment on his body and soul to the full accomplishment of the statutes. The statutes also established duties peculiar to women, a policy that indicates the active participation of women which, in the Statutes of St. Mary fraternity at Arezzo (1262), was mostly encouraged.[57]

The development of medieval fraternities confirms anew the strong links between the sacred and the profane in medieval society without implying a complete identification between the popular religion and the institutionalized Church or, furthermore, between the Church and medieval society as a whole. On the contrary, despite their religious essence, the fraternities were not usually placed under the Church's supervision, although they looked for the special tutelage of a patron saint. The cult of the saints thus represents one of the most common expressions of popular religion: saints served as concrete mediators between the believer and God and provided an escape from an abstract monotheism and a terrifying reality. Different needs in space and time encouraged the compilation of a detailed catalogue of saints, each one acquiring his own characteristics. St. Hubert shared with St. Eustace the patronage of huntsmen and was also invoked against hydrophobia. St. Roch was especially invoked against the plague, while one single glance at St. Christopher, painted or carved, was sufficient for protection from a fatal end. St. Foy safeguarded war captives in the Iberian peninsula and worked for their release. St. Olaf became the patron saint of Norway. Exclusive symbols were also attributed to each saint in order to facilitate their recognition. St. Achatius wore a crown of thorns, St. Giles was accompanied by a hind, St. George by a dragon, and St. Christopher was of gigantic stature. St. Blaise was depicted in a den of wild beasts, St. Cyriac with a chained devil, St. Denis carrying his head under his arm, St. Erasmus being disembowelled by means of a windlass, St. Eustace was shown with a stag carrying a cross between its antlers, St. Pantaleon with a lion, St. Vitus in a cauldron, St. Barbara with a tower, St. Katherine with her wheel and sword, St. Margaret with a dragon, and so on. The visual representation of saints enriched the medieval world of images and created an international language of nonverbal communication, the basic concepts of which were comprehensible to all believers, literate and illiterate alike. On the other hand, the cult of saints enriched the world of Church communication with a flexible tool, which could be adapted in accordance with special requests, professional or social status. The cult of saints thus represents the completeness of the Church appeal, which—whether by legates, Ecumenical Councils, or synods; by the mediation of bishops, priests, and saints; by oral, written appeals, or symbolic representations—was a universal message both in its geographical and social dimensions.

All channels discussed hitherto, although aimed at strengthening the inner unity of Christian society, relied on the common Christian basic premises of a united Christendom. The convergence of Jews and Moslems into medieval Europe encouraged the Church to develop additional channels, more suitable for the *propagatio fidei* among non-Christians. As early as 1142–1143, Peter the Venerable, abbot of Cluny, initiated the translations of the Koran and other Islamic treatises and wrote a refutation of the Moslem doctrine in his *Liber contra sectam sive haeresim Saracenorum*. He argued that his main aim was to strengthen the faithful against Moslem propaganda and to provide Christians with a weapon for future ideological conflicts. He claimed further, "I do not attack you (the Moslems) as our people often do by arms, but by words; not by force but by reason; not in hatred but in love."[58] This peculiar kind of "love" increased during the thirteenth century, when the overall strategy of Christendom with regard to Islam underwent modification. As claimed by Chenu, the battle now was "not only military but doctrinal, through a dialogue of controversy."[59] This "dialogue of controversy" involved Moslems and Jews alike, while the degree of dialogue or coercion changed according to the political circumstances. The Church's proselytizing efforts reached their peak in the thirteenth century, when the Jews of southern France were forced to attend the sermons of the friars in the area. King Louis IX ordered all Jews to attend the preaching of Friar Paul Christian and "to respond fully, without calumny and subterfuge, on those matters which relate to their law. . . . whether in sermons, in synagogues, or elsewhere"(1269).[60] In those areas where Jews were not the only religious minority, the royal order also included the Moslems. Jayme I of Aragon also established that "whenever the archbishop, bishops, or Dominican or Franciscan friars visit cities or locales where Saracens or Jews dwell, and wish to preach the word of God to the said Jews or Saracens, these shall gather at their call and patiently listen to their preaching. If they do not wish to come, our officers shall compel them to do so, rejecting all excuses" (1242).[61] In addition, both Dominicans and Franciscans encouraged proselytizing missions in the Moslem world while fostering secret conversions via commercial, religious, or other contacts. Their strategy also involved infiltration through a metaphysical dialogue with whatever Islamic scholars came to hand—diplomatic maneuvers toward converting a potentate in whose footsteps many subjects could drift into Christianity or public proselytizing in territories conquered from Islamic rule.

Another important communication event used by the Church "to lead wanderers onto the path of truth" was provided by the *religious controversies* with Jews and Moslems, which aimed at bringing about a massive conversion. This appeared as a most desirable goal, which, from the Christian perspective, legitimized the confrontation between the representatives of different religions. Controversies on religious issues were of ancient origin and preceded Christianity (see plate 9). As early as in the third century BC, public disputations were performed before the most important political personalities of the time. The spread of both Christianity and Islam in a common monotheistic framework further encouraged controversy between them and Judaism, as the three religions appealed to more or less the same audiences and developed in a somewhat circumscribed geographical area. Early controversies centered around the articles of faith and the issue of divine choice. Christians often criticized the Jews for the materialism of the Mosaic Law and their misunderstanding of the Holy Scriptures, which had led to their exile from the Holy Land. The Jews argued the open contradiction between the Trinitarian faith and monotheism, and between the divine and human nature of Christ. Christian sources suggest that in debates held in Jerusalem between Moslems and Christians during the seventh and eighth centuries, Jews played an auxiliary role on behalf of the Moslems.[62] In Western Europe, and in the Spanish kingdoms in particular, the convergence of Christians, Jews, and Moslems encouraged religious controversy and, in parallel, the diffusion of polemical literature from both sides.[63] However, the terms "controversy," "debate," or "disputation" became quite inaccurate in the climate of opinion of the Central Middle Ages, when the polemic tradition of ancient times was replaced by inquisitorial investigation. The term "religious controversy" thus often reflects a situation formally initiated by the Church hierarchy which established the topics as well as the ways of discussion and, quite naturally, forced its own conclusions upon those representing the religious minority.

The Debate of Paris (1240) has been considered as the turning point in the polemic tradition between Christians and Jews in the Middle Ages; from then on, the Christian offensive focused on Talmudic literature, both for proselytizing among Jews and for developing the main principles of Hatred of the Jew, which crystallized at the time.[64] The confrontation with Jewish rabbis was used as a "case study" to check the validity of the Christian argumentation in light of the large-scale missionary offensive of the times. From the perspective of Church propaganda, the utilization of rabbinical sources to prove funda-

PLATE 9. Ibn Wasiti's illumination of al-Harirri's *Maqamet.* Two Moslem gentlemen in a lively dispute.

mental Christian principles had considerable advantages over the earlier patterns of Christian predication. The reliance upon texts with which the Jewish audience was already familiar ensured a favorable impact, as the Jewish public heard its own sources used for corroborating the principles of the Christian faith. This impact was further strengthened by the fact that the Christian side was usually represented by a converted Jew who volunteered to prove the blasphemies implied in the Talmud and to convince his former coreligionists of the irrefutable truth inherent in the Christian dogma. The political and religious elite also played an active role in the deliberations. In the Paris controversy, for example, the Queen Mother Blanche of Castile presided over the meetings, her assessors being the ecclesiastical elite of the times. In the debate of Barcelona (1263), King Jayme I attended to the deliberations held in the royal palace, together with the ministers provincial of the Dominicans and the Franciscans. In the debate of Tortosa (1413–1414) the antipope Benedict XIII himself presided over the meetings, suitably surrounded by his College of Cardinals. The Jewish side was represented by local rabbis who had usually been forced to attend. According to the Hebrew account of the Paris controversy, the Jewish rabbis faced their chief antagonists and answered them directly. Yet the accuracy of this report is rather doubtful in light of the inquisitorial patterns suggested by the Latin protocol, according to which the Jews played the passive role of defendants and were not even allowed to see their accusers.[65] The rabbis were further compelled to testify under oath; they had been interrogated separately and had not been allowed to consult each other. The Controversy of Barcelona followed similar patterns twenty-three years later, while contemporary sources emphasize the large attendance of Jews and Christians from the surrounding areas. Other public disputations took place at Pamplona (1373), Burgos (1375), Avila (1375), and Granada (1430), all of them following the inquisitorial patterns recorded in Paris.

The scarcity of sources and their apologetic nature make it quite difficult to know for sure what actually happened during the public confrontations. Nevertheless, the massive attendance of different social strata at the meeting place indicates the importance of the event, not only with regard to those "infidels" whom the controversy apparently aimed at converting, but also to the many Christians who attended and therefore could have been favorably influenced by the patronizing attitudes of the Christian communicators. Besides, wide social strata could easily find a common denominator in their xenophobic outbursts, while the latent hatred of strangers was justified on religious principles. The public controversies thus appeared as additional propagandistic means at the Church's disposal for advancing its interests in the internal order, as they increased, foremost, the self-confidence and the group solidarity among the Christian population itself. Jews and Moslems merely served as groups of reference which, in the sociopolitical conditions of medieval society in general, and the critical situation of the Spanish society in particular, could be easily sacrificed in the search for self-definition. The "propagation of the faith" was therefore mainly directed at Christendom itself, reducing the importance to be ascribed to members of other monotheistic religions, whether Jews or Moslems. From the point of view of their declared goals, moreover, it is quite doubtful whether the religious controversies achieved a significant success, since available data do not indicate massive conversions of Jews or Moslems following the impressive performance of the Christian propagandists.

Proselytizing missions and public controversies corroborated anew the universal scope of the Church communication in the Central Middle Ages. The lack of an integrated communication system had forced the Church to exploit an endless number of alternative channels while turning its administrative framework into a means of communication. Archbishops, bishops, and priests thus adapted and communicated the papal message for their various audiences. The vitality of the ecclesiastical communication is further substantiated by the increasing use of written reports, which reflect more advanced communication practices and a growing awareness of their importance. The Fourth Council of Lateran heralded, in this regard, the standardization of Church practices through written reports, regular assemblies, and the universal duty of confession. The primitive means of communication, the bad conditions of the roads, and the lack of safety in large areas of medieval Europe, further testify to the success of the Church, with the papacy at its head, in bringing its message to all components of Christendom and, foremost, to the ecclesiastical order itself. One can further argue that the success of the Gregorian Reform was significantly influenced by the development of the Church communication system, one that allowed the Vicar of God to voice the *Vox Dei* notwithstanding the isolation and autarky inherent in the feudal system. Faithfully assisted by prelates throughout Christendom, their message further legitimized by the will of God, the medieval popes thus developed their administrative skills to the full while bringing about the development of a communication system which, in time, was imitated by the secular rulers of Christendom.

4

The Catechism of Fear and the Cult of Death

The development of the Church in the Central Middle Ages embodies an essential paradox: the ecclesiastical order reached its maximal influence at a time when Western society gradually evolved from corporate frameworks into more developed socioeconomic systems which by their very nature opposed the Church's monopoly. This paradox raises questions as to the Church's capability to maintain its predominance while adapting its message and communication channels to the changing circumstances. This chapter deals with death and the world to come as examples of the Church's dynamism and its ability to convey new messages while advancing its own interests. Both death and the world to come were connected with ancestral traditions and deeply rooted beliefs, which the Church tried and ultimately succeeded in replacing by its own Catechism of Fear in general and the Cult of Death in particular. The importance to be adjudicated to the Church's attitude in this regard results, therefore, from the extreme propagandistic effort implied therein, since its fundamental contempt for earthly reality created a crucial antagonism between the ecclesiastical message and the increasing capability of medieval man to enjoy this world.

From its earlier stages, Christianity projected an unambiguous attitude with regard to this world and the world to come; its emphasis on the ephemeral, sinful essence of this world, ensured the complete preference of the next. This new scale of values affected not only the ideas but also the style of life of ancient Christian societies, which were encouraged to adopt a new code of behavior. In their attempts to alienate the new Christians from the pleasures of this world and to inculcate instead the *contemptus mundi*, the Fathers of the Church vigorously opposed mourning ceremonies, which were a very common practice in the late Roman Empire. Roman rituals were thus condemned as remnants of paganism that had to be rooted out, as they conferred ceremonial legitimacy on the pain of leaving this world. St. John Chrysostom (c. 347–407) severely attacked the custom of mourners paid to encourage external signs of grief, and threatened with excommunication all those who would persist in such a sinful custom. His opposition did not focus on the salaried aspects inherent in such a use, but rather on the inconsistency of mourning with the Christian faith of redemption and the expected rejoicing of the faithful towards death.[1] The Fathers of the Church forbade also the use of musical instruments in funeral processions, such as flutes, ciatharas, and kettledrums, which were often played to give maximum resonance to the mourning at death marches. St. Basil (c. 330–379) justified this prohibition in light of the inner contradiction between mourning rituals and Christ's victory over death with its resulting hopes for salvation to the whole community of the faithful. The Church's tendency to chant the Alleluia at funeral offices reflected the belief in the world to come and the resulting joy expected from all the faithful towards their release from this ephemeral world.[2]

Ancient Christian traditions of welcoming death were complemented in the Early Middle Ages by what can be called the solidarity of the dead, namely, the perception of doomsday as a collective experience of all humanity which, as such, reduced to some degree the fears and anxiety of each individual. Between the sixth and tenth centuries, Western society had not yet acknowledged individual responsibility in the face of doomsday, which was further postponed to an unknown date: the end of time. Both the solidarity of the dead and the postponement of the Last Judgment brought about new burial customs: the dead were buried in the Church graveyard, thus waiting in safety for the second coming of Christ while enjoying ecclesiastical tutelage. Sigfrid, Bishop of Le Mans, for example, entered a monastery a few days before finishing what a contemporary chronicler described as "a miserable and sinful life," and was buried there.[3] Those who had trusted the Church mediation could therefore have full confidence that they would be properly rewarded in the Heavenly Jerusalem. This new and fearless concept of death found full expression in the Gregorian Chant. It reflects the greater reconciliation of the faithful with death through its polyphonic adaptations in the *Requiem* and the *Missa pro defunctis*.[4]

The reconciliation with death in the Early Middle Ages resulted from the corporate conception of death and the after-death. Both were perceived as collective experiences, which, as such, alleviated the anxiety of the individual and further undermined the very justification of fears. The denial of religious legitimization of the fear of death was indirectly expressed in the memorial by which Dante Alighieri welcomed all sinners condemned to the *Inferno*: "*Lasciate ogni speranza, voi ch'entrate*," abandon every hope, ye that enter.[5] At first glance, this inscription might suggest the hopeless condition of all those condemned to hell but, as pointed out by Sachs, it rather expresses the Church's claim that "no one would suffer damnation who had not despaired, for the abandonment of hope is the very condition of entrance into the devil's domain."[6] Until the eleventh century, therefore, whether as a result of its struggle against the pagan heritage or out of its belief that "this world is but a pilgrimage," the Church had not yet developed the Catechism of Fear nor the Cult of Death. Both the poetry and the burial customs of the time suggest a reconciliation with death which aimed at reducing fears either individually or in the corporate framework of medieval society.

The gradual improvement of socioeconomic structures, however, encouraged a deeper emotional involvement of medieval people with this world and a resulting change of attitudes in the perception of death. As Aries pointed out: "*La mort est devenue le lieu où l'homme a pris le mieux conscience de lui même*."[7] The solidarity of the dead, characteristic of the Early Middle Ages, was thus gradually minimized and replaced by a new, more individualistic concept of death. Death continued to be regarded as the common fate of all flesh and blood, since everyone is mortal by nature, but death still became an individual experience to be undergone by everyone, an experience between oneself and his Creator alone. The individualization of death became a central factor mainly among the townsmen, whose approach to death resulted from the emotional aspects of the accelerated tempo of change of the historical process. A major sector of townsmen had been forced to leave their native land and to divorce themselves from the ancestral solidarity of the family framework. The distance from home often brought about an individualization of death in actual practice, as many had to be buried in strange territory, far from their forefathers and the family burial place. in addition, most inhabitants of medieval cities had in fact abandoned the corporate-feudal structures of the countryside and integrated into more individualistic frameworks in which their status was highly influenced by their own initiative and their capability of adaptation. Moreover, most townsmen were deeply involved in the new reality they had helped to foster and, quite naturally, were less ready to disregard this sinful, ephemeral world in favor of a more spiritual one.

The socioeconomic development of Western society in the Central Middle Ages left its mark on the increasing ability of contemporaries to improve former ways of life. Vinay refers quite enthusiastically to this process and its resulting implications in the *Umwelt* of medieval men: "If there was a joyful century in the Middle Ages it was this one, when Western civilization exploded with astonishing vitality, energy, and desire for renewal. . . . The twelfth century is typically the century of liberation, when men cast aside all that had been born rotten for more than a millenium. . . . " He refrains, however, from leaving such an optimistic perspective, as he further recognizes that "the men of the Middle Ages really began to suffer in their happiest period, just when they had begun to breathe deeply, when for the first time, it seemed, they had become aware that all the future lay ahead, when history took dimensions it had never before assumed."[8] The increasing interest of medieval people in their environment and their resulting readiness to enjoy the new possibilities they discovered thus encouraged a new awareness of the ephemeral nature of life, which became better known and, as such, more loved. Julien of Vézelay (twelfth century) was well aware of this dichotomy between life and death and confessed his distress quite vividly:

> Three things terrify me, and at their mere mention my inner being trembles with fear: death, hell and the judgment to come. I am therefore frightened by the approach of death, which will take me out of my body and away from the pleasing light shared by everyone, into I know not what region reserved for

> the spirits of the faithful. . . . After me the history of mankind will unfold without me. . . . Farewell, hospitable earth, on which I have long worn myself out for futile purposes, on which I have inhabited a house of mud which, though it be mere mud, I leave unwillingly. . . . It is unwillingly (that I go) and I will leave only if I am driven out. Pale death will burst in upon my doubts and drag me, in spite of my resistance, to the door. . . . Along with the world, we leave behind everything that is of the world. The glory of the world is left behind on that sad day: farewell to honors, wealth, and property, to broad and charming meadows, to the marble floors and painted ceilings of sumptuous houses! And what about the silks and squirrel furs, the variegated coats, the silver cups and fine, neighing horses on which the wealthy who gave themselves airs once paraded! But even this is as nothing, for we must also abandon a wife so sweet to look at, we must abandon our children, and we must leave behind our own bodies, for which we would willingly pay gold to redeem them from this seizure.[9]

Julien of Vézelay's reasonings mirror the pain of the individual faced with his own expected departure from this beloved world, further aggravated by his fears of the unknown. The void the deceased left behind became a central motif in the prevailing attitudes of medieval people regarding death, together with the shocking spectacle of the deterioration of human beauty and the equality of humanity before death, further emphasized by the Dance of Death.[10] In addition, both in the countryside and in the nascent towns, death was an ubiquitous element of daily life. The high frequency of plagues and natural disasters in medieval society reached its peak toward the end of the period, when natural catastrophes such as the Black Death, and long-lasting military confrontations such as the Hundred Years War, turned the fourteenth century into one of the most critical periods in the history of Western society. Available data clearly reflect the demographic implications. In the village of Givry in Burgundy, for example, within three and a half months, between August and November 1348, 643 inhabitants died out of a population of 1,500. In a group of villages in Savoy, 52 percent of the taxpayers died between 1347 and 1349, while in the district of Aix-en-Provence, the mortality rate ranged from 29 to 43 percent and 75 percent between the years 1346 and 1351. This state of affairs was clearly reflected by the popular maxim: "For we are here today, and gone tomorrow."[11] All these resulted in irrational fears that were increased by ancestral traditions. Folk tales tell about Lazarus, who, after his resurrection from the Kingdom of Death, lived in awful anxiety, for he knew that he would have to reenter the gates of death. Grief at the thought of leaving a dear and familiar world, distress and anxiety in the face of death and the unknown were but different facets of one of the oldest existential questions in Western culture on which the medieval Church bestowed the perspectives of a cult, the Cult of Death.[12]

As a result of actual reality or due to the Church's consistent tendency to consolidate its own position by the Catechism of Fear, death became, by the Central Middle Ages, an integral part of daily life and, as such, served as the most suitable instrument for manipulation by the Church. The new message it developed embodied the whole range of human affections: it legitimized happiness at leaving a world which was essentially sinful, but it also fostered anxiety in the face of the forthcoming divine judgment and its unknown sequences. Between both extreme emotions, the ecclesiastical message embodied the quietude promised to those righteous who committed themselves to the mercy of God while submitting to His representatives. Yet the concept of quietude was too abstract to neutralize the weight of the diverging emotions raised by death, nor was it completely defined at this stage.[13] Furthermore, although happiness, anxiety, and quietude could stand together as progressive stages along the way of the faithful to God, the changing emphasis on each suggests the process undergone by the ecclesiastical establishment through which it actually reconsidered its own scale of priorities.

The Cult of Death appears as the Church's response to the socioeconomic process of change and the individualization of death, which could threaten its leading role in contemporary society. The gradual decline of the feudal regime, indeed, had widened the maneuvering scope of medieval people while creating the first breaches in the Church's hegemony. The improvement in living conditions, the growth of cities, and the increase in literacy, although not undermining the essential religious nature of medieval people, still restrained their former complete dependence on the Church and the clergy. On the other hand, the individualization of death and the increasing fears and distress in face of leaving behind a more beloved world, created a fruitful background to strengthen the Church's status while encouraging the reception of its message. Moreover, the gradual integration of medieval man in more anonymous social structures fostered his exposure to the Church's manipulation, which was hardly practicable in the corporate framework. The need to prevent the harmful consequences of the socioeconomic process while profiting from the new awareness toward death brought the Church to convey new concepts on death and the world to come, summarized in the Catechism of Fear.

Encouraging the search for the most effective ways to ensure the achievement of fear in the long term,[14] the ritual of fear became an indispensable tool in the hands of the ecclesiastical communicators with regard to both the laity and the clergy. Manuals written for the assistance of preachers mentioned the intimidation of the audience as a goal that had to be sought, although Arnold of Liège, for instance, recommended a double caution in the manipulation of fears.[15] The *Liber exemplorum* (1275–1279) encouraged preachers to provide their public with a detailed list of the punishments intended for the sinners after death in order to bring about a "healthy weeping, as they must fear of the punishments which await those who deviated from the right path."[16] What can be called the Cult of Death, however, was not conditioned by circumstantial preaching but was rather an integral element, if not the most crucial aspect, of Church communication in the Central Middle Ages. Priests often read to the faithful a biblical or a patriarchal source that focused on the issue of death. They explained it and, when necessary, translated it from Latin to the vernacular. Afterward, the whole congregation accompanied the priest in prayer and, often chanting, begged for God's mercy.

In order to promote the Catechism of Fear, the Church recruited all communication channels at its disposal, including the cathedrals' monumental sculpture, poems and prose works, ballads, paintings on wood, stone, or canvas, stained glass windows, embroidery, tapestry, metal works, engravings on stone or metal, and woodcuts. Furthermore, the Church communicators, aware of the heterogeneous nature of medieval society, adjusted their message to the various sociocultural strata, while differentiating not only between clergy and laity, but also between the different sectors among the laity and the ecclesiastical order. In their appeal to the lower strata, Church spokesmen frequently used short stories that were aimed at ensuring the intimidation of the audience, a goal evaluated as the safest criterion to measure the successful reception of the Church message.[17] In order to put the fear of God in the faithful, preachers usually focused on the individual judgment after death and the decay of the flesh; this became a central motif toward the end of the period. In its appeal to the intellectual elite, the emphasis of the Church message turned to different aspects of the Cult of Death, either because of the disregard of the body in monastic circles,[18] or because of the more sophisticated perception of the intellectuals. In their appeal to this sector, the Church communicators focused on the vacuum left behind the dead, *ubi sunt*?, or on problems related to the sudden death, *mors repentina*, which left no time for the extreme unction to be performed. Yet the emptiness the dead left behind had anthropocentric dimensions too wide to be accurately used by Church propaganda. Quite different was the Church's approach to the sudden death, which became the leitmotif of the ecclesiastical message, "*memento mori*!" remember that thou shalt die! solemn words that for medieval people, laity and clergy alike, were a call for repentance.[19]

The individual's divine judgment after death, an issue to touch the hearts of the lower social strata and the cultural elite as well, thus became the very focus of the Cult of Death. This individual perspective of death weakened the foundations of the solidarity of the dead and served the Church's purpose. Left alone before the judgment of God, it was rather expected that the faithful would try to improve their chances, through strengthening their links with God's representatives. The perspective of a personal judgment after death heightened the helpless situation of each individual and his resulting dependence on the Church *mediation*. It might also weaken the growing identification of medieval people with this world and turn the focus of attention on grounds more favorable to the reception of the Church message. The search for the most suitable premises to develop ecclesiastical propaganda thus encouraged the return to the former contempt for earthly reality, while the Church was faced with the challenge of disregarding five centuries of historical development. Hence the Church message again turned this world into a "dark road," a "house in ruins," and a "mass of sin." The clergy renewed the Pauline patterns of the *contemptus mundi* and emphasized anew a complete disregard for actual reality.[20] The ephemeral essence of earthly pleasures was set against the eternity of God's mercy and presented as a serious obstacle in the path of salvation. Such an attitude was clearly reflected in the sarcastic diatribe of an anonymous fourteenth-century preacher who demolished, in one stroke, the very value of this life: "*Why rememberest thou no thyng of thy riches that thou browzttiste into this worlde? Thou wottyste well i-nowze that nowzte thou browzttiste into this worlde and nowzt thou schalt be bere hens. Now thou hast wordly riches, and seyste all is thyne. Why doyste not thou take it wythe thee?.*"[21]

Disregard for this world emphasized the crucial importance of the Day of Judgment, described in its most depressing, terrifying aspects (see plates 10, 11, 12). To stress human helplessness against divine omnipotence, Church communicators united God and Satan in their search for truth. Satan became the head prosecutor of sinners together with the testimonies of their conscience and the sins perpetuated by each one. Although Satan was not actually present at the trial, he assumed a dominant role therein, his opinions being faithfully reported by angels sent by God to hell for this specific task. Dedicated to the constant growth of the underworld, Satan's sharp eyes missed nothing, and discovered the misdeeds of everyone as through an open book presented before God. Only divine mercy could restrain the power of Satan, and yet no one could rely upon it completely as God "punishes ummercifully."[22] Only in exceptional cases could sinners receive a second opportunity to return to life and expiate their sins, thus improving their chances in a second trial. In most cases, however, God's verdict was final, leaving no hope of further appeals (see plate 13). At this stage, Church communicators provided a most detailed description of hell, which, according to Holy Scripture was identified with fire (1 Cor. 3:13–15). But the torments ascribed to hell went beyond the scriptural sources and were heightened by all the terrifying power the preachers' imagination could provide. Nevertheless, the many sufferings ascribed to the underworld appeared many times as being drawn from the ancestral stories of the torments applied to the Christian martyrs at an earlier date (see plate 14). Yet medieval communicators still managed to establish two distinctive features of hell: the everlastingness characterizing punishment there, which left no room for further hope, and the utter chaos of hell, which stood in complete opposition to the harmony characteristic of heaven.[23]

The clouds of the unknown that shrouded doomsday and the resulting fears of its implications increased the anguish of individuals and fostered endless feelings of guilt as well. The complete impotence of humanity before the omnipotence of God, the lack of mercy ascribed to His judgment, sometimes brought about the outbreak of popular movements that mirror the inability of the clergy to alleviate all psychological tensions encouraged by its own Catechism of Fear. Inevitably, there were some who could not suffer for long the feelings of uncertainty, anxiety, and guilt and committed suicide. The data available do not allow us to draw accurate conclusions with regard to suicide rates in the Central Middle Ages, but they suggest that the clergy itself was not immune to this way of escape. From a sociological perspective, the

PLATES 10, 11, 12. Reims Cathedral, north transept, *Judgment portal,* c. 1230. In the apex of the tympanum appears the strikingly large figure of the judge, with arms outspread and, on either side, the kneeling figures of Mary and John the Baptist. Beyond them two angels kneel with the cross, spear, and crown of thorns. Two zones

PLATES 10, 11, 12. (*continued*) immediately below the judge are filled with the resurrected. The two bottom bands show scenes of paradise and hell. At the bottom left Abraham is shown with souls in his bosom, brought to him by angels on either side, flanked by angels escorting the blessed. In the zone above are seated figures of the blessed: an archbishop with pallium and rationale, an abbot with staff, a royal couple, a virgin with chaplet. On the right is an angel with censer and incense boat, and next to him a tree of paradise. At the bottom right the angel with the flaming sword drove the damned toward the cauldron of hell, toward which the devil at their heels is pulling them on a chain. The procession includes a king, a bishop, a monk, and, at the rear, a miser with his money bag. Above were groups of the damned.

mere act of suicide represents the failure of the Church to perpetuate its control over medieval people who, in the most extreme cases, chose in their despair to dissociate themselves forever from the community of the faithful. Such failure was interpreted by the Church communicators in cosmic dimensions. Suicide was presented as the victory of Satan, a challenge against God and the outcome of despair, sadness, melancholy, or anger. The uncompromising condemnation of suicide further justified the punishment of the homicide who was denied Christian burial. Yet the futility of such a policy was clear enough so that more effective ways were used to induce the fear of God into medieval society as a whole. In most cases, the Church encouraged the secular

PLATE 13. Tympanum and lintel, west portal, Autun Cathedral, *Last Judgment,* c. 1125–1135 (11′4″ × 21′). The composition is dominated by the central figure of Christ with outstretched hands, enthroned within a mandorla supported by four angels. The upper zone is heaven. On Christ's left are the damned in hell and the *Psychostasis,* the weighing of souls. On his right, the elect are led to heaven by St. Peter and an angel. On the lintel below hell are the resurrected destined for hell and chased there by an angel; below heaven is a procession of the resurrected about to be rewarded, including ecclesiastics and pilgrims.

authorities to confiscate the property of the deceased, a practice that very often represented a death sentence for all his family after him.[24]

The phenomenon of suicide hints at the difficulties that plagued both laity and clergy when undergoing the Catechism of Fear. One can argue that the Church was not the sole factor that influenced the world of ideas and beliefs, and neither did the Church enjoy total control over actual reality. Still, the crucial influence of Church propaganda in medieval society should not be underrated. With an almost complete monopoly over education, the Church enjoyed most of the communication channels at its disposal, and thus an overriding influence on the climate of opinion at the times. Nor was its influence restricted to the Cult of Death: it painted the whole world picture of medieval people with a polarity that left no room for nuances. This world and the world to come, as the inner existence of every human being, similarly underwent a bipolar categorization of heaven and hell, good and bad, piety

PLATE 14. S. Maria Assunta Cathedral, Orvieto, c. 1310–1330. Scenes of the *Last Judgment* carved by the Siennese sculptor Lorenzo Maitani.

and sin, mercy and punishment. All the alternatives of human behavior were thus expressed in absolute, eternal terms, which left little room for further considerations.

The absoluteness and infinity implied in the very essence of hell, the belief that no prayers, charity, or sacraments could change the verdict of God,[25] however, proved to be inoperative in the long run. Such concepts might further undermine the Church's role as mediator between humanity and God. An intermediary field between hell and paradise thus came into being, one that would widen the freedom of action for God's representatives and would strengthen their status on earth. The new concept of death paved the way for the "discovery" of purgatory toward the second half of the twelfth century. Purgatory was conceived as an intermediary other world in which some of the dead were subjected to trials which could, however, be mitigated by prayers or the spiritual aid of the living. At the Second Council of Lyons, Pope Clement IV defined the concept of purgatory as follows:

> Those who fall into sin after baptism may not be rebaptized but, through genuine penitence, they can obtain pardon for their sins. If, truly penitent, they die in charity before having rendered satisfaction through penance for what they have done by commission or omission, their souls. . . . are purged after their death, by purgatorial or purificatory penalties. . . . For the alleviation of these penalties, they are served by the suffrages of the living faithful, to wit, the sacrifice of the mass prayers, alms and other works of piety that the faithful customarily offer on behalf of others of the faithful, according to the institutions of the Church.[26]

By means of prayers, masses, or whichever way indicated by the Church, the suffering of the deceased could thus be alleviated while reducing their term in purgatory. A fearful image of purgatory, therefore, had to be created. Many of the punishments ascribed to purgatory were indeed similar to those reputedly inflicted by Satan in the underworld, a feature that created a common bond between purgatory and hell, while emphasizing the gap between them both and paradise. The essential difference between hell and purgatory, however, focused upon the "fact" that the suffering in purgatory was limited in time, as the Church could control the hands of the clock. Purgatory, therefore, became the obvious outcome of the Catechism of Fear by which the Church tried and often succeeded in overcoming the dangers inherent in either complete happiness or despair. Against the alternatives of leaving the Church's tutelage, either by the nonconformist way of heresy or by the most extreme escape of suicide, the Church offered an alternative solution that was also presented as the most desirable in the eyes of God and His saints. By offering a second chance to sinners and their family, the idea of purgatory actually expanded the power of the ecclesiastical establishment. It turned the clergy into an effective channel for divine justice, one that was empowered to alleviate God's verdict by teaching the way of repentance to the individual himself or, if he should die before achieving full penance, to his relatives.

The implications of this new concept of purgatory were vividly reinforced by means of the *exempla*. They faithfully indicate the weight ascribed in heaven to all aspects of human behavior while delineating the terrifying facets of divine judgment. In an edifying story told by Peter Damian, one of the leading Italian hermits, he saw "Saint Agnes, Saint Agatha and Saint Cecilia herself in a choir of numerous resplendent holy virgins. They were preparing a magnificent throne which stood on a plane higher than those around it, and there the Holy Virgin Mary with Peter, Paul and David, surrounded by a brilliant assembly of martyrs and saints, came to take their place." In this particular case, the appeal by a woman who described the deceased's charity toward the poor and his devotion to the holy places, changed the sentence passed on a patrician who "was dragged in, bound and chained, by a horde of demons. Whereupon Our Lady ordered that he be freed, and he went to swell the ranks of the saints. But she ordered that the bonds from which he had just been set free be kept for another man, still living." Peter Damian ends this striking account with the lesson to be learn: "Thus does divine clemency instruct the living by means of the dead."[27] This *exemplum* faithfully reflects the nature of purgatory and the socioreligious advantages it portrayed from the Church's perspective. From the moment the Church became the exclusive mediator between the living and the dead and its monopoly was finally legitimized by the idea of purgatory, both sacraments and masses acquired the highest validity, since they could sway the balance of God's judgment. By giving the living the chance to help their relatives and dear ones after death, the Church satisfied not only its own interests but also the need inherent in medieval people for some kind of solidarity with the suffering of the dead. Jean Gobi also tells the story of a man who sailed to the Holy Land in order to carry out a vow that his father, now deceased, had made. He was rewarded for his praiseworthy action as his father appeared in his dream and thanked him for his liberation from purgatory.[28]

Belief in purgatory further encouraged masses for the soul of the deceased, a well-known practice from early times. In sixth-century Gaul, it had been customary to recite mass for the dead every day of the year immediately following the death; some 365 masses were thus performed *in memoriam* of those who could afford it. Yet these figures shrank in comparison with the vast number of masses performed toward the end of the Middle Ages. A fourteenth-century French nobleman, Bernard d'Escoussans, for example, required 25,000 masses to be prayed for his soul, while he carefully detailed in his will the frequency of their performance. Requests of this kind became more the rule than the exception. Many members of the social elite of Southern France asked for between 2,000 and 50,000 masses to be said *pro remedio animae*, for the salvation of their souls, the exact number varying according to the financial ability to bear the expenses. These impressive numbers, confirmed in medieval wills and parish records,[29] clearly indicate the reception of the Church's message and its translation into daily practice. For the socioeconomic elite, at least, the proliferation of masses became the most effective way to reduce its fears and anxieties.

The clergy also benefited from both the rise in status and a growing source of income whose importance cannot be ignored. The material aspects of such a practice were clarified by Archbishop of Canterbury John Peckham, who, aware of the situation, asserted: "Let no man think that one mass said with pure intention for a thousand men might be considered equal to a thousand masses also said with pure intention."[30] All monastic Orders also publicized themselves on the grounds that they could provide their benefactors with thousands of masses properly performed *in memoriam* through their different houses both in the Holy Land and Europe. A pamphlet composed on behalf of the Hospitallers in the thirteenth century encouraged monetary contributions to the Order on the grounds that this convenient arrangement enabled one side to provide the money, while the other advanced the salvation of the soul. In other words, the relationship between medieval people and their Creator gradually acquired unforeseen mathematical and economic dimensions. The Catechism of Fear thus achieved its goals: it translated irrational fears into most convenient practices from the Church perspective, which alleviated the anxiety and distress promoted by its own message. The Order of Cluny also participated in this process, organizing the masses for all the dead on All Souls' Day (November 2).

The judgment after death with its implications of hell, purgatory, or heaven represents the more abstract aspects of the Church message. More material facets of the Cult of Death focused on the deterioration of the corpse in all its macabre aspects, which became a central issue in the Church's appeal to the lower social strata. The wide reception of the ecclesiastical message in these circles was further facilitated by the equalizing message implied therein: death creates a common bond between the rich and the poor, the beautiful and the ugly, the strong and the weak. As pointed out by Geoffroy Babion (d. 1112): "Listen all you people, and the ignorant shall better know. Go to the graves and contemplate the lesson the dead have left for us, the living. There lie just bones. Man is dead and his trial is waiting. Among the living, he resembled us in his vanity. . . . but now he lies in his grave, has deteriorated into dust; his flesh which he fed with delicacies, is gone, the nerves have been disconnected from their systems. Only his bones remain, providing a lesson for the living."[31] An anonymous thirteenth-century sermon also emphasized in the vernacular the common end of all creatures: all corpses deteriorate without there being any difference between "the rich and the poor, those who took revenge and their victims as well; everyone goes in one way, everyone dies and becomes decay."[32] Flowers, pretty women, and handsome men, all go the same way, since they all "*devienge carugne*."[33] The didactic elements indicating the ephemeral nature of human beauty were expressed in the letter of St. Anselm to Gunhild, a nun who had fled her convent to become the mistress of Count Alain le Roux. Anselm strongly emphasized the temporal nature of human beauty as well as the punishments intended for all those who indulged themselves in carnal pleasures. Both "facts" justified full repentance and, as a result, the immediate return of Gunhild to her nunnery.[34]

The morbid tendency of medieval people to focus attention on the deterioration of the body appears, according to Huizinga, as "a kind of spasmodic reaction against an excessive sensuality, while it expressed a very materialistic sentiment"[35] (see plates 15, 16). Materialism and spirituality, existing fears and Catechism of Fear, all combined to bring about a significant change at burial ceremonies, while the clergy completed its circle of power by acquiring a most dominant role. The priest would entreat God in the eulogies that the holy angels be ordered to receive the soul of the departed and to conduct it *ad patriam paradisi*, heaven thus becoming the common fatherland of all Christians. Sometimes, musical eulogies, the *lamentatio* or *planctus*, were also written *in memoriam* for the more important personages. The "*Fortz chausa*

PLATE 15. Sacro speco, Monastery of S. Benedetto, Subiaco, Rome. Fresco painted at the end of the fourteenth century showing the triumph of death.

est," a musical piece in the troubadour style, was dedicated to Richard Coeur de Lion. King Sancho III of Castile served as the source of inspiration for the "*Plange Castella misera*," while the "*Rex obit*" was dedicated to Alfonso VIII of Castile and the "*Monialis concio*" to Doña María a Gonzales de Aguero, the mother superior in a Castilian convent.[36] Toward the end of the thirteenth century, the polyphonic *planctus* replaced former monophonic pieces, as it was considered more suitable to express the acceptance of death by the faithful.

Burial procedures also became more sophisticated, as they had to be concomitant with the socioeconomic status of the dead. The tombstone was prepared in advance and represented the deceased in his life; it depicted a more realistic portrait which aimed at moving people to pray for his soul (see plate 17). The funeral procession reflected the former status of the deceased, becoming an important milestone along the believers' path to God. The idea of purgatory and the mediatory role ascribed to the clergy, moreover, encouraged the renewal of earlier burial practices. An increasing number of knights, nobles, and princes asked again to be buried in the graveyard of monasteries and churches, near the holy relics. In this way, they could enjoy a double privilege: the prayers said at the house of God and the special grace accorded to the remains of saints. There were also some who asked to enjoy this special grace while they were still alive, thus increasing the number of those who took

PLATE 16. Tomb of Frances I de La Sarra (died 1362), Chapel La Sarraz. Watched over by members of his family in full dress, the dead is represented as he was supposed to look after several years in the grave: long worms slither in and out of the body, and the face is covered by toads in such a manner that two of their heads replace the eyes, offering a macabre spectacle.

PLATE 17. Tomb Slab of Archbishop Siegfried III of Eppenstein (died 1249), Mainz Cathedral. The archbishop's head peacefully resting on a pillow appears trampling the basilisk and the dragon while simultaneously placing crowns on the heads of two kings, Heinrich Raspe of Thuringia and William of Holland.

holy orders, spending their last years in monasteries. By beginning the process of penance in advance, they hoped to shorten any further stay in purgatory. Such a practice developed in the medieval kingdoms of Christian Spain. Many nobles joined the Military Orders to which they contributed their possessions and lands.[37] With the development of the Crusader kingdoms overseas, many nobles and knights entered the Military Orders in the Holy Land as well, and devoted their last years to the Holy War against the Infidel.[38]

The gradual ritualization of death through Church practices thus helped to restore the former reconciliation with death, one which the Catechism of Fear had crucially undermined. This reconciliation with death, however, could in fact be achieved only via the Church, which, in this way, finally consolidated its mediatory role between man and his Creator. Yet the overwhelming fears of a sudden death still had to be overcome, as this kind of death exposed the faithful to divine judgment without being properly armed with the Church's blessing. According to popular traditions, the most pious and righteous people were always blessed with a warning of their coming death and were thus given time to prepare themselves to die in peace with God. This became the most desirable death, one which had been perpetuated in the *Chanson de Roland*, Tristan, and the Legends of the Round Table. It never surprises the hero but rather warns him of its forthcoming fate, either by cryptic or unnatural signs. Such a "special" death, known in advance, however, is reserved only for the few, the chosen spiritual men. Ordinary men are most vulnerable to the more common, sudden death. This frightening possibility became one of the pillars of the Church's message, particularly among the intellectual elite. In a society where death was a kind of masterpiece, a sudden death without receiving the last sacraments was considered a clear expression of God's will, one that condemned the deceased with the hopeless loss of his soul.[39] This lay behind the Church's emphasis on the need to perform the sacraments, which accounts even for babies in danger of death to be properly baptized in order to safeguard their souls and assure them of eternal life. On the other hand, there were doubts regarding the burial to be given to those adults who died suddenly. A number of penitential books credited the suddenly deceased with the benefit of the doubt and accorded them Christian burial. This privilege, however, did not apply to those who died suddenly in the course of a duel.[40]

Thirteenth-century *exempla* faithfully reflect the fear mechanism that the sudden death motif aimed to develop. Very often, the sudden death was an immediate result of sin, as was the case of a wealthy man who made a false oath and was found dead on the spot—an unmistakeable, didactic moral for all those who swear falsely.[41] Even children were not spared by divine justice, a lesson which implies the universality of the Church message regardless of age or social status.[42] Sometimes, Satan himself or one of his demons informed the living, usually the most pious, about his taking over the souls of sinners. In other cases, the loss of the soul was suggested by unnatural forces, conveying a message from the underworld in an uncompromising form. This was the case of the fire that started in a grave and turned the corpse into dust, signaling the hopeless fate of the sinner buried there.[43] Stories of this kind aimed at encouraging penitence not merely as a sacrament to be taken circumstantially but rather as a way of life, so that death would not catch the Christian unprepared. A permanent form of penitential life was the most effective answer to the Cult of Death and the only one to ensure the expiation of sins, even though an unfortunate incident might prevent the fulfilling of the last sacraments in actual practice. This was the message behind the story of a man who prayed all his life that he would not die without receiving extreme unction. His prayers were apparently not heeded, since he died a sudden death before receiving the last sacraments. Thanks to his righteous way of life, however, he was given a second oppportunity. He was allowed to return to life in order to receive the extreme unction, thus ensuring eternal life for his soul.[44] The moral here was quite simple, referring, as it does, to the crucial role of the Church in this world and the world to come: even divine justice has need of the sacraments in order to reward the pious after death. Appeals to Saint Barbara were also believed to be particularly effective against sudden death.

Research on the Catechism of Fear and the Cult of Death in the Middle Ages suggests the main stages in a complex process of communication in which the Church message underwent a process of continuous change in order to ensure its universal reception. The Catechism of Fear and the resulting Cult of Death thus present a faithful example of the Church's ability to adapt its message to new conjunctures while safeguarding its own predominance in medieval society. The Church actually succeeded in neutralizing the harmful implications of the individualization of death while utilizing the new sociopolitical structures to its own advantage. Church propaganda, hardly practicable in the corporate framework of the village or the manor, found most fertile ground in the growing towns. Alone before an unknown reality and the frightening possibility of the after death, medieval people thus became more receptive to the Church message and more vulnerable to its manipulation. The degree of efficacy of this campaign is faithfully reflected through its results in the long term: the Catechism of Fear and the Cult of Death strengthened the Church's predominance in this world and turned repentance into the leitmotif of everyday life. Church supremacy, moreover, spread beyond the limits of earthly life while the idea of purgatory bestowed upon the clergy an everlasting influence over the world to come as well. The prevailing approach to death as an integral component of life, then, loses its primary paradoxality and the Cult of Death becomes an additional and crucial aspect of ecclesiastical communication in the Central Middle Ages. The influence of the Church's message and its propagandistic success in actual practice was further corroborated by the Crusader movement; this appears as yet another expression of the power inherent in Church propaganda during the Middle Ages.

5

Ecclesiastical Propaganda in Practice: The Crusades

The idea of the Crusades embodies one of the most deeply rooted concepts in Western culture. General Eisenhower's call for "a Crusade in Europe," to mention just one example, suggests the long-lasting impact of the Crusades on Western society and the constant need of Western man to justify his wars by well-known concepts.[1] This need was also common to medieval man when he departed overseas "to liberate the Holy Sepulchre and the Eastern Christians from the yoke of the Infidel." The awareness of the papacy, the promoter, and main supporter of the Crusades, as to the need to legitimate the Christian enterprise in the Levant brought about an unprecedented propaganda campaign, which is the subject matter of this chapter. The Crusades present one of the earliest examples of what has since come to be known as the use of mass media, whose impact in medieval society is hardly questionable. Yet the unprecedented reception of Crusade propaganda in wide social strata eventually challenged the papal plan of a selective military enterprise overseas while it distorted the original Crusade message. The massive enlistment in the First Crusade required a reconsideration of the former papal schemes in order to advance the goals which had originally initiated the call for the Crusade. The Crusades provide, therefore, a suitable case study for the use of mass media in medieval society and the ability of the papacy to adapt its policy according to the reception of its message. Research into Crusade propaganda further suggests the manipulative use of old concepts and stereotypes enlisted in the service of the Vicar of God on earth. This chapter deals with the papacy's view of the Crusade and, in parallel, the message it publicized and the channels of communication it utilized. In addition, the reception of the Crusade message and the papal reconsideration of former plans would be analyzed as complementary factors; these together depict the various stages of the most spectacular propaganda campaign in the Middle Ages.

Urban II's call for a Crusade at the Council of Clermont (November 27, 1095) raises questions that have not yet been satisfactorily answered. If the imperative to help the Eastern Christians had indeed encouraged the idea of a selective military enterprise overseas, then the propaganda campaign promoted by the pope throughout Christendom appears rather inaccurate. At the end of the eleventh century, feudal society had already developed recruit systems that could suitably fulfill these needs without requiring the pope's personal involvement or that of the ecclesiastical order as a whole. Besides, the crystallization of the Crusader kingdoms in the Levant spread far beyond the supposed assistance to the Eastern Christians. These developments could not be merely regarded as a reflection of the customary gap between original expectations and actual practice inherent in historical processes. They seem rather an outcome of the consistent policy of the Reformer papacy to advance its own interests, a goal that though not publicized, stood beyond the papal uncompromising support of the Crusade throughout the Middle Ages.

It is the thesis of this study that the Crusades suited the tendency of the Reformer papacy to turn Western society into the *Societas Christiana*, namely, to consolidate the complete subordination of the ecclesiastical order to the papal monarchy and, through the mediation of the clergy, the whole of Christendom as well. This perspective, clearly indicated by the basic premises of the Gregorian Reform, suggests the scheme of priorities of Urban II and the role ascribed by the pope to the European support of the Eastern Christians. From the papal point of view, a military victory in the Levant would not only help the Eastern churches but would also indirectly solve the painful problem of the Orthodox schism; it would also strengthen the papal leadership throughout Christendom as a whole. The Crusades thus resulted from the Reform policy of the papacy and its constant struggle to strengthen its position at home. This interpretation does not corroborate the thesis of Erdmann, who regarded the assistance to the Eastern Churches as the final goal of the Crusades (*Kriegsziel*).[2] The papal commitment to the Eastern churches and to the Byzantine Empire was linked to the challenge of consolidating the papacy and, from the papal scheme of priorities, probably subordinated to it. The strong links between the Crusade projects of the papacy and the Gregorian Reform elucidate the former plans of Pope Gregory VII to carry out a military journey eastward,[3] as an integral part of his program to consolidate the papal monarchy in the West. The correlation between Gregory VII and Urban II, that is to say between the Gregorian Reform and the Crusades, turned the assistance to the Eastern Christians and the liberation of the holy places into an important means manipulated in the service of the Reformer papacy and its interests in Christendom.[4] This perspective could also solve the paradox inherent in Urban's reluctance to allow the massive incorporation of the clergy in the Crusades. Since the Crusades were meant to support the papal monarchy in Europe, the clergy's involvement was essential at home where they were expected to play an active role in consolidating the status of the papacy. The goal of stressing the papal monarchy in Christendom, however, could hardly have found consistent support in medieval society nor could it foster the reception of the papal message. Neither the schism nor the imperative to help the Eastern Christians might, in 1095, have encouraged a mass movement eastward as it had not brought about any massive support in 1453, in light of the Turkish threat over Constantinople. Urban II successfully applied the lessons learned from the failures of his predecessors and avoided publicizing any plans closely connected with the hierarchic views of the Reformer papacy.[5] The lack of expedience in publicizing the needs of the Reformer papacy and its political goals in the East logically turned the focus of papal propaganda into the pan-Christian pilgrimage to Jerusalem. The idea of a pilgrimage to Jerusalem embodied both the desirable assistance to the Eastern Christians and the consolidation of the papal monarchy, since the pope was the sole initiator of the Crusade and, as such, its undisputable leader. By amalgamating the knightly ideals of a "Just War" with the moral code of pilgrims, Urban II ensured his own supremacy,[6] having an unquestioned monopoly over the remission of sins and the distribution of indulgences.

One major methodological problem in the research of the Crusades concerns the message publicized by the pope at the initial stages of the movement. The speech of Pope Urban II at the Council of Clermont, considered the first step toward the development of the Crusade, has not been preserved and it is quite impossible today to reconstruct its original content. Different versions of Urban's speech were recorded about a decade later, thus giving voice to the events in the Levant, rather anachronistic to the climate of opinion in Clermont. Munro has pointed out some common features among the various versions of the pope's speech[7] which reflect the major themes of Crusade propaganda at the initial stage: *The Eastern Christians' call for help* against the Moslem offensive had been argued by Urban II as the main justification for the Crusade. Urban emphasized the desecration of the holy places and the distress of Eastern Christians and pilgrims, suffering under the yoke of the nonbelievers. The report given in this regard by Fulcher of Chartres is quite pathetic and leaves no room for doubt as to the importance ascribed by the pope to the immediate improvement in this shameful situation:

> For as many of you have already been told, the Turks, a Persian race, have overrun them right up to the Mediterranean Sea. . . . they have conquered those who have already been overcome seven times by warlike invasion, slaughtering and capturing many, destroying churches and laying waste the kingdom of God. So, if you forsake them much longer, they will further grind under their heels the faithful of God.[8]

The future enterprise, however, went far beyond the limits of the Byzantine Empire, focusing on *the liberation of Jerusalem and the Holy Sepulchre.* According to Robert of Reims, in order to make this imperative clear, Urban II emphasized the unique role of Jerusalem in Christian tradition: "May you be especially moved by the Holy Sepulchre of Our Lord and Saviour, which is in the hands of unclean races, and by the holy places, which are now treated dishonourably and are polluted irreverently by their unclean practices."[9] An appeal of this sort transformed the journey overseas into a *pilgrimage*, which rewarded its participants with plenary indulgences in compensation for their total commitment to God. The canons of Clermont, indeed, pointed at the complete dedication to God as a prerequisite for obtaining the desirable expiation of sins: "If any man sets out from pure devotion, not for reputation or monetary gain, to liberate the Church of God at Jerusalem, his journey shall be reckoned in place of all penance."[10]

The journey eastward was therefore regarded as a faithful expression of God's will. Since *God would lead the Christian army*, no earthly power would stand in its way. This concept was behind the different versions of the pope's speech, its nuances only increasing the intensity of the papal message. The chroniclers attributed to Urban II the use of phrases such as "With the Lord as their leader,"[11] "God acting through you," "With God working with you"[12] or the biblical quotation which served as the final clause to the pope's speech in the version of Robert of Reims: "Whosoever doth not carry his cross and come after me is not worthy of Me."[13] These all hint at the leitmotif behind the papal call: being God's Vicar and His spokesman upon earth, Urban II was responsible before the whole of Christendom for the divine sanction of the journey overseas. As an outcome, the journey eastward became a *Bellum Justum*, a "Just War," and as such, stood in total contrast to the bloodshed between Christians that had hitherto prevented the salvation of many. Guibert of Nogent ascribed to the pope the following lesson:

> Until now you have fought unjust wars; you have often savagely brandished your spears at each other in mutual carnage, only out of greed and pride, for which you deserve eternal destruction and the certain ruin of damnation! Now we are proposing that you should fight wars which contain the glorious reward of martyrdom, in which you can gain the crown of present and eternal glory.[14]

In his letter "to all the faithful in Flanders" (end of December, 1095) Urban himself corroborates the main motifs reported by the chroniclers:

> We believe that you, brethren, learned long ago from many reports the deplorable news that the barbarians in their frenzy have invaded and ravaged the churches of God in the Eastern regions. Worse still, they have seized the Holy City of Christ, embellished by his passion and resurrection and, it is a blasphemy to say it, they have sold her and her churches into abominable slavery. Thinking devoutly about this disaster and grieved by it, we visited Gaul and urged most fervently the lords and subjects of that land to liberate the Eastern churches. At a council held in Auvergne, as is widely known, we imposed on them the obligation to undertake such a military enterprise for the remission of all their sins and we appointed in our place as leader of this journey and labour our dearest son Adhemar, Bishop of Le Puy. . . . If God calls any men among you to take this vow, they should know that it will depart, with God's help, on the Feast of the Assumption of the Blessed Mary (August 15) and that they can join the company on that day.[15]

Beyond the pope's appeal on behalf of the Crusade, this letter, preserved in its original form, acquires maximal importance from the perspective of medieval communication: it indicates both the media utilized at the time and their degree of efficacy in ensuring the transmission of news. The pope assumed that the deteriorating situation of the Eastern Christians was already known in Europe, as he acknowledged the existence of many *reports* on this matter. Besides, only one month after the Council of Clermont, Urban also laid down the widespread knowledge of its decisions, especially of the plans for the journey overseas. The papal letter further reflects the kind of audience appealed to by the pope, namely, knights—particularly the French knights—who were expected to ensure the military success of the enterprise. Urban's preference for French knights was emphasized by Robert the Monk, whose version of the pope's speech bestowed on them the flattering epithets of "race chosen and beloved by God. . . . distinguished from all other nations as much by the situation of your lands and your Catholic faith as by the honour you show to the Holy Church."[16] The pope's emphasis on a military journey, as confirmed by other sources, however, raises questions concerning the possibility of a massive expedition eastward, as recorded by Fulcher of Chartres. According to his version of Urban's speech, the pope had encouraged rich and poor alike, without any distinction of class or military training, to take up the cross.[17]

Nevertheless, the audience addressed by Urban II discouraged any massive participation by clerks or monks. The pope's reluctance to allow the clergy's personal involvement in the Crusade was clearly expressed in his letter to the religious of the Congregation of Vallombrosa (October 7, 1096):

> We have heard that some of you want to set out with the knights who are making for Jerusalem with the good intention of liberating Christianity. This is the right kind of sacrifice, but it is planned by the wrong kind of person. For we were encouraging knights to go on this expedition, since they might be able to restrain the savagery of the Saracens by their arms and restore the Christians to their former freedom: we do not want those who have abandoned the world and have sworn themselves to spiritual warfare either to bear arms or to go on this journey; we go so far as to forbid them to do so. And we forbid religious men—clerics or monks—to set out in this company without the permission of their bishops or abbots, in accordance with the rule of the holy canons.[18]

The resolute papal decision to prevent any mass recruitment within the ecclesiastical order was expressed anew in his letter to his supporters in Bologne (September 19, 1096) when the pope extended some of his reservations to the laity as well:

> We do not allow either clerics or monks to go unless they have permission from their bishops and abbots. Bishops should also be careful not to allow their parishioners to go without the advice and foreknowledge of the clergy. You must also care about young married men that would not rashly set out on such a long journey without the agreement of their wives.[19]

These statements faithfully indicate the kind of public addressed by the pope, both the initiator of the Crusade and, at the time, its main propagandist. The call for the Crusade appealed to all Christian knights and particularly the French, while the Church hierarchy was excluded from the camp of those heading eastward. Besides, the pope took care to prevent any state of anarchy expressing full confidence in the clergy's capability to deal with the organization of the journey and the careful selection of its participants.

In the first stages of the Crusade, Urban II appears therefore not only as its promoter but also as its main propagandist.[20] He tried first to enlist the support of the ecclesiastical order in the propaganda campaign of the journey eastward. About 300 prelates listened in Clermont to the pope's passionate preaching: "You brothers and fellow-bishops, you fellow priests and fellow-heirs of Christ proclaim the message in the churches committed to your care and give your whole voice to preaching manfully the journey to Jerusalem."[21] Yet contemporary data do not record a large response from the clergy to the pope's appeal. The prevailing apathy of prelates to the papal propaganda campaign is suggested both by the few reports that remain from the Council of Clermont,[22] and the lack of evidence with regard to the preaching of the Crusade by bishops. Beside his appeal to the clergy, the pope concentrated his publicity campaign in the kingdom of France, in areas surrounding the Capetian domain. He preached the Crusade at Limoges on Christmas, 1095, at Angers and Le Mans in February, 1096, at Tours in March, and at Nimes in July, but he might also have preached elsewhere. By the time he left France, the publicizing of the Crusade had already begun. Urban II also elaborated the communication patterns that served the Crusade communicators in the next two centuries, namely, the gathering of synods, the mission of legates, and a flow of correspondence addressed to both clergy and laity. The communication channels of the Crusade were thus established; throughout the Crusade period, *synods*, *legates*, and *written messages* became the main channels at the disposal of the medieval papacy. They were flexible enough to be adapted both to the changing needs of the Crusades and to the ever-changing European public opinion as well.[23]

The use of *legates* began with the very first attempts to carry out the Crusade propaganda. At the Council of Clermont, Urban II appointed Adhemar, bishop of Le Puy, as his legate in the journey overseas. Until his death in August, 1098, Adhemar became the undisputed leader of the Crusaders, both lay and clerical, and enjoyed their deferential esteem.[24] Although legates still participated in further journeys, their main role remained in Europe, recruiting volunteers and encouraging monetary contributions to the Crusade. In February, 1096, Urban II commissioned Robert of Arbrissel and Gerento, abbot of St. Benigne of Dijon, to preach the cross in the Loire Valley, Normandy, and England. In 1100, Pope Paschal II authorized the Cardinals John of St. Anastasia and Benedict of St. Eudoxia to summon synods in Valence, Poitiers, and Limoges, in order to reinforce the first wave of Crusaders, which had left for the East in 1096. From the second half of the twelfth century, the use of legates became quite customary, though their mission underwent further changes. On the eve of the Second Crusade, Pope Eugene III appointed Teodwin, cardinal bishop of Santa Rufina, and Guido, cardinal priest of San Chrysogono, as his legates *a latere* in the kingdom of France. Both cardinals were joined by local prelates in order to overcome their communication problems: as one came originally from Germany and the other from Florence, their knowledge of French was nil. This state of affairs supports the premises advanced in former chapters regarding the practical problems inherent in the use of legates: the pursuit of widest support required not only an elementary knowledge of the local language but also a fairly good acquaintance with the local customs. The search for a practical solution encouraged Pope Innocent III to further support the participation of the ecclesiastical order in the Crusade campaign. During the thirteenth century, the local clergy became responsible for recruiting Crusaders in the parish while they were reinforced by the friars, mainly the Dominicans.

The participation of the friars further underlined the importance of *preaching,* which remained the most important media for spreading the Crusade message. Crusade sermons gradually became an integral part of liturgy and ceremonial and were therefore given in the church buildings. New prayers for the good of the Holy Land and the success of the Christian enterprise overseas were added to the customary Mass and were followed by the psalm *Deus venerunt.* At the end of the sermon, all the congregation joined the priest in singing hymns such as "*Veni, Sancte Spiritus*," "*Veni, Creator Spiritus*," "*Vexilla regis*," or "*Salve, crux sancta.*" On such occasions, the preacher would rise from his pulpit and address the audience: "Now, who is hoping for the blessing of God? Who would join the congregation of saints? All those ready to come to me and receive the cross will be rewarded by God with all of these."[25] Sometimes, the impact of preaching was enforced by particularly dramatic circumstances: in July, 1099, for instance, the Crusaders surrounded the walls of Jerusalem seven times, waiting for the same miracle performed by God at the gates of Jericho. Yet the walls of Jerusalem still stood firm before the Christian army, which, lacking a more portentous tool, was most receptive to the mystic preaching of a hermit from the Mount of Olives, who gave the last impetus to the final conquest of the holy city.[26] A similar scene unfolded in October, 1147, when, before the walls of Moslem Lisbon, the Crusader army also listened to preaching; on this occasion, the preaching was suitably heightened by a glimpse of a fragment of the true cross.[27] Both cases portray the mechanism of what would now be called a "demonstration," followed by a "teach-in."

According to Humbert de Romans, the preacher himself had to take the Crusader vow in order to set a personal example. He also required preachers to become familiar with the Christian doctrine of indulgences and its implications in actual practice. But Humbert went even further and, in addition to rhetorical skills, encouraged a wide knowledge of the biblical sources concerning the Holy Land, its geography, or, at least, its exact place on the earth's surface and the nearby countries. Humbert also advocated the study of Moslem history and the basic principles of Islam. He even enclosed much reading matter available at the time.[28] Although his advice was hardly followed, a thirteenth-century satirical poem elaborated on the preachers' dominant influence on their audience, and warned against the consequences of rushing heedlessly to take the Crusader vow:

> If you go to hear the preachers,
> Do beware of clever teachers,
> Who can with their style and gloss
> Make you captive of the cross.[29]

These words were illustrated by an *exemplum* depicting a wife who, having heard some words of a Crusade sermon, locked her husband at home to prevent his further involvement in the journey overseas. But, being forewarned by providence, the husband succeeded in listening to the preacher, escaped by jumping from the roof, and eventually took the vow.[30]

Aware of the weight ascribed to the personal contact between communicators and their audiences, the popes also used to enlist all those known for their *rhetorical skills* in the Crusade campaign. One of the most famous preachers of the Crusade was the twelfth-century Cistercian monk, Bernard de Clairvaux who, according to Prawer, turned the second Crusade into "a journey for the salvation of the soul."[31] Encouraged to preach the Crusade by his former pupil, Pope Eugene III, Bernard began to preach at Easter, 1146, at Vézelay, where he succeeded in enlisting Louis VII and a large number of French noblemen. In the winter of 1146–1147, he preached in the Lowlands and Germany, receiving the enthusiastic support of Conrad III and the German princes.[32] Between June and July, 1147, he returned to Southern France, where he encouraged military journeys against the Moslems in Spain. Bernard's messengers also reached remote places in the Lowlands and the Rhine valley. His letters were received by the Bretons and the Diet of Regensburg and Bohemia by February, 1147. Henry of Olmutz carried a personal letter from Bernard to Abbot Gerlach of Rein, who preached the Crusade in the Steiemark and Carinthia.[33] The influence of Bernard de Clairvaux on medieval audiences thus reached its zenith, when his letters were read before audiences anxiously awaiting to receive his message.

The imperative to operate a long-term communication system encouraged the papacy to implement the parallel use of additional channels. Papal *encyclicals* were read before the whole congregation of the faithful in a particular area. Pope Alexander III ordered the prelates to read in public his Crusade letters in all churches, and his message to be explained to the faithful (1181).[34] We have satisfactory evidence that Eugene III's bull calling for the Second Crusade ("*Quantum predecessores*") reached England, Denmark, Tournai, the monastery of Lobbas in the Lowlands, the count of Flanders, and Bishop Arnulf of Lisieux very close to the papal call.[35] By the synchronic mobilization of both the legates and the local clergy, either through written and oral messages, in Latin or in the vernacular, Crusade propaganda thus spread throughout all social strata in Christendom. The ecclesiastical order, too, functioned as a most accurate source of information about current events overseas. After the conquest of Jerusalem, the archbishop of Reims, for example, announced the good news to his suffragans. He asked all bishops in his province to hold thanksgiving masses and to encourage fasting and the giving of alms to the poor, thus bringing about the personal involvement of all the faithful in the Christian enterprise.

The length of time required for the *transmission of information*, however, remained a critical problem during the whole Crusade period. The need to achieve a maximal reception of the Crusade message in a minimal period of time was often dictated by the accelerated tempo of events in the East, and the delay caused by the long distances separating the Crusader kingdoms from Western Christendom. On the eve of the Second Crusade, the news about the fall of Edessa in December, 1144 was formally delivered to the pope almost one year later, by messengers sent from Antioch to the papal court. The arrival of the Crusader delegation moved Eugene III to an immediate reaction, and the bulls calling for a new Crusade, were dated December 1, 1145. Nevertheless, the Crusader armies left Europe in April, 1147. In other words, almost eighteen months were required to complete the preparations for a new Crusade, actually delaying the European response by two and a half years after the Moslem conquest of Edessa. Nor did such a state of affairs improve in the fourteenth century when Western society still continued to foster plans for a forthcoming Crusade, which, ultimately, never materialized. For example, the Carmelite prior in Albi, received a copy of the papal bull of July, 1333, authorizing the tithe only after March 23, 1334. Similarly, the preceptor of Lodève was notified of the papal instructions on August 18, 1334, with a delay of thirteen months.[36]

Quite paradoxically, 200 years earlier, the time schedule of the First Crusade had been much shorter. At the end of November, 1095, Urban II preached the Crusade at Clermont and planned the departure of the Christian armies for August 15, 1096, thus leaving only nine months between his announcement and the actual implementation of the journey. Nevertheless, the waiting time announced by the pope seemed to have been too long, for as early as in April, 1096, the first armies had left Germany, namely, five months after the papal call. The general readiness to sail immediately overseas was often related to the social environment of the First Crusade, most of its participants coming from the lower social strata and thus having fewer material interests to risk. Besides, dramatic natural phenomena that took place close to the council gave further validity to the papal call while bestowing it with an apocalyptical aura: a shower of meteors (April, 1095), lunar eclipses (February and August, 1096), signs in the sun together with an aurora (March, 1096), comets (autumn, 1097), the sky glowing red (February, 1098), great lights in the sky (autumn, 1098), eclipse of the sun (December, 1098), and another red aurora (February, 1099) filled the western skies. These phenomena brought about the proliferation of masses, and litanies were also said in churches.[37] Meanwhile, a severe drought that had lasted for several years ended abruptly in 1096 with a wet spring followed by a magnificent harvest. All these were perceived as further signs of the approaching doomsday, while at the same time supplied a convenient stimulus for the universal reception of the Crusade message. Nobles committed themselves promptly to the First Crusade as well. The delegates of Count Raymond of Saint Gilles arrived at Clermont on December 1, only four days after the papal announcement. They formally notified the pope of the count's readiness to undertake the Crusader vow, and his hopes to be rewarded with the leading of the Christian army. The willingness of Count Raymond was not exceptional at the time. Philip I, king of France, his brother Hugh of Vermandois, and other members of the French nobility discussed the papal plans on February, 1096, namely, three months after Clermont, and some of them joined the Crusader army shortly thereafter.

The gap between the First Crusade and later journeys could not be explained on the grounds of technical improvements, but, rather, of the different degrees of motivation. At the end of the eleventh century, Urban II found a most propitious ground to propagandize, as his contemporaries' knowledge about the Holy Land and the Moslems was almost nil. This information vacuum left much room for maneuvering, manipulation of information, and the selective transmission of messages. Most contemporaries were aware of the unprecedented speed of transmission of the Crusade message, and there were some who tried to find the explanation in the eschatological climate of the times:

> In order that it should be clear to all the faithful that this journey was decreed by God and not by man, it happened. . . . that on the very day on which these things were said and done, the reverberating report of that great decree shook the whole world, even in the maritime islands in the Ocean. It was spread abroad that the Jerusalem journey had been thus established at the council. It is clear, therefore, that this was the work of no human voice, but of the Spirit of the Lord who fills the whole earth.[38]

Guibert of Nogent was particularly impressed by this state of affairs and reported in full detail the unprecedented promptness in the communication of the Crusade message:

> It was not necessary for any ecclesiastical person to make orations in churches to stir up the people, since one man told another, both by word and by example, at home and elsewhere about the plan to set out. . . . A great rumor spread into every part of France, and everyone to whom the advancing report first brought the pope's command, then approached all his neighbors and family about undertaking the Way of God, for so, metaphorically, it was called.[39]

These reports suggest the crucial role played by *informal channels of communication* in the publicizing of the First Crusade, when those receiving the message shared the wonderful news with their relatives while giving to the spreading rumors a maximal credibility. Besides, the journeys of Urban II in the kingdom of France encouraged the diffusion of the Crusade beyond the areas in which the pope actually preached. The desecration of the holy places by the Moslems, the pilgrimage to Golden Jerusalem, the Just War against the heathens, were messages accepted without hesitation among medieval audiences, whose basic world of ideas was nourished by a Christian scale of values. Moreover, Urban II had ensured a priori the positive response from the feudal nobility with whom he had probably held conversations at the time of Clermont.[40] In addition to Raymond of Saint Gilles, it seems quite probable that papal legates had shared the project with the most prominent members of the French aristocracy, such as Robert of Flanders and Robert of Normandy. Urban's care to associate the feudal nobility in his plans reinforces the military essence of the Crusade from the papal point of view. Even the personal participation of kings was not formally encouraged in the initial stages.[41]

The spontaneous intervention of *wandering preachers* added yet another factor to the amorphous nature of the First Crusade and also underlined the inability of the clergy to channel the general euphoria into more organized patterns. Peter the Hermit was one of the most prominent figures, largely responsible for the mass recruitment among the poor. A small man, thin, with long grey hair, he fitted the image of a biblical prophet and enjoyed complete control over his audiences. Large masses of people seemed to hear the word of God through him and made intense efforts to obtain some of his personal belongings as holy relics.[42] In order to enhance his prophet-like image, one of the popular stories told about Peter refers to his revelation of the Crusade by the Holy Ghost in Jerusalem, before the pope actually issued his call in Clermont. Peter contributed to the veracity of this story by carrying everywhere a heavenly letter, in which the conditions of his mission were carefully stipulated by the Holy Ghost. He began to preach the Crusade at Berry at the end of 1095. In February-March, 1096, he preached in Champagne, and later in Lorraine. From there he passed from the Orléans area, through the cities of the Meuse Valley and Aachen, to Cologne, where he spent Easter with a crowd estimated at about 15,000 people; he was joined later by an additional 5,000 volunteers recruited by his messengers. Most of them sought the Holy Land of milk and honey to which Peter, as a new Moses, would lead them. Psychologically they were ready to fight the armies of Antichrist on their way to a Golden Jerusalem and to stand the historical trial ordained by God. The concept of the Crusade as a sacred mission reserved for the chosen people found a most fruitful arena within the lower social strata; among the social elite as well, there were also some who followed in the steps of Peter, such as Count Hugh of Tubingen, Count Henry of Schwarzenberg, Walter of Teck, and the three sons of the count of Zimmern. The thousands who followed Peter were not an isolated phenomenon and, albeit in smaller measure, one can point at similar spontaneous groups elsewhere. Armies of Saxons and Bohemians were also recruited by two priests, Folkmar and Gottschalk. In addition, a large army of Rhineland, Swabian, French, English, Flemish, and Lorrainer Crusaders were led by Count Henrich of Leiningen. As time went by, and both the Levant lost its initial mystique and the defeats of the Crusaders in the battlefield increased, the ability of the ecclesiastical order to restrain the popular elements was further weakened. The Children's Crusade in 1212 and the Shepherds' Crusade in 1251, to mention just two examples, were symptomatic of the Church's inability to fulfill both the papal expectations and those of medieval men, while it failed in channeling the latter into more organized patterns.[43]

In addition to the active participation of preachers, the Crusades were also considerably publicized by *amateur communicators*, who, though lacking professional skills, still played an important role at the information level. Some of the Crusaders who returned home full of glorious anecdotes of the Holy Land became spontaneous spokesmen on behalf of the journey eastward. Following his return from Jerusalem at the beginning of 1106, Bohemond of Taranto made a "tour de France" and donated treasured relics to churches and monasteries. He shared his marvelous experiences with local audiences eager to hear news of the Crusader kingdoms from an eyewitness. In the eyes of their contemporaries, Bohemond and his peers personified the biblical heroes and as such, enjoyed the highest prestige, exemplified by the readiness of the feudal aristocracy to appoint them godfathers of their children. The admiration reserved for those who had actually participated in the journey eastward thus facilitated the communication process with medieval audiences, which, having rarely crossed the familiar borders of the village or the manor, were largely receptive to those who had become acquainted with the wide world.[44]

The papal expectations for a selective military journey were not therefore fulfilled in the First Crusade nor in the Second, when the massive patterns of participation of 1095–1099 were repeated.[45] The prevailing atmosphere of the approaching Kingdom of God was too sweeping, the effectiveness of the diocesan system too weak for the papal plans of a selective military journey to be accomplished. Medieval sources mirror the diversity of factors that brought about the massive participation in the Crusade. On the eve of the Second Crusade, an anonymous chronicler disapproved of the fact that:

> Some, eager for novelty, went for the sake of learning about strange lands; others, driven by want and suffering from hardship at home, were ready to fight not only against the enemies of the cross of Christ but also against Christian friends, if there seemed a chance of relieving their poverty. Others, who were weighed down by debt or who thought to evade the service that they owed their lords or who were even dreading the well-merited penalties of their crimes, simulated a holy zeal, and hastened (to the Crusade) chiefly to escape such inconveniences and anxieties. With difficulty, however, there were found a few who had not bowed the knee to Baal, who were indeed guided by a sacred and salutary purpose and were kindled by love of the divine majesty to fight manfully, even to shed their blood for the sake of the holy of holies.[46]

Despite the decreasing number of Crusaders in the thirteenth century, the passenger list of a pilgrims' ship in 1250 further testifies as to the social heterogeneity characteristic of those moving eastward.[47]

Research into the European response to the papal call indicates, therefore, a significant gap between the First and Second Crusades and those that came afterwards. In the initial stages, the Crusade message achieved an unprecedented and unexpected success. Although we lack accurate data regarding the rate of positive response to Urban's call, medieval reports indicate an immediate massive reaction which, in fact, nullified the projects of a selective military journey. The chronicler Ekkehard claims that about 100,000 men took up the cross at Clermont,[48] an exaggerated estimate in light of the fact that only some 300 prelates were actually present. Even if one takes into account the customary presence of the laity and the possibility that the pope made his speech outside the church building, the number of 100,000 men still lacks authenticity, but does indicate an unexpectedly large response. Sigebert de Gembloux emphasized the wide geographical scope of the Crusade as those who took the vow included "Western people. . . . whose number can hardly be estimated. . . . from Spain, Provence, Aquitaine, Bretagne, Scotland, England, Normandy, France, Lotharingia, Burgundy, Germany, Lombardy, Apulia, and other Christian kingdoms."[49] As to more accurate figures, Riley-Smith estimates the number of warriors in the siege of Jerusalem in 1099 at 43,000, a number that does not include the great mob of noncombatants who had joined the Crusade or those who had died on the long journey to the Holy Land. Despite the pope's reservations, the clergy, too, was present among the multitude moving eastward. At least ten bishops from Italy and France participated in the First Crusade, together with two archdeacons and about thirty priests, most of whom were probably members of the lower clergy. Besides, contemporary sources indicate a varying number of monks, five abbots, and a nun, fleeing from their monasteries to join the army.

This state of affairs actually made a mockery of Urban II's warnings regarding both the clergy's exclusion from a personal involvement in the Crusade and their need to obtain ecclesiastical authorization in advance. The massive response to the papal call could not therefore be merely regarded as a propaganda success, as it embodies a gap between the pope's declared intentions of a selective military journey and the recruitment of thousands whose military skills, if any, were rather low. Besides, the wide identification with the Crusade in its earlier stages requires further inquiry regarding the issues that ensured the universal acceptability of the papal message and its further distortion in actual practice. The participation of the clergy as well as the massive recruitment among people from various social strata, moreover, raises questions with regard to the degree of efficacy of the diocesan system in order to fulfill the papal expectations. One should also question whether this mass recruitment was due to a process that surpassed the original papal plan or whether it reflects a new scale of priorities elaborated by the papacy in light of the enthusiastic reception of the Crusade message. A partial answer to these questions is provided by a deeper insight into Crusade ideology and the rewards designed by the eleventh-century papacy in order to foster the acceptability of its propaganda campaign.

The papal awareness of the climate of opinion in medieval society paved the way for the ideological foundations of the Crusades and provided their juridical basis. The Crusade was depicted as an expedition authorized by the pope, its leading participants took vows and, consequently, they enjoyed the privileges of protection at home as well as a *plenary indulgence* of all their sins.[50] The indulgence granted via the Church, on God's behalf, the remission of all or part of the penalties that resulted from sin. It was based on the concept of a "Treasury of Merits," an inexhaustible amount of virtues stored up by Christ and the saints on which the Church could draw on behalf of a repentant sinner. Besides, the vow to take up the cross turned the Crusaders into an integral part of the Church, who could thus enjoy its privileges. This state of affairs was faithfully reflected in the letter of Gregory VIII to Hinco of Scrotin who, following his intention to sail overseas (1187), was rewarded with the pope's commitment to "take under the protection of St. Peter and ourselves, your person with your dependants and those goods which you reasonably possess at present. . . . stating that they all should be kept undiminished and together from the time of your departure on pilgrimage overseas until your return or death is most certainly known."[51] A detailed scale of compensations was carefully drawn up, further encouraging the feeling of uniqueness within the Crusaders. The *sign of the cross* on their outer garment became their distinctive symbol, one that further strengthened their collective consciousness as the chosen people elected by God to release His Sepulchre from the yoke of the infidel. The cross became the first badge worn by an army in post-classical times and, according to Cohn, "the first step towards modern military uniforms."[52] The distinctive use of the cross bestowed upon all the Crusaders a unique and unifying identity, transcending the lines of national borders or social class. The use of the cross thus complemented the message of Crusade propaganda: in a direct, visual way, it reflected the link between the Crusaders, ready to become martyrs in the path of the Holy Land, and Jesus Christ, Who had sacrificed Himself on the same ground for the salvation of humanity. The emphasis on the symbol of the cross in medieval sources, and the large number of miracles ascribed to its use, clearly indicate its large promotional success.[53]

The massive reception of the Crusade propaganda was further conditioned by the ability of the Church communicators to link the Crusade with deeply rooted ideas at both the popular and the intellectual levels of perception. The classification of the Crusade into the category of a "*Just War*" was an important tool mobilized in the service of Crusade propaganda, one that embodied both the German war epic[54] and the feudal knighthood ideals.[55] It provided a most effective channel through which to reach the wide support of knights, the main target of Crusade propaganda. Yet it faced the Church communicators with the imperative to give a new interpretation to the evangelical condemnation of bloodshed. In the confrontation between evangelical principles and political reality, nevertheless, the Crusade served as a stimulus to legitimate wars; it also encouraged the integration of the papacy into militant political systems, thus culminating the process of compromise whose genesis reached back to the fourth century.

The gradual spread of Christianity and its final recognition as the formal religion of the Roman Empire (AD 380) had weakened the original polarization between Christianity and warfare.[56] It further encouraged a reexamination of warfare, no longer within the framework of a persecuted sect but as the recognized religion of the Roman State. The gradual justification of war paved the way for a new social ideal of the warrior-saint, such as Demetrius, Theodore, Sergius, and George, the cult of whom suggests the theological compromise reached by the Church. The Christianization process of the Germans encouraged a similar evolution in medieval Europe as well. Already in the sixth century, Gregory of Tours had presented the wars of the Frankish King Clovis as part of the divine plan to annihilate any seed of the Arian heresy upon earth, thus giving divine justification to the aggressive policy of the first German king who embraced the Catholic faith.[57] The popes had also made significant steps toward legitimizing warfare when they applied to secular rulers for military assistance. Popes Gregory I, Leo IV, John VIII, Leo IX, and Alexander II went even further, and promised everlasting life to all those who fell in combat while defending the Church against the Moslems or the Normands.[58] The military needs of the papacy in the Early Middle Ages thus fostered the legitimization of warfare in the context of a "Just War," which, by definition, was a war of defense. This process both weakened the strict limitations of warfare established by the Fathers of the Church and gave the papacy a more flexible ideological platform, one that fitted the changing needs of the papal institution.[59] The concept of a "Just War" became an effective propagandistic tool in the service of papal security, while it paved the way for the papal enterprise *Outremer*, as the journey overseas was called in Crusade historiography.

The Church's recognition of warfare and its integration in the concept of the "Just War" left its mark in the elaboration of a complicated ceremonial, which hallowed knighthood with divine sanction. By the eleventh century, the use of blessed banners and special consecrations of weapons suggests the Church's tendency to patronize medieval knights while channeling the prevailing violence in more acceptable directions. The *Oratio super militantes*, a blessing conferred upon knights at the ceremony of ordination, asked the Lord to "Bless your servants who bend their heads before you. Pour on them your stable grace. Preserve them in health and good fortune in the warfare in which they are to be tested. Wherever and whyever they ask for your help, be speedly present to protect and defend them."[60] In addition, also in medieval Europe, the Church encouraged the cult of the warrior-saint, in patterns similar to those found in the Byzantine Empire. Saints who had been soldiers, such as Martin, Maurice, and Sebastian, became ideal guides for medieval knights, as well as St. George, who, coming originally from the East, became a popular figure during the Crusade period.[61] The popularity of the Archangel Michael also reached its peak in the eleventh century; his image was engraved on the shields of Henry I and Otto the Great. Considered the prince of heaven, to him was attributed the leading of soldiers into victory in the battlefield. He might also help monks in their endless fight against the devil and his temptations (see plate 18). These developments reflect one essential aspect

PLATE 18. Missal of Henry de Midel, Hildesheim, c. 1159 (285 × 190 mm). Michael and the battle between angels and devils supported by the dragon.

that further encouraged the reception of the Church's message among various social strata. Faced with developments that undermined the unity of Christian society and also harmed its own leadership, the papacy refrained from entering a struggle condemned to failure. The popes opted, instead, for a way of compromise, thus improving their chances of becoming an integral part of medieval society while advancing their own interests. Against the anarchical and adventurous image of medieval knights, the Church communicators placed the challenge of the warrior-saint as one who had been distinguished by his military skills but who, nevertheless, implemented them for salutary purposes. All ecclesiastical media were enlisted for this goal, including the especial blessing of knights, the ceremony of ordination to knighthood, the consecration of banners and weapons, and the cult of the warrior-saints.

The implementation of the Crusades, however, faced the Church with the imperative to fit the movement into the category of a "Just War," either by giving the Crusade the defensive nature inherent in such definition or by adapting the concept of a "Just War" to the new circumstances. Gregory VII had made a crucial contribution to the obliteration of defensive criteria. He defined a "Just War" as any war made in the service of the Holy See which, per se, rewarded its participants with the expiation of their sins.[62] Before the Crusade actually materialized, a convenient ideological ground was therefore carefully formulated in order to allow the integration of any papal enterprise into the patterns of a "Just War." As such, it was led by the pope, and granted its participants the expiation of their sins.[63] The call of Urban II to defend the Eastern Christians against Islam was now completely legitimated, as it fitted the patterns established by his predecessors with regard to the "Just War." Furthermore, the Moslem profanation of the holy places where Christ had lived and been crucified was regarded as usurpation and conquest, two "facts" that turned the Crusade into a defensive war. This was the conclusion reached by Odo of Chateauroux when he claimed: "But someone says, 'the Moslems have not hurt me at all, why should I take up the cross against them?'—But if he thought well about it, he would understand that the Moslems have done, indeed, great injury to every Christian."[64]

Yet the problem of identifying the enemy was still to be solved, since, on the eve of the First Crusade, the medieval West suffered from an almost absolute lack of information about the Moslems.[65] The inability to reach a clear definition of the enemy brought about its generalization as "heathens," which, as such, reject any belief in Christ the Savior. The ill-defined perception of the adversary and the lack of information about the Moslems and Islam facilitated a stereotyped picture of the enemy. The Moslems further served as a group of reference that advanced the stereotyping of the Crusaders as well. Those who took up the cross were thus identified with the mythical figures of Joshua, Gideon, David, and Judas Maccabeus. The geographical goals of the Crusades also played an important role in the integration of the biblical heritage since both the Crusaders and the children of Israel fought on the ground of the Holy Land. Besides, both the biblical heroes and the Crusaders were fighting a Holy War against the nonbelievers, a battle desirable in the eyes of the Lord.

These common denominators became the leitmotif of Crusade propaganda as it developed after the call of Clermont:

> If the Maccabees in the old days were renowned for their piety because they had fought for the rituals and the Temple, then you too, Christian soldiers, may justly defend the freedom of the fatherland by the exercise of arms. . . . If the children of Israel oppose this by referring me to the miracles which the Lord performed for them in the past, I will furnish them with an opened sea crowded with gentiles. To them I demonstrate, for the pillar (of fire), the cloud of divine fear by day, the light of divine hope by night. Christ Himself, the pillar of rectitude and strength, gave to the Crusaders instances of inspiration; he strengthened them without any earthly hope, only with the food of the word of God, as it were with heavenly manna.[66]

Fulcher of Chartres saw in the Maccabees and Gideon a prefiguration of this renewed alliance between God and the Crusaders in his own days.[67] Moreover, the struggle against the nonbelievers, common to the children of Israel and the Crusaders, would ensure the desired victory on the battlefield since the Crusaders were following in the steps of Israel; they had left Egypt and were guided by God through the desert. "The children of Israel," claims Baldric of Bourgueil, "who were led out of Egypt and after crossing the Red Sea appropriated this land for themselves by force, prefigured you, with Jesus as their leader."[68] After the Battle of Dorylaeum (July, 1097), Robert of Reims made the Crusaders sing a triumphant hymn, based on the biblical hymn of gratitude in Exodus, following the destruction of the Egyptians in the Red Sea.[69] "Although for its sins it sustained the whiplash of its God," wrote Raymond of Aguilers, "the army of God stood victorious over all paganism through his mercy."[70] Describing the Crusaders' conquest of Antioch, Petrus Tudebodus began his list of local kings with the prestigious figure of Judas Maccabeus.[71] The conquest of Jerusalem (July, 1099) gave further impetus to the eschatological expectations, as the heavenly city was expected to follow the steps of the earthly one.[72] Robert of Reims found much inspiration in the prophetic books that fitted "this deliverance made in our times. We see *in re* what the Lord promised to Isaiah; what Isaiah said of the spiritual Church is now fulfilled in reality: the gates of Jerusalem are open, which used to be shut day and night."[73] The Crusaders' successes were believed to have been divinely predestined,[74] and the Crusaders, "a blessed people chosen by God," were chosen by providence expressly in order to accomplish this task.[75]

The use of biblical images and metaphors was not unique to the propagandists of the First Crusade, but rather served as a source of inspiration throughout the Middle Ages. Bernard de Clairvaux saw in the Knights Templar the personification of the Maccabees and the children of Israel in their readiness to fight the Holy War.[76] The links between God and the former chosen people also explained the defeats suffered by the Crusaders in the battlefield while turning their military reverses into a divine trial by which God tests the endurance of His new chosen people. It follows that a historical moral had to be learned, namely, "in the same way we read in Holy Scripture that the children of Israel were frequently afflicted and defeated in war by the Philistines and the Edomites and the Midianites and other neighboring peoples, to force them to return to God and to persevere in keeping his commandments."[77] This common suffering served as a unifying force between the children of Israel and the Crusaders, who followed the same path of pain. Reporting the troubles of the Christian army at Damietta (1219) and the resultant resentment among the Crusaders, two commentators ascribed to the Crusaders the former complaint of the children of Israel to Moses: "Because there were no graves in Egypt, hast thou taken us away to die in the wilderness? wherefore hast thou dealt thus with us, to carry us forth out of Egypt?" (Ex. 14:11).[78] There were, nevertheless, some authors who pointed at a gap between the Crusaders and the Israelites, suggesting, rather surprisingly, the moral superiority of the latter: "But as they looked after the failure of the Crusade, unlike the Israelites who did not in their hearts return to Egypt, the Crusaders failed to sustain their faith, and so the Lord's mercy could not prejudice His justice. Like them, they should return a second and a third time to the task."[79]

These admonitions, however, did not undermine the spiritual essence ascribed to the Crusaders as the "Pilgrim Church of God."[80] The focusing of the journey's goal (*Marschziel*) upon Jerusalem through the liberation of the Holy Sepulchre provided one of the most creative ideas in the history of mankind. It offered an irresistible combination of spiritual and temporal rewards, the former by the donation of the pope, the latter by his sanction. An ordinary Christian living in the eleventh century might not have known the name of his nearest city, and it is doubtful whether he knew the name of his king's residence; but he knew of the existence of Jerusalem, of which he had heard during Mass, its image glowing in the stained-glass windows of the Church. Jerusalem had become a fairly tangible place for him, the place where Jesus the Savior was crucified, where God will shape the final battle which heralds doomsday.[81] The mythical perception of Jerusalem found a practical expression in medieval cartography, which located Jerusalem in the center of the world (see plates 19, 20). It was perceived as the site of the biblical paradise where the first man had been overcome by Satan. Jerusalem was pictured as the holiest relic, one that incorporated the holiness of the prophets, Christ, and the Apostles, and, as such, one that had been chosen by God as the arena for the crucial events in the history of humanity.[82] This state of affairs brought about much confusion of earthly Jerusalem with the heavenly city. Earthly Jerusalem was incorporated into the kingdom of purity since it had been blessed with both spiritual and material beauty. Even theologians regarded earthly Jerusalem as a symbol of the heavenly city "like unto a stone most precious," which, according to the Book of Revelations, was to replace the earthly city at the end of time.[83]

The use of clichés, such as *Exercitus Dei*, the army of God, *Fideles Christi*,

PLATE 19. Map of Psalms, c. 1250. The map is a supplement of the Book of Psalms; it depicts Jerusalem at the center of the world with the inscription: God working salvation in the midst of the world.

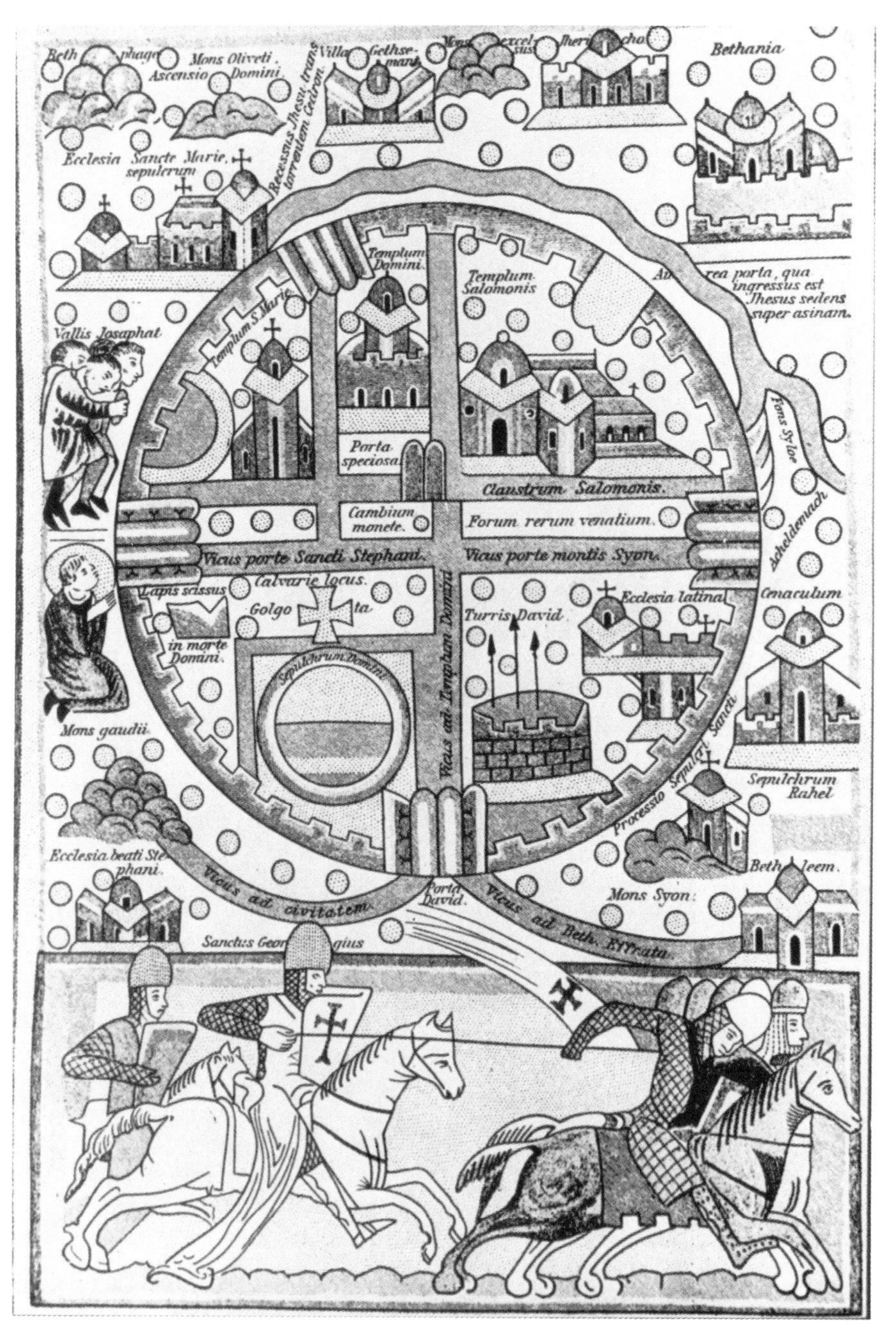

PLATE 20. Crusader map of Jerusalem from about 1170. The map represents the city and its vicinity, from Bethlehem to Mont Joie (Nebi Samwil). The city is shown schematically, as a circle divided into four quarters with the east at the top. The Dome of the Rock, *Temple of the Lord,* Al-Aqsa, *Temple of Solomon,* Golgotha and the Tower of David are prominent. Below is depicted the victory of a Crusader over infidels.

the loyal army of Christ, and *Gens sancta*, holy people,[84] suggest the crucial influence of the Crusade in fostering a collective identity among those departing overseas. Quite naturally, this search for self-definition focused on the Christian common denominator while strengthening the concept of a *Societas Christiana*, a more desirable goal from the perspective of the Reformer papacy. As a result, the enviable status ascribed to the Crusaders as active partners in the divine plan acquired a new light, providence being responsible for their journey, and even "physically" accompanying it. These "facts" were emphasized time and again in Crusade propaganda under the motto of *Deo duce*, under the leadership of God, or *Deus lo vult*, it is God's will. Before the term *Crusade* came into existence,[85] the Church propagandists depicted the Christian enterprise overseas through a metaphorical spectrum hinting at this deepest spiritual experience. The journey eastward was described as *Via Dei*, the way of God, *Via Jesu Christi*, the way of Jesus Christ, *Expeditio Dei*, God's expedition, *Iter Dominicum*, the way of the Lord, or even *Gesta Dei*, the deeds of God. All communication media spread the same message through the use of edifying anecdotes, which emphasized the divine reward awaiting the Crusaders in the world to come,[86] or through folk poetry, which voiced the same message in a rather convincing manner:

He who embarks for the Holy Land,
He who dies in this campaign,
Shall enter into heaven's bliss
And with the saints there shall he dwell.[87]

The convincing argumentation of the Crusade campaigners, however, does not indicate the degree of acceptability of their message. Still, the reception of Crusade propaganda is further confirmed by a participant in the Third Crusade, Conon of Bethune, who claimed:

God is besieged in His inheritance
And we shall see how people will respond
Whom He redeemed from darkness and the grave
By dying on the cross the Turks now hold.
You know, they are dishonoured who refuse,
Unless they are too poor or old or sick.
Those who are healthy, and are young and rich
Cannot without disgrace remain behind.[88]

The same concepts were voiced by Prevostin, chancelor of the University of Paris in the thirteenth century, who, in his sermon at Advent, argued somewhat metaphorically that "the Holy Sepulchre knew glorious days, but because of our sins had lost its sacred halo. The demons jumped on it, singing: 'Where is the God of the Christians?' The Saracens, at least, did not lose their God. The God of the Jews is asleep. But the God of the Christians is dead. Our sins have brought about a shame such as this."[89]

Feudal vocabulary was used as well, a tendency that caused the French medievalist Duby to emphasize the influence of the "feudal spirit" that had actually turned the Crusaders into the "loyal vassals of a jealous God."[90] The Dominican Etienne de Bourbon encouraged preachers to approach the Holy Land as the royal domain of Christ, which had been unjustly conquered by the Moslems. It follows that there is an imperative duty for the subjects of Christ to release His land from the nonbelievers.[91] Thirteenth-century popes elaborated on these concepts and reached the inevitable conclusions: though Christ is indeed the ruler of universe, He is linked to the Holy Land with special bonds.[92] The Moslem usurpation of the Kingdom of Christ allowed and even forced His Vicars to initiate the liberation of the holy places while recruiting all the faithful in the sacred enterprise.[93] The Christian kings had to be pioneers in the liberation of Christ's lands, since they had to serve as examples to be imitated by their subjects.[94] Pope Innocent III exemplified this idea through an interesting analogy:

> Let us say that a temporal king had been expelled from his kingdom by his enemies; if his subjects do not commit themselves and their property to him, when he reconquers his kingdom he will regard them as traitors; thus will the King of Kings, the Lord Christ, Our Savior, if you hesitated to come to His help when He was driven out of His kingdom, which He bought at the price of His blood.[95]

Humbert de Romans summed up the whole concept: "As the royal majesty makes it necessary to have an army always at the king's command, so does the majesty of Christ, the King of Kings, makes it necessary that He should have a Christian army, always ready to fight against the enemies of His cross."[96]

The harmonious blending of the ideals of a "Just War," a pilgrimage to Jerusalem, and feudal terminology borrowing the concept of *auxilium*, thus gave an unprecedented impetus to the reception of the Crusade message. The mass commitment to depart for the East transcended the lines dividing clergy and laity and found an additional expression in medieval historiography. One can hardly find another event that received such publicity in medieval times: whether in chronicles, reports, poetry, written or told folk tales, either by eyewitnesses or by those who remained in Europe and nourished themselves on their own imagination or the reports of others, the Crusades caught the attention of medieval audiences by all available means. The unprecedented success of the Crusade thus faced the papacy with the challenge of adapting its former schemes while reaping the maximal profit from this state of affairs that undoubtedly opposed the concept of a selective military journey to the Levant. This imperative allows a reconsideration of the Crusade from the three main premises of this study, namely, the papal view of the Crusade, the reception process of Crusade propaganda, and its resulting influence on the papal policy in the long term.

At the beginning of this chapter, I proposed to evaluate the success of the Crusade from the scheme of priorities dictated by the Reformer papacy, the promoter of the Crusade and, in its initial stages, its main propagandist. This perspective turned the Crusade into one of the main tools at the disposal of the papal monarchy and ascribed only marginal importance to the declared goals of the assistance to the Eastern Christians and the liberation of the holy places. The degree of reception of the papal message, therefore, had to be evaluated accordingly from these different but still complementary angles, namely, the declared goals in the East and the papal interests in the West. From a historical perspective, one can easily conclude that the declared goals of the Crusades in the East were not achieved. The Crusades failure to strengthen the Byzantine Empire did not help the Eastern Christians, but rather deepened the enmity between the Western Catholic and the Eastern Orthodox Churches. Nor did the Crusades bring about the liberation of the holy places, and the chances of a successful mission among the Moslems weakened.[97] This state of affairs was already evident to Roger Bacon, who, in the aftermath of Louis IX's defeat at Mansurah, questioned the value of further Crusades: "If the Christians are victorious, no one stays behind to defend the occupied lands. Nor are non-believers converted in this way but killed and sent to hell. Those who survive the wars together with their children are more and more embittered against the Christian faith because of this violence. They are indefinitely alienated from Christ and rather inflamed to do all the harm possible to Christians."[98] This opinion was also voiced by Humbert de Romans, who reported that people

> are asking what is the purpose of this attack upon the Saracens. For, by this killing they are not aroused to conversion, but rather are provoked against the Christian faith. Moreover, when we are victorious and have killed them, we send them to hell, which seems to be against the law of charity. When we gain their lands, we do not occupy them as colonists. . . . because our countrymen do not want to stay in those regions and so, there seem to be no spiritual, corporeal or temporal benefits from this sort of attack.[99]

These reasonings mirror the decline of the Crusade through the thirteenth century, when the Holy Land lost its former mythical attraction and the Christian defeats in the battlefield reduced the readiness of European society to fight a war condemned to failure. Moreover, the identification with biblical stereotypes, which had contributed to the feelings of mission among the Crusaders while strengthening the negative image of the enemy, was not further practicable in light of the better knowledge of the political situation and its resulting implications on the image of Moslems as well. The resulting sense of defeat sometimes brought about the nonfulfillment of the Crusader vow as well as attempts to desert the army. Already in December, 1099, Pope Paschal II threatened excommunication for those who did not fulfill their vows, and a papal bull in this spirit was transmitted to the bishops. In the provincial synod held at Lyons, the papal order was read out before many prelates while those who had deserted the Crusade were strongly encouraged to return eastward in order to complete their pilgrimage.[100] During the thirteenth century, as the original enthusiasm decreased, the number of desertions increased, and brought about a crisis that Pope Innocent III tried, unsuccessfully, to solve. The constitution *Ad liberandum* of the Fourth Council of Lateran stressed the link between participation in the Crusade and divine reward, while the pope, acting as Vicar of God, allowed himself the prerogative of finding the most appropriate timing for punishing the defaulters: "But to those declining to take part (in the Crusades), if indeed there be by chance such men ungrateful to the Lord our God, we firmly state on behalf of the Apostle that they should know that they will have to reply to us on this matter in the presence of the Implacable Judge on the Final Judgment."[101]

The original attraction of Crusade propaganda, therefore, was not immune to change, nor could it be evaluated as a static phenomenon. The critical approaches of Roger Bacon and Humbert de Romans suggest the changes undergone by European public opinion during the thirteenth century and the resulting inefficiency of Crusade propaganda when it was not further backed by victories in the battlefield. The former complaints of Orderic Vitalis, Iohannes de Tulbia, and Bernard de Clairvaux (see above, notes 77, 78, 79) further suggest the demoralization process which was part and parcel of those who no longer enjoyed the glory of victory. Yet the decline of the Crusade in actual practice and the final failure of Crusader Acre in 1291 did not blur the crucial contribution of the Crusade in advancing the interests of the Reformer papacy, its promoter and main supporter. On the contrary, the papal Crusade policy mirrors the ability of the papacy to adapt its schemes to changing circumstances. From the perspective of the goals formally declared by Urban II at Clermont, it soon became evident that medieval society as a whole was not receptive to the selective concept of a military journey eastward. The papal plan of the Crusade was distorted by the unorganized multitudes traveling to the Levant in the footsteps of mythical leaders, looking for the land of milk and honey. The massive response of contemporary society was extrinsic to the original papal plan and, as such, could not be merely regarded as a propaganda success. Although knights were the main target of the papal message and for them the concept of the Just War had been elaborated, Crusade propaganda was not yet employing clear class-distinctive patterns, but its spread remained rather universal. The resulting massive involvement of the laity undermined the monopoly of the clergy in the organization of the Crusade, and revealed the failure of the diocesan system to fulfill the papal expectations in actual practice. On the other hand, the spontaneous involvement of popular elements in the First and Second Crusades, although extrinsic to the papal original plans, nevertheless served the papal non-declared goals. It provided the Reformer papacy with a channel of communication to Christendom as a whole for the first time in the Church's history.

The mass reception of the Crusade message thus faced the papacy with the imperative to amend its original plans while making the maximal profit from the enlistment of large social sectors. This state of affairs encouraged the papacy to bestow apostolic sanction on the historical process and acknowledge the pan-Christian essence of the Crusade while abandoning its former selective patterns. Christian knights were still required to participate in later

Crusades, but under the supervision of Christian kings, who were encouraged to lead the journeys overseas. The Crusade thus provided the papacy with a formidable weapon at the time of the Investiture Contest, when both kings and priests were fighting an unconditional struggle over their hegemony in Christian society. From the point of view of the Reformer papacy, the lack of military expedience in bearing amorphous multitudes eastward was therefore suitably balanced by its positive implications with regard to the status of the Holy See in Christendom. The papal initiative in the Crusade accompanied by plenary indulgences became the most faithful implementation of the papal plenitude of power, since it subordinated both clergy and laity to the decisions of the Vicar of Christ on earth. The elaboration on the concept of a "Just War" further bestowed on the papacy the military support it needed, not only against Moslems, but against all those who might challenge the papal authority in Europe as well, be they heretics, German emperors, or the Italian city-states. The massive recruitment in both the First and Second Crusades thus mirror the universal acceptability of the call from Clermont as understood by contemporaries, who were not yet able to distinguish the military and political considerations behind the papal initiative. The disregard for the papal policy, however, did not include the feudal elite, whose members were ready to become involved in the implementation of a selective military journey overseas. Such a state of affairs indicates the different approaches to the Crusade which is further corroborated by contemporary reports (see note 46). Nevertheless, the variety of attitudes could not weaken the universal appeal inherent in the concept of Crusade. The Crusade provided the different social strata with a challenge suitable to their own scale of values while the papacy successfully adapted its media and channels of communication to the changing needs. In this regard, the Crusades contribute the most spectacular example of medieval political communication in actual practice. Despite the large distances, the lack of a sophisticated communication system, and the many obstacles presented by particular languages and customs, the papal message succeeded in reaching Christendom as a whole. The use of well-known feudal vocabulary enlivened by biblical stereotypes, the elaboration upon the concept of the Just War, the distinctive use of the cross by those sailing *Outremer*, the collective feeling of identification as a chosen people, all these were powerful factors in overcoming the particularism and isolation inherent in feudal society. They paved the way for a communication process of unprecedented dimensions that further reflect the vitality of the Reformer papacy and its capability of adaptation in changing circumstances.

II

The Monarchy

6

The Foundations of the Medieval Monarchy

The ascendancy of the monarchy in the Middle Ages was deeply influenced by the success of royal propaganda in presenting the kings as a personification of the three major streams of thought that nurtured contemporary society—namely, the biblical-Christian, the German, and the Roman. The conceptualization of monarchy through well-known traditions encouraged its acceptability while it became an essential component of the world view of medieval men. The longings for social justice and universal peace, for inexhaustible courage on the battlefield and irreproachable behavior, were combined in the idea of monarchy. All questions that occupied medieval men and even eschatological trends focused on the monarchy; kings also took on a leading role in the desirable changes heralding doomsday. Yet from a historical perspective, the medieval monarchy embodies an essential contradiction. The gap between ideas and reality, the existing and the desirable, achieved a maximum scope in the prevailing attitudes toward the idea of monarchy on the one hand, and toward the actual policy of monarchs on the other. Research into the foundations of medieval monarchy, therefore, requires a clear distinction between the idea of monarchy and the role actually played by monarchs within feudal society.

The gap between theory and practice reached its peak in the Early Middle Ages, when, against the gradual fragmentation of political power, the Christ-centered concept of monarchy was elaborated. Between the twelfth and thirteenth centuries, however, the Western kings aimed at and, to a certain extent, succeeded in bridging the gap between theory and practice while paving the way for centralizing monarchies. Yet the centralizing policy of monarchs created a further contradiction between actual reality and the biblical-Christian and German traditions, which did not acknowledge the king's status as sovereign nor were they acquainted with the concept of sovereignty. This state of affairs did not lead kings to relinquish the use of well-known concepts whose propaganda efficacy had been satisfactorily proved. Yet it created a precarious situation following the further use of concepts whose implementation was actually avoided by the royal centralizing policy. This process provides an interesting field of research as to the elements selected in order to foster the acceptability of a new policy whose very essence was in open contradiction to the ancestral traditions that had given it birth. The canonization of Louis IX in 1297 mirrors the completion of this long process through which the kings of France managed to personify the longings for justice and peace, being blessed with God's grace by the sanction of His Vicar. On the other hand, the coronation oath of Edward II in 1308 bears testimony to the struggle between the centralizing tendencies of the kings of England and the no less constant tendencies of the community of the realm's representatives to share in the art of government. The attempts to forge a link between the lofty ideals of monarchy and everyday reality, between God's will and its further translation into the political language of the times, form the subject matter of this chapter.

The German tradition provided two essential features of the medieval monarchy: the personal loyalty to the kings, corraborated by the oath of fealty, and the divine attributes of the rulers. The founding of the political body on personal loyalty, combined with the geographical instability of the German tribes, fostered the state of anarchy that followed the collapse of the Western Roman Empire (476). The German kingdoms that flourished in Western Europe by the fifth and sixth centuries lacked continuity in time and stability in space. Some were so short-lived that they were described by the name of their ruler, as the "Kingdom of Samo," which flourished for a brief period in Eastern Germany. Others, which lasted longer, moved fantastically in space, as did the West Goths, who jumped from the Baltic to the Black Sea and later on to the Bay of Biscay. No regularly functioning institutions nor premonitions of sovereignty could be found in such a society. The king existed to deal with emergencies, not to head a stable legal or administrative system. The result was the emergence of a limited monarchy that lacked a clear juridical basis, a state of affairs faithfully depicted by the Roman historian Tacitus:

> They choose their kings by birth, their generals for merit. These kings have not unlimited or arbitrary power, and the generals do more by example than by authority. If they are energetic, if they are conspicuous, if they fight in the front, they lead because they are admired. But to reprimand, to imprison, even to flog, is permitted to the priests alone, and that not as a punishment, or at the general's bidding but, as it were, by the mandate of the god whom they believe to inspire the warrior.[1]

The lack of an ethnic basis did not improve the shaky status of German kings. In spite of the fact that they often acquired ethnic titles such as "King of the Franks" (*Rex Francorum*) or "King of the English" (*Rex Anglorum*), they actually ruled over several tribes that spoke different languages, observed different customs, and had not yet settled permanently in a defined territory. The usual pattern was a dominant warrior group, drawn from several German peoples, ruling a subject population which was Roman, Celtic or Slav. Within this heterogeneous society, the king's success was assessed according to his ability to carry a victorious sword and provide endless booty for his followers. These two achievements bestowed on him his people's unconditional loyalty, which was formally established through the oath of fealty.

> When they go into battle, it is a disgrace for the chief to be surpassed in valour, a disgrace for his followers not to equal the valour of the chief. And it is an infamy and a reproach for life to have survived the chief, and returned from the field. To defend, to protect him, to ascribe one's own brave deeds to his renown, is the height of loyalty. The chief fights for victory; his vassals fight for their chief.[2]

The personal links between the leader and his warriors, enforced by the oath of fealty, provided the political nucleus of the German kingdoms. They may have fluctuated in size and in the territories they occupied, but as long as a sizable number of people recognized a certain man as their king, a kingdom existed. The transition into sedentary patterns strengthened the authority conferred to the kings. Having led their people to victory, they became not only rulers over the conquered territory, but also their very owners "by the right of conquest." The sedentary style of life also fostered new expectations from the kings, upon whom were bestowed the magical attributes that had formerly been the sole prerogative of priesthood. Kings were thought to possess a certain power over nature, which they gracefully shared with their subjects, providing successful harvests in times of peace and victory on the battlefield. The divine capacity of kings bridged the gap between the royal house and the god or gods that carried out the royal requests. The historian Jordanes reported that the Goths regarded their kings as demi-gods, the god Woden being considered the forefather of the ruling dynasty.[3] The traditional long hair, an attribute of the Frankish dynasty, had certainly originally been a symbol of the kings' supernatural nature. Hair that had never been cut was thought of as the seat of the miraculous power inherent in the *reges criniti*, the kings having long hair, as had been prefigured by the biblical Samson.

The growing prestige of the royal house became a crucial factor in the capacity of the German kingdoms to survive. It provided regularity in space, the king being the leading force behind permanent settlement in a fixed area, and continuity in time, due to the emergence of dynastic patterns. The French and Anglo-Saxon kingdoms established themselves around the tenth and eleventh centuries in a fixed area for a respectable length of time, with a king at their head. Interests and loyalties were still primarily local, and limited to the family, the neighborhood, or the county. The fragmentation of political power, its treatment as a private possession, and the tendency to local autonomy heralded the emergence of feudalism, which appeared whenever the strain of preserving a relatively large political unit proved to be beyond the socioeconomic resources of society. Early feudalism relieved the strain by simplifying institutions and personalizing loyalties. The emergence of feudalism, however, neither diminished nor harmed the special privileged status accorded to the monarchy in the field of ideas. Even the feudal aristocracy, the main opponent of monarchy, was not immune to the sacred halo of kings nor aimed to bring about the abrogation of the monarchy in actual practice. Following the death of the last Carolingian, Louis V, in 987, the Frankish nobility did not use the occasion to encourage the abolition of the monarchy, but rather chose the strongest candidate, Hugh Capet, whose dynasty was to rule France for the next 340 years.[4] This was hardly an exceptional phenomenon. The peculiar political interplay inherent in the feudal regime was further indicated by the readiness of the French barons to respond to the summons of Blanche de Castile in 1230, notwithstanding the fact that they were in open rebellion against the queen. Moreover, the forces of Henry III, against which they had been eventually summoned, intended to support their struggle.[5]

Though the development of feudalism restrained the prerogatives of monarchs in actual practice, it still acknowledged the unique status of the monarchy in the interplay between lord and vassal. This provided the juridical basis for the development of the monarchy in the Central Middle Ages, when the socioeconomic process of change required greater political stability. By the mid-twelfth century, medieval political systems were becoming more sophisticated and, as such, more reliable. The feudal monarchies gradually acquired the essential elements of the modern state: political entities, each with its own basic core of people and lands, gained legitimacy by enduring through generations. This process of state-building was essentially *sui generis* but resulted in the large categories of a "unitary state," as was developing in medieval England, or a "mosaic state," as was emerging in France. In the unitary state, provincial liberties were not particularly significant, while in the mosaic state, the king slowly extended his authority over the provinces, a process that allowed the development of particular laws and institutions. The French model was preeminent in Europe for the purposes of state-building, in which the two essential areas of royal rule were justice and finance.

The advancement of the monarchy in actual practice encouraged a parallel development in the realm of ideas in order to perpetuate the growing prestige of kings while justifying their centralizing tendencies. The king was no longer regarded as a lord among his peers or, in the well-known Crusader phrase, a *primus inter pares*. The feudal system ultimately accorded the king the status of a suzerain,[6] a claim that provided the motto of the Capetian chancellery from the twelfth century onward. Philip August (1180–1223), for instance, stated quite plainly that the king of France should not pay homage to anyone.[7] This claim achieved a formal juridical basis in the parliament declaration of 1314 that: "it has never been accustomed that the kings of France could pay homage as subjects do."[8] In the kingdom of England, as well, attention was called to the fact that the king's right overrides all other feudal rights. A twelfth-century lawbook asserted that "the lord king can have no equal, much less a superior."[9] German traditions were amalgamated into the biblical-Christian and the Roman heritages while bringing about an ideological wholeness, which, in time, acquired the dimensions of a cult, the Cult of Monarchy.

The intellectual elite voiced the basic principles of the royal message. According to Hugh of Fleury:

> There are some who affirm that kings had their origin not from God, but from those who, ignorant of God, at the beginning of the world through the agitation of the devil strove in blind greed and unspeakable presumption and temerity to dominate their fellow-men by pride, rapine, perfidy, murders, and nearly every kind of crime. How foolish this opinion is, is evident by the teaching of the Apostle, who says, "For there is no power but of God: the powers that be are ordained of God" (Rom. 13:1). By this statement, it is certain that the royal authority was ordained or disposed on earth not by men, but by God.[10]

The words of Hugh de Fleury mirror the conflict of opinion which characterized medieval society at the time of the Investiture Contest. The tendency to countervail the Church's claims on the political sphere had encouraged a historical awareness as to the ancient origins of German monarchs that antedated the conversion to Christianity. Yet recognizing the ancient origins of monarchy did not bring about a denial of its divine source, but rather the opposite. Belief in the divine source of monarchy, as voiced by Hugh of Fleury, became an essential trend in the Cult of Monarchy around which the many facets of the royal magistracy were elaborated. The Christian concept of the king ruling by the grace of God (I Cor. 15:10) thus perpetuated the German belief of the king being the son of gods, a demigod on earth.

The evangelical heritage also provided a convenient basis for the behavior norms with regard to the king as head of the whole political system. According to Christian dogma, man suffers the state because he is under judgment, and the state is a part of the curse which lies upon him for his sins. All four Gospels insist on the fact that Jesus was not crucified for any political offense. On the contrary, He had considered political power as a fact of life and offered nothing except submission to this fact. He placed no limitation on the exercise of political power except the sovereignty of God, which is exercised on earth by His agents. Individual men and nations alike act as such agents of God's power to save or to judge, and they are morally neutral in so far as they are agents. It makes no difference whether they are good or bad, for God can use either. God makes use of the state as long as this age endures and Christians do not have to oppose the institution of the state as such, but rather to acknowledge its existence: "Whosoever therefore resisteth the power, resisteth the ordinance of God: and they that resist shall receive to themselves damnation" (Rom. 13:2). The New Testament examines neither the credentials of Rome nor its use of power. As pointed out by McKenzie, "It would make no more sense than the examination of the credentials of an earthquake. Government. . . . is to be obeyed because it exists."[11] The axiomatic recognition of the state added further validity to the Pauline doctrine of subjection, which called for absolute subordination to the ruler. John Chrysostom claimed that "It is the divine wisdom and not mere fortuity which has ordained that there should be rulership, that some should order and others should obey."[12] This concept was shared by St. Augustine, who also stated that "the Christian is to be led by the weight of authority," or, conversely, that obedience to the command of the superior authority was his hallmark.[13] This resulted in the perception of the believer as a *fidelis Christianus* who, as such, took no active part whatsoever nor shared in the art of government.

The divine origin of government found a most fruitful arena in medieval society while royal communicators developed its practical outcomes. The *Dialogue of the Exchequer*, a political treatise written in the reign of Henry II by the treasurer FitzNigel, stated:

> It is necessary to submit ourselves in all fear and to obey the powers ordained by God. All power, in fact, comes from our Lord God. . . . But although abundant riches may often come to kings not by some well-attested right, but perhaps by ancestral customs, or perhaps by the secret counsels of their own hearts, or even through the arbitrary decisions of their own will, their deeds, however, must not be discussed or condemned by their inferiors. For their hearts and the workings of their hearts are in the hands of God, and the cause of those to whom the care of subjects has been entrusted by God Himself depends on divine judgment and not on human judgment.[14]

The special status of the king by the grace of God, Who had entrusted to him the guidance of the congregation of the faithful, became the main theme in the royal message. Yet it faced royal communicators with the need both to release kings from the tutelage of the Church while elevating them over feudal practices. In order to spread the premises of monarchy more effectively, the most suitable media had to be used, while royal communicators incorporated the German and Christian traditions into the Cult of Monarchy. By the use of ancestral symbols, royal propaganda answered the prevailing expectations of contemporaries and fostered the emergence of a favorable public opinion to the kings' policy. The ceremony of royal unction, the prayers said in the king's honor, the symbolism implied in the royal insignia and minted in coins, as well as the throne and the magnificent burial monuments reserved for kings, all those were turned into nonverbal communication channels, which transmitted the Cult of Monarchy to medieval men. These audio-visual means perpetuated the message of monarchy in terms receptible for each generation, and their impact on both the learned and the unlettered was maximal.

The *anointing of kings* with holy oil represents one of the first stages of royal propaganda. This had been a common practice among the Visigoths from early times, but initially appeared in the Frankish Kingdom in 751: Pepin III, the first Carolingian king, tried by means of the anointing to blur his usurpation of the crown, which had been possessed by the Merovingians for almost 300 years. The endurance of the Merovingian dynasty and the common belief that its members enjoyed the special protection of the forest gods paved the way for the anointing of the Carolingians. It appeared as the most effective way to bestow on the new dynasty the legitimacy it needed. The ceremony of anointing brought medieval kings closer to the privileged position enjoyed by priests and bishops, who underwent a similar ritual when receiving holy orders. It also fostered the adoption of the biblical status of the king-priest, personified by Melchizedech, of whom it was written that he was both "King of Salem. . . . and the priest of the most high God" (Gen. 14:18). The rule of monarchs became linked with the divine government of Christ, while the royal unction indicated the synchronic harmony between earth and heaven, between the world visible to the human eye and that which is concealed.[15] Through his consecration with holy oil, the king became closer to King David, the Anointed of God, the *Christus Domini*. To use the biblical expression, he turned into the "Lord's Anointed," protected by the divine word from all machinations of the wicked, for God himself had said: "Touch not mine anointed" (Ps. 105:15). These claims actually transformed any opposition to royalty into sacrilege, since kings had been exalted by God far above the common crowd. As claimed by Barrows Dunham, "Divinity is, upon the whole, the most ingenious device that political theory has ever discovered. It protects the ruler in life and in power, but more particularly, it protects the office against the human frailties of the incumbent."[16]

The special holiness implied in the royal unction was emphasized throughout the *laudes regiae*, prayers for the king which, after the Carolingian period, became an integral part of the coronation ceremony. The *laudes* were among the earliest attempts to establish in the political and the ecclesiastical spheres a likeness of the City of God, as the newly ordained ruler was entrusted with the execution of God's will. By means of the *laudes*, God and His saints became the allies of the monarch, who was to be "glorified without end together with the Redeemer and Saviour Christ," Whose name and place he was believed to represent. Jesus Christ was begged by the prelate to "confirm the king as mediator between the clergy and the people in the throne of this present kingdom and grant him to rule together with Him in the eternal kingdom." From this came the idea of *condominium* of the Anointed with God, with Whom the king was to share His throne in the world to come.[17] The anointing of kings with holy oil and the praying of the *laudes* at the coronation ceremony paved the way for the Christ-centered concept of the monarchy. The Norman Anonymous exemplifies this philosophy in his treatise *De consecratione pontificum et regum* (c. 1100): "The power of the king is the power of God. This power, namely, is God's by nature, and the king's by grace. Hence, the king, too, is God and Christ, but by grace; and whatsoever he does, he does not simply as a man, but as one who has become God and Christ by grace."[18] Although the Christ-centered view of the monarchy did not exert a decisive influence on the political philosophy of the Central Middle Ages, it left its mark on the use of *special royal insignia*, which further spread the Cult of Monarchy. The kings of France were anointed from the *sainte ampoule*, the sacred vessel which from the time of Clovis had supposedly been brought by a dove from heaven; their standard was the Oriflamme, a banderole of red silk on a lance, which had also descended from heaven. The motto, *Christus vincit, Christus regnat, Christus imperat*, (Christ the victorious, Christ the ruler, Christ the emperor), which from the twelfth century onward appeared on French *coins*, became another symbol of the sacred prerogatives of the crown, together with the *Golden Lilies*. The French kings displayed that device on their coins, such as those magnificent *floreni ad cathedram*, the so-called *chaises d'or*, which had been minted in France since the end of the thirteenth century and showed the king enthroned on a high Gothic chair (see plate 21). The universal use of the *throne* from the eighth century onward became a further expression of the theocratic concept of the monarchy. The royal throne had symbolized the power of biblical kings (I Kings 10:9, 1:35, 46), of Roman magistrates, and, from the third century onward the ecclesiastical authority bestowed on bishops.[19] The use of the throne by medieval kings further indicated the gap between the crowned and all other mortals. It brought the images of king and Christ together as closely as possible, as Christ was also represented in medieval paintings sitting on an high throne. The throne thus complemented in a fairly visible, unambiguous way, the message of the coronation ceremony while perpetuating the theocratic ideas implied thereby.

Similar aims were pursued by the selection of *tombstones* reserved for the king and the royal family, through a meticulous choice of the resting place for the royal remains. In contrast with the simplicity that characterized burial practices in the Early Middle Ages, by the eleventh and thirteenth centuries, tombstones had become more sophisticated, following the patterns elaborated on Byzantium and the Holy Roman Empire. Henry I and Stephen favored the erection of magnificent ecclesiastical establishments at Reading and Faversham, respectively, which, in time, became their burial places. In this way they ensured for themselves personalized prayers and suitable burial churches, which perpetuated the prestige of their reigns for generations. A similar role was bestowed on the monastery of Fontevrault in Anjou, chosen as the burial place of Henry II, Richard I, and Eleanor of Aquitaine. All three were honored with a series of magnificent tombs, probably dating from the early thirteenth century, and forming part of a very early group of *gisants* or recumbent effigies, produced by the monumental sculptors of the region[20] (see plate 22). Not satisfied with the splendid burial monument erected in Westminster for his wife, Eleanor of Castile, Edward I perpetuated her funeral procession through the "Eleanor crosses," a series of stone crosses commemorating the overnight halts of her funeral cortege between Lincoln and Westminster Abbey. Several years earlier, a similar series of monumental crosses was erected in France to mark the resting places of the funeral procession of Louis IX on its way from Tunis to St. Denis. These burial arrangements perpetuated the message of the monarchy in a fairly concrete form, since the crosses erected on the main roads of France and England became a constant reminder of the glorious status of monarchy perpetuated forever.[21]

The formal recognition of St. Denis and Westminster Abbey as royal mausoleums by the thirteenth century complemented the continuing tendency

PLATE 21. *Ecu d'or.* Gold coin minted by Philip VI of Valois in 1337. The king is shown in war dress with a sword sitting on a gothic throne. On the reverse the motto of the *Christus vincit* in a floral cross.

PLATE 22. Tombs of Henry II (died 1189) and Eleanor of Aquitaine (died 1204). Abbey Church, Fontevrault. The earliest attempt at making recumbency visually explicit in the normal gisant. The effigies repose on a draped *lit de parade,* and their garments are carefully arranged so as to conform to the idea of bodies lying in state. Henry II holds his scepter, and the hands of Eleanor are brought close together in holding a small, open prayer book.

to give a more material expression to the kings' heavenly glory. It was also meant to solve the problems inherent in the endless struggle among ecclesiastical establishments to be honored with some part of the royal remains, especially the heart. This state of affairs suggests the almost mythical attraction exercised by medieval kings and the tendency of their contemporaries to share the royal glory through the *worship of the royal remains*, which had come to acquire the prestige of holy relics. At this stage one can follow the development of a dialectic process: kings who in the Early Middle Ages had been most in need of ecclesiastical legitimacy had become, by the Central Middle Ages, a source of prestige for the same Church that had elevated them above the average man. Those prelates and monks who vigorously claimed their right to store the royal body were well aware of the popularity of monarchs and, consequently, of the possibility of turning their burial place into a center of pilgrimage, after the royal corpse had begun to produce suitable signs of its supernatural power. Besides the heavenly grace conferred by the deceased, those ecclesiastical establishments could count on the generosity of the royal family, expressed through a long list of economic benefits. The monks of St. Denis profited most from this state of affairs after they had managed to secure the privileged position of their patron saint as the special benefactor of the French royal family (see plate 23). Beyond his unquestioned support of the Capetians in the battlefield, St. Denis was credited with caring for the good health of all members of the ruling dynasty, two "facts" that were strongly

PLATE 23. Tomb of Louis de France (died 1260), Abbey Church, St. Denis. Commissioned by Louis IX c. 1263–1264, it has a frieze around the tomb-chest consisting of figures in high relief showing the funeral procession.

emphasized by Capetian propagandists.[22] The special links of a particular saint with the royal family acquired high significance in the process of state-building. The first stirrings of national sentiment among the young states of Europe were often expressed and given specific content through the choice of a protective saint whose special responsibility was to oversee the destinies of his people and to preserve the realm from any threat. This resulted in the advantages implied in the immediate appeal to a particular saint rather than to God, Who, though almighty, cannot prefer one nation above all others.

The religious aura ascribed to the royal remains further encouraged the widespread belief in the kings' *thaumaturgic powers*, either as a direct result of the unction or as an indirect outcome of German traditions. Peter of Blois argued: "I would have you know that to attend upon the king is something sacred, for the king himself is holy. He is the Anointed of the Lord (and) it is not in vain that he has received the sacrament of royal unction, whose efficacy—if someone should chance to be ignorant of it or doubt it—would be amply proved by the disappearance of that plague affecting the grain and by the healing of scrofula."[23] For many years, medieval kings used to "touch for scrofula," which was a virtually endemic kind of tuberculosis adnitis, characterized by glandular swellings, due to the tuberculosis bacillus. Both the Capetians and the Norman kings of England claimed to be able, simply by their touch, to cure people suffering from this disease, which was consequently called in France "*le mal du roi,*" and in England "*the king's evil.*" The kings of England also used to distribute the so-called cramp rings, which, by virtue of their consecration at the hands of the monarch, were thought to have acquired the power to restore health to the epileptic and to alleviate all kind of muscular pain.[24] Medieval society, well acquainted with sickness and plagues, became a most fertile ground for the belief in the kings' supernatural power, a belief which was systematically encouraged by royal propaganda. At the beginning of the eleventh century Robert the Pious claimed that "We continuously transfer divine grace to all mortals."[25] This was a claim of high propaganda value in view of the readiness of kings to gracefully share their power with all their subjects. Edward I blessed 1,736 people during the eighteenth year of his reign, 1,219 in the thirty-second year, and 983 in the twenty-eighth. The Capetians as well touched the sick with their hands and performed on them the sign of the cross, two acts claimed by ancestral customs as being the most secure ways to overcome all manner of sicknesses.

The thaumaturgical power attributed to kings could be, and was actually used effectively, to advance political interests. According to a report of the Dominican Francis, bishop of Bisaccia, at the outbreak of the Hundred Years War, Edward III proposed to Philip VI a most suitable way to secure divine arbitration between them both while preventing further bloodshed between Christians: combat in the lists, true judgment of God, either in the form of a duel between the two claimants themselves, or a contest on a larger scale between two groups of from six to eight loyal supporters; alternatively, one or another of the following trials: "If Philip of Valois is, as he affirms, the true king of France, let him prove the fact by exposing himself to hungry lions, for lions never attack a true king; or let him perform the miraculous healing of the sick, as all other true kings are wont to do," a sentence that suggests the readiness of the true king of France, namely, Edward III, to face the challenge.[26] Although the reliability of this account has not been satisfactorily proved and, besides, Philip VI opted for more conventional ways to settle the conflict, it nevertheless indicates the widespread belief in the supernatural powers of monarchs and the resulting attempts of the political elite to profit from them.

At the other end of the social spectrum, the quasi-religious image of the monarchy encouraged the development of *messianic expectations* centered on the royal dynasty. On the eve of the Second Crusade, Louis VII was regarded by many as the "Emperor of the Last Days," who would lead the Christian armies in their fight against the Antichrist. Louis VII's son, Philip August, assumed the role of the priest-king before the Battle of Bouvines (1214) and, like Charlemagne in the *Chanson de Roland*, blessed his army as a host fighting for the true faith. In those same days, some sectarians in Paris saw in the Dauphin, the future Louis VIII, a messiah who, under the dispensation of the Holy Spirit, would reign forever over a united and purified Christendom. By the mid-fourteenth century, Jean de Roquetaillade attributed to the king of France the conquest and rule of the whole world, including the kingdoms of Asia, Africa, and Europe. After the death of both pope and emperor, the ever-victorious descendant of Charlemagne would establish a reign of peace for a thousand years.[27] The very fact that messianic expectations were elaborated around the monarchy underlines anew the crucial role ascribed to the monarchs in medieval society while even the desirable changes in actual reality were in fact conditioned by the royal initiative.

According to Cohn, the masses of the poor became the ideal arena for the elaboration of eschatological expectations that focused in the monarchy since they were most in need of a radical change. Yet the Cult of Monarchy went far beyond class differences and became the common denominator of the various social strata. The monarch was considered the representative of the powers governing the cosmos, an incarnation of the moral law and divine intention, a guarantor of the order and rightness of the world and, as such, the loving father of all his subjects whose welfare had been entrusted to him by God. The king had been sent by God to serve as a model to be imitated by all ordinary mortals. The idea of "*rex imago aequitatis*," the king as the reflection of equity, enhanced the crucial role played by the implementation of royal justice. According to Isidore's etymological interpretation of the Latin term *rex*, the king is the one who rules justly.[28] This opinion was voiced by popular French poetry in the vernacular, as "*Rois, tu ies rois pour droit roïier.*"[29] St. Augustine went even further, arguing that the implementation of justice ultimately provides the state with its *raison d'être*, as without justice the State becomes a "big crowd of robbers." As picturesquely argued by a pirate before Alexander the Great: "As I plunder people on a small boat I am called a pirate, but you plunder the whole world and, consequently, are addressed as Emperor!"[30]

Hugh of Fleury saw in the king's responsibility for justice a clear expression of God's will to secure the normal management of human society:

> The ministry of a king is to correct his subjects and call them back from error to the path of equity and justice. Wherefore in the Book of Judges one also finds that, before the children of Israel had a king, Jonathan, the grandson of Manasseh, says, "In those days there was no king in Israel, but every man did that which was right in his own eyes" (Judg. 17:6). It is evident, therefore, that where there is no king to rule and to draw the people away from arrogance, the whole body of the kingdom totters. Deservedly, is he called king who knows how to rule their ways fittingly and to control those subjected to him. For this reason, I say, the omnipotent God has set a king over men, who lives and dies like them, in order that he may coerce the people subject to him by his terror and that he may subdue them with laws for right living. . . . . For the people is easily corrected by fear of the king. But the king is deterred from the path of injustice by nothing except the fear of God and dread of hell. . . . Also, all who are placed in power are to be honored by those over whom they preside, not because of themselves, but because of the order and rank which they have received from God.[31]

Even the king's disregard of justice could not dim his sacred halo nor justify further criticism of his deeds. The tendency to safeguard the prestige of monarchy often encouraged blaming the "bad" counselors for the misdeeds perpetuated by kings.[32] Being God's messenger upon earth and, as such, a reflection of the heavenly order, kings could not therefore come under the judgment of flesh and blood. The author of the *Dialogue of the Exchequer* denied, in a fairly dogmatic fashion, any right to criticize the king's deeds nor to judge him.[33] Being directly dependent on God, the king became immune to human judgment, without this justifying any further development of an arbitrary rule. Already in the eleventh century, the thin line separating the "king by the grace of God" and the "tyrant" was carefully defined. Manegold of Lautenbach emphasized this distinction through a peculiar metaphor:

> It is necessary that he who is to bear the charge of all and govern all, should shine above others in greater grace of the virtues and should strive to administer with the utmost balance of equity the authority allotted to him. . . . Yet when he who has been chosen for the correction of the wicked and the defense of the upright begins to foster evil against them. . . . is it not clear that he deservedly falls from the dignity entrusted to him and that the people stand free of his lordship and subjection, when he has evidently been the first to break the contract for whose sake he was appointed? Nor can anyone justly and rationally accuse them of lack of faith, since it is quite evident that he first broke faith. For, to draw an example from baser things, if someone should entrust his pigs to be pastured to someone for a fair wage, and afterwards learned that the latter was not pasturing them, but was stealing, slaughtering, and losing them, would he not remove him with reproaches from the care of the pigs, retaining also the promised wage? If, I say, this principle is maintained in regard to base things, that he is not considered, indeed, a swineherd who seeks not to pasture the pigs, but to scatter them, so much the more fittingly, by just and probable reason in proportion as the condition of men is distinct from the nature of pigs, is he who attempts not to rule men, but to drive them into confusion, deprived of all the authority and dignity which he has received over men.[34]

The impulse to deny private responsibility for royal deeds performed for public ends undoubtedly existed, but, following St. Augustine, theologians did not hesitate to point out the final responsibility of rulers for all their acts and their special obligation to adhere to the highest standards of conduct. John of Salisbury related the principle of justice to the absolute authority of kings, but he subordinated princes to the enforcement of law: "The prince, although not bound by the ties of law, is yet the servant of law as well as that of equity;. . . . he bears a public person, and. . . . he sheds blood without guilt." From this point of view, the prince is therefore freed from human judgment, but he is also *legibus solutus*, because he is expected to act on the basis of his innate sense of justice. The king is bound *ex officio* to venerate law and equity for the love of justice itself and not for the fear of punishment.[35] Similar concepts were elaborated on the Sicilian constitutions published in Malfi by Frederick II in 1231: "The Caesar. . . . must be at once the father and the son of justice, her lord and her minister: father and lord in creating justice and protecting what has been created; and in like fashion he shall be, in her veneration, the son of justice and, in ministering her plenty, her minister." Henry Bracton, a contemporary of Frederick II, claimed in a similar manner: "The king's power refers to making law and not injury. And since he is the author of the law, an opportunity to injustice should not be nascent at the very place where the laws are born."[36] The subordination of the monarchy to the implementation of justice left its mark on the political interplay of the times. In the *Mise of Amiens*, the arbitration verdict on the struggle between Henry III and the barons (1264), Louis IX combined the recognition of royal power with the liability of kings to the customs and laws of their realm. He acknowledged the *plena potestas* bestowed on Henry III as "the said king shall have full power and unrestricted rule within his kingdom and such status and such full power as he enjoyed before the time aforesaid." On the other hand, this "full power" could not "derogate from royal privileges, charters, liberties, establishments, and praiseworthy customs of the kingdom of England which existed before the time of the same provisions (of Oxford)." As pointed out by Wood: "If such limitations, and notably the Magna Carta, were to continue to have force even while the king enjoyed full power, then it follows that this power, far from being absolute, continued . . . . to be constrained and defined by custom and by law."[37]

The basic readiness of both Henry III and the insurgent barons to ask for the arbitration of Louis IX hints at the prestigious position the king of France achieved in his lifetime. Either in medieval reports and in modern historiography, Louis IX came to personify the image of the ideal monarch. The French historian Fustel de Coulanges underlined his unconditional love of God and his piety as his most essential virtue. In the words of the king's biographer, Jean de Joinville: "He loved God with all his heart." Louis's nature was also marked by his charity and love of all mankind. Joinville testifies that "the king daily gave countless generous alms to poor religious, to poor hospitals, to poor sick people, to other poor convents, to poor gentlemen and gentlewomen and girls, to fallen women, to poor widows and women in childbed, and to poor craftsmen who because of old age or sickness were unable to work or follow their trade."[38] The three faults most common among ordinary people—jealousy, pride, and greed—were absolutely absent from the king's character, nor did he ever fall victim to such weaknesses. Louis vigorously opposed bloodshed and the territorial expansion by Christian princes at the expense of their neighbors. On the other hand, he regarded the war against Moslems as a personal duty of every member of Christendom. In fulfilling the royal ministry, Fustel de Coulanges concluded, Louis exemplified the "just king," as he both consulted the magnates of his realm three times a year and respected the customs and charters of towns[39] (see plate 24).

The prevailing expectations from the ideal king were further emphasized by the many hymns written after Louis's canonization. His deep knowledge of Holy Scriptures, his moral and spiritual perfection, his unquestioning devotion to the implementation of justice, and his withdrawal from earthly pleasures, together with the many fasts, self-discipline, and ascetism, were re-

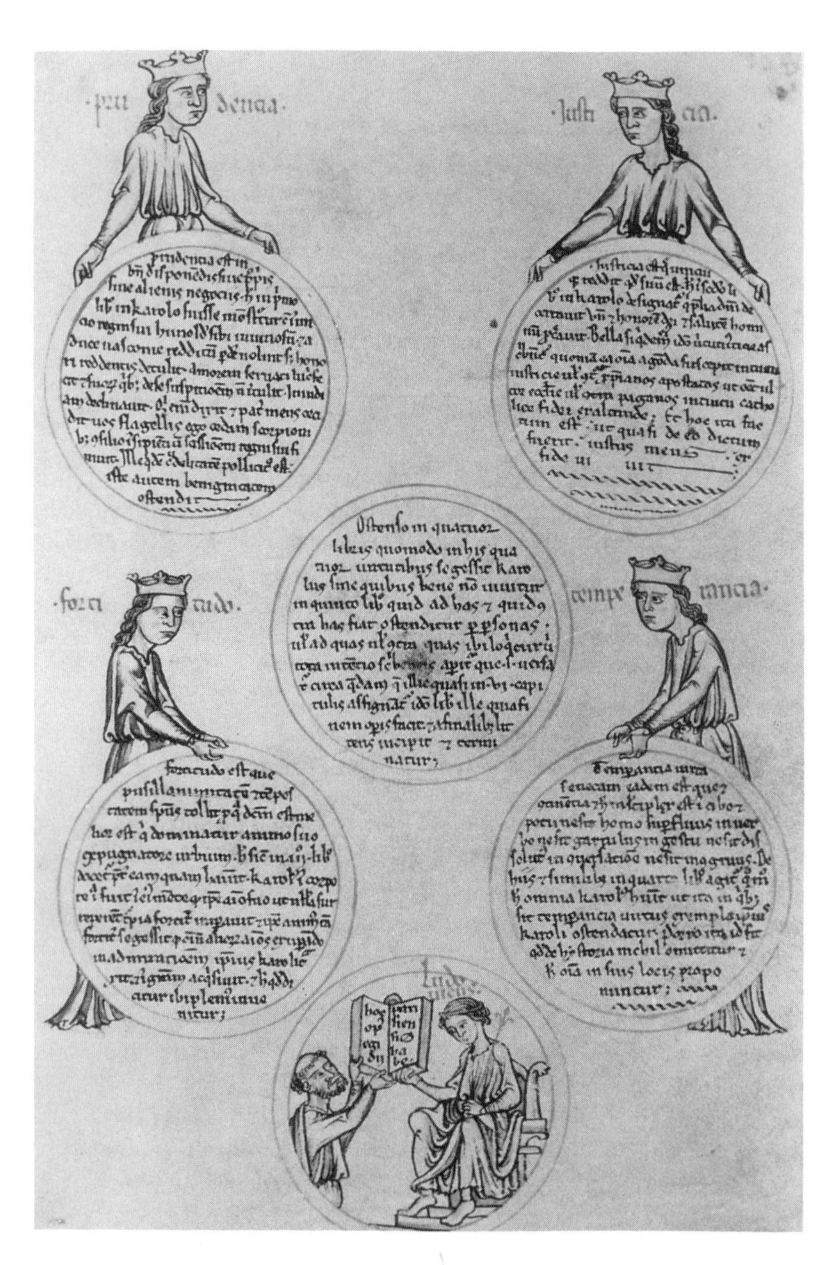

PLATE 24. The virtues and Prince Louis, Gilles de Paris, *Miroir des Princes*. Manuscript painted in Paris during the reign of St. Louis.

garded as the basic features that typified Louis's way of life. His biography was presented as the most faithful reflection of the misfortunes and achievements that were part and parcel of the biblical heroes: like Jacob, he traveled and met God far away; like Joseph, he was released from captivity in Egypt; like Jonah, he miraculously escaped drowning in the sea.[40] Without being marked by personal or national allegiances, contemporary reports also contributed to the mythification of the king's image. The English chronicler Matthew Paris described the reign of St. Louis as the pinnacle of the kings on earth.[41] The Italian Franciscan Salimbene considered Louis as holy, long before he was actually canonized.[42] Wide support of the king and his policy was further voiced by French chroniclers who, struck by his virtues as a judge, referred to him as *Ludovicus justus*, Louis the Just.[43] Such attitudes were categorically summed up by a chronicler from Limoges, who stated that there had never been a better prince than St. Louis.[44] The pan-Christian admiration of St. Louis reflects the essential contradiction between the use of religious and, as such, universal concepts in advancing the Cult of Monarchy and the kings' interest to foster a distinctive political solidarity between their subjects. This contradiction, however, did not lead to royal communicators abandoning the use of religious symbols. They opted, rather, for bestowing on them a national meaning, thus creating the basis for propaganda campaigns, which will be analyzed in further detail in the following chapters.

The unquestioned admiration for St. Louis reflects his success in fulfilling most expectations bestowed on the monarchy for generations. Jonas of Orléans had seen the three pillars on which the Christian monarchy should be established in the implementation of justice, piety, and compassion. The king had to be mostly obligated toward the implementation of justice, first with regard to God, but toward all the inhabitants of his kingdom alike. This implies the king's duty to avoid any discrimination in trials; to protect the churches as well as all those most in need of royal assistance, such as widows, orphans and foreigners; to prevent theft and adultery; to support the poor; to keep only experienced counselors who do not fall readily to prejudice, and to protect the kingdom from all enemies. All these were summarized and given divine sanction by the primary requisite of Christian kings "to live in accordance with the Catholic faith."[45] It was therefore the threefold duty of a king to maintain peace, to administer justice with equity, and to put down iniquity in all classes of society. The "good king" was henceforth not merely the brave and generous German leader but, rather, the just, wise, and merciful Christian king by the grace of God, who took the great kings of the Scripture as his model.

The king's duty to implement justice acquired a more categorical basis through the *oath* he pronounced at his coronation ceremony. French monarchs were traditionally crowned in the Cathedral of Reims, where Clovis, the first Frankish king, had converted to Christianity and received baptism from the hands of St. Remigius. The *Ordines ad consecrandum et coronandum regem*, a treatise written around the 1270s, reports in detail the dialogue between the king and the prelate responsible for the royal unction:

> —I assure you I will observe the Canonical privileges of each one of you and your churches, the law and justice, as far as I can, so help me God. I promise to protect the bishops and their churches as (properly has to) a king in his kingdom. . . . In the name of Jesus Christ, I promise the Christian people committed to my rule to keep true peace for all of them and the churches of God. I promise to prevent any injustice or theft of any kind. I promise to maintain justice and royal grace in trial, so help me God through his grace. . . . I promise to devote myself with all my might and in faith to the removal of any seed of heresy, declared as such by the Holy Church, from all the lands conferred to my rule. . . . I undertake all these duties under oath, so help me God and these Holy Gospels.

The king swore on the Gospels and all the assembly escorted him, chanting the *Te Deum* ("Thee, God, we praise").[46] Following these proceedings, the newly-crowned ruler was awarded the *acclamations* of the whole congregation, laity and clergy alike, a reminder of old Roman and German traditions.

The contractual character of the royal oath before the magnates of the realm, however, was in fact counterbalanced by the hereditary claims of the Capetians to the throne. The development of dynastic patterns in the kingdom of France was significantly fostered by the Capetian policy to enable the heir to share the rule of his father. This practice began only three months after Hugh Capet reached the throne in 987, and lasted until 1223, when the strengthening of the monarchy allowed Philip August to relinquish this practice. It turned the royal oath into a ceremonial act whose implementation was in fact conditional upon the king's goodwill. The use of the Oriflamme from the reign of Louis VI onward reflected the dynastic stability of the kings of France as it linked the Capetians with Charlemagne, the first king who had been entrusted with its use.[47] The links with the Carolingians thus further strengthened the French monarchy since they had been the first dynasty formally blessed with the papal sanction. After anointing Pepin III and his two sons in 754, Pope Stephen II had indeed "forbidden all Franks, under the threat of excommunication, to choose a king other than of this dynasty, which was elevated to the throne by the grace of God, and approved and consecrated by the holy Apostles, through His vicar, the pope."[48] The dynastic principle reinforced the holiness of the Capetians while conferring on the royal house the eternal blessing God had bestowed on the house of Jesse (II Sam. 7:8) (see plate 25).

Quite different was the status of the monarchy in England. Both the Norman conquest in 1066 and the rise of the Angevin dynasty in 1154 had created breaches in the hereditary principle. William the Conqueror was well aware of his weak dynastic position. Following his victory at the Battle of Hastings, he tried to justify his coronation on the grounds that Edward the Confessor had promised him the crown of England. He also claimed kin-right, somewhat dubiously, due to the fact that the Confessor's mother, Emma, had been the daughter of his own great-grandfather, Richard I of Normandy. He further argued that God had demonstrated the justice of his cause in battle

PLATE 25. Stained glass window of Jesse's family tree, Chapel of the Virgin, St. Denis, twelfth century. On the right, Suger, Abbot of St. Denis (1122–1151), at the center, Jesse, King David's father.

and that he had been elected by the Witan (Assembly of magnates) of England and subsequently crowned.[49] Endowed with weak hereditary rights, the Norman monarchy was forced to rely on the coronation ceremony. The coronation oath acquired a formal contractual effect between the king and the representatives of the community of the realm, who zealously protected their right to control the royal policy and, accordingly, to react against any disregard or injury incurred by the monarch. In 1308, the Earl of Lincoln openly challenged royal authority by drawing his sword before Edward II. He claimed that his duty of loyalty focused essentially on the royal institution rather than on the monarch himself, unless the king actually put into practice the rights of the crown. If the king injured the ancestral rights of the crown, or incurred an action considered by the Earl of Lincoln as harmful to the crown, it would be therefore completely justifiable to oppose the monarch in protecting the monarchy.[50]

The declaration of the Earl of Lincoln clarifies the peculiar status of the English monarchy while delineating a clear line between the king and the crown, through a depersonalization of the monarchy. This state of affairs facilitated the removal of kings, a practice unknown in Capetian France. During the fourteenth and fifteenth centuries, kings could be removed on the grounds of their being a "bad king," namely, one who had violated the country's laws, customs, or morality or, a "useless king," the incompetent executive and inept politician who had not mastered the art of government.[51] The removal of Edward II, Richard II, Henry VI, Richard III, and Edward V in the course of 150 years clearly indicates the vulnerability of the dynastic principle and the importance ascribed in England to the king's undertakings at his coronation. Though exposing the monarchs to criticism and eventually to their removal from office, the weakness of the dynastic principle also weakened the religious nature of monarchy. Yet backed by a substantial income from their Continental domain and enjoying the advantages of the centralizing Norman tradition, the Angevin kings of England could strengthen their position without enjoying the quasi-religious prestige of the Capetians. The very fact that the kings of England succeeded first in advancing a centralizing monarchy in medieval Europe suggests that the lack of religious legitimacy did not avoid per se the strengthening of the monarchy. On the contrary, it released royal propaganda from the imperative of facing universal Christian concepts, while facilitating the reception of the royal message in a national context.

In England, as in France, the coronation oath originally embodied three main royal duties, namely, to secure the peace of the Church and all the Christian people, to forbid robbery and crime, and to impose justice and mercy. There were, nevertheless, royal attempts to restrain the practical consequences of these far-reaching engagements. In the addendum introduced by Henry II in 1154, the king committed himself to protect the interests of the crown as well, thus turning his oath into a monarchical program.[52] By 1308, however, the declarations forced upon Edward II mirror the prevailing tendency to turn the coronation oath into a royal formal undertaking against any possible mismanagements in the future:

> Sire, are you willing to grant and preserve and by your oath confirm to the people of England the laws and customs granted to them by former kings of England, your righteous and godly predecessors, and particularly the laws, customs and liberties granted to the clergy and the people by the glorious king St. Edward, your predecessor?
> —I grant and promise them.
> —Sire, will you for God and holy Church and for the clergy and for the people keep peace and accord in God, to the best of your ability, intact?
> —I will.
> —Sire, will you in all your judgments have impartial and proper justice and discretion done in compassion and truth to the best of your ability?
> —I will.
> —Sire, do you agree to maintain and preserve the laws and rightful customs which the community of your realm shall have chosen and will you defend and enforce them to the honour of God to the best of your ability?
> —I agree and promise.[53]

The attempts to restrain the freedom of action of the monarchs, though favoring constitutional developments, did not ultimately harm the essential eminence of kingship in medieval society as a whole. The king remained the most tangible symbol of social justice, universal peace, and unreserved devotion to the faith and the Church. Contemporary communicators used an organic analogy to transmit this message. The state was represented as a body whose members had to obey the heart while working together for the common welfare. In the words of Hugh of Fleury:

> God indeed set the first man in the world, already furnished with the primordial dowry of wisdom, above all the creatures of the world. And thus He subtly intimated to him that there is one King and Lord of the whole creation, Whom that celestial court which is above us rightly serves and obeys. And that we may recognize this equally in the form or our body, we see that all the members of our body are subject to the head. It is apparent, I say, that all the members of the human body are subject and subordinate to the head both in position and in rank. Whence it is quite clear to us that the omnipotent God differentiated not only between the various members of the human body, but also the distinct ranks and powers of the whole world, corresponding to the distinctions which we know to exist in the celestial court in which God, the Father omnipotent, alone holds the kingly dignity and in which after Him, as we know, the angels, archangels, thrones, and dominations, and other powers stand one above the other in a wonderful and seemly variety of powers. . . .[54]

The *Rex pacificus*, a political treatise written in the University of Paris at the end of the thirteenth century, saw in the human head the symbol of the spiritual realm, while the heart exemplified the temporal rule. The writers accepted Aristotle's approach, which antedated the creation of the heart above all other members of the body. As the heart sends blood to the whole body, becoming the source of life, the political leader is the keystone of an ordered society. The heart is essential for the existence of all members of the body, including the head, as ceasing its operation causes death. This leads to the inevitable conclusion that the monarchy is indispensible for the very existence of the state and its proper operation.[55]

The recognition of the monarchy as the axis of a well-ordered society fostered a growing willingness to entrust in the monarch the total control over his subjects, their lives and their property as well. The king's power thus went far beyond the traditional links that bound the individual with his family, his place of birth, and the universal Church. According to Jean de Blanot (c. 1250), if a vassal of the duke of Burgundy is called for military service to join in the duke's war against the duke of Lorraine and, at the same time, is called to serve the king to protect the kingdom, he should obey the king first. This primary duty to the king was justified on the grounds that "the duke's summons is only for his own private utility, but the king's is for the public utility, which must be placed above the private. Only if the king summons them for any target which does not concern the public utility, should the vassals obey their overlord, the duke." This theory was shared by the French legist, Jacques de Révigny (c. 1270), who added the historical justification that among the Romans, the defense of castles prevailed over their love of their own children.[56] The loyalty to the king thus gradually paved the way for the revival of the Roman concept of "love of the fatherland" (*amor patriae*). It included the territory where the individual was born, the individual himself as subject of that love, and the monarch as the personification of the fatherland.

The former personal loyalty to the king underwent a process of change while surpassing feudal barriers and local attachments. This process brought about the revival of classical concepts, such as Cicero's "the fatherland is dearer to me than life." In his continuation of Aquinas' *De regimen principum*, Tolomeo of Lucca stated that "love of the fatherland is founded in the roots of charity which puts not one's own things before those we hold in common, but the common things before one's own. . . . Deservedly the virtue of charity precedes all other virtues because the merit of any virtue depends upon that of charity. Therefore, the love of the fatherland deserves a rank of honor above all other virtues."[57] Bracton associated the defense of the fatherland with the superior right of the king to defend his kingdom. Canonists and Decretalists alike were unanimous in their opinion that the most heinous acts when performed in the defense of the fatherland or for the benefit of the community—including the murder of a son by his father or, conversely, of the father by his son—should not be regarded as crimes, since the fundamental principle of giving priority to the common good was above personal interests. As Post has pointed out, reason of state should not be regarded as an invention of the Renaissance; it had already been in existence in the Middle Ages.[58] The bridging of the monarchy and the love of the fatherland with a potential ideology of reason of state indicates the long way undergone by the monarchy since the process of settlement in the Western Roman Empire. The German and biblical-Christian traditions had been amalgamated with the Roman heritage while contributing to the preeminence of monarchy and the reception of its message in the framework of ancestral norms. Most successful kings, such as St. Louis, achieved wide support when presented their policies as dictated by their essential respect for ancestral customs. Royal attempts to change accepted norms, and there were indeed such attempts in the Central Middle Ages, still encountered varying degrees of opposition, with the secular and ecclesiastical aristocracy at its forefront. Reluctance to relinquish the status quo encouraged, therefore, a wide use of propaganda in an attempt to weaken any substantial opposition to the kings' centralizing policy at home.

Despite the fact that the papacy generally supported the monarchy and that papal policy often served as a most reliable shield for royal prerogatives, royal communicators found in the papal institution a most convenient target for developing the first definitions of sovereignty. Without relinquishing the holiness of monarchy, royal propagandists asked for a more accurate division of labor between king and priest, an important step toward the laicization process of Western society. John of Paris pointed out the inalienable links between kings and their subjects, which did not depend upon the papal sanction nor could they have been, as "royal power preceded papal authority, and there were kings in France before there were Christians there. It follows, therefore, that royal power. . . . does not flow from papal authority but from God and the people who personally elect the monarch or the dynasty."[59] The marginal position accorded to the pope in the political field stressed the close links between the anointed king and his people who had chosen him or the royal family without dimming the religious halo of the medieval monarchy. The propagandistic advantages implied in such concepts reached their climax in fourteenth-century France when royal communicators succeeded in creating a united national body under the idea of "a people (which) shall dwell alone and shall not be reckoned among the nations" (Num. 23:9). Pope Boniface VIII represented, then, all the forces of evil threatening royal sovereignty and, ultimately, the welfare of French society as a whole.[60] This development did not yet bridge the gap between theory and reality, and the idea of an absolute monarchy remained extrinsic to medieval society. Nevertheless, by uniting the king, his subjects, and the Catholic faith together in a common territory,[61] and by leaving little room for outside interference, the foundations of the medieval monarchy actually encouraged the emergence of the national state later on.

7

Political Communication

The development of political communication in the Central Middle Ages mirrors a changing scale of priorities while its degree of efficiency indicates the power achieved by kings and the political audiences to which they appealed or were able to reach. In the Early Middle Ages, the main targets of political communication were foreign princes and the papal curia. The isolation inherent in the feudal regime limited, domestically, the political audience of the royal court to the king's direct vassals, the tenants-in-chief. Although medieval kings did not totally relinquish communication with their subjects, its scope was minimal and operated through the feudal channels. The development of a nation wide communication system toward the end of the period heralds a more advanced communication stage, when the king's increasing power led to a more regular and immediate contact with his subjects. The elaboration of such a system appeared as a most suitable means to strengthen the status of kings, while encouraging a public opinion favoring their policy. It appears both as a cause and an effect in the centralizing policy of monarchs, who, being the promoters of the communication system, were also responsible for its development and operation. This development was concomitant to the emergence of royal administration and was actually subordinated to it. Lacking more specialized channels, medieval kings integrated their officers into their communication system, as they could supply reliable information and propagandize the royal policy throughout the countryside. Royal administrators thus acted as communication channels between the rulers and their subjects.

This chapter deals with the channels of communication developed by the monarchy, particularly in the two major powers of medieval Europe: France and England. Between the eleventh and fourteenth centuries, the royal policy in both countries was characterized by a marked tendency to strengthen the kings' position at home, counteracting the sociopolitial influence of the feudal aristocracy. These centralizing tendencies not only affected the prerogatives of the political elite, but also differed from well-known, accepted norms, two facts which led to an unprecedented use of propaganda. The need to legitimize the kings' policy brought about a widening of the political audience, whose support the royal communicators were seeking. The feudal elite, personified by the aristocracy and clergy, could hardly remain the main target for royal propaganda as it represented those sectors most affected by the kings' centralizing policy. The knights and the bourgeoisie were thus integrated in the political audience while becoming active partners of the royal policy. This process changed the sociopolitical balance of power and created a convenient background for the emergence of a national consciousness. The development of political communication systems in the Central Middle Ages indicates, therefore, the gradual enlargement of the political society, which was evolving from feudal patterns into the representative framework of parliament, a process in which the monarchy played a leading role.[1] The attempt to look for parallel developments in England and France, however, raises methodological problems, as the emergence of political communication systems was a process *sui generis* and, as such, prevents synchronic research. The Norman Conquest in 1066 and the reign of Philip August (1180–1223) have been regarded as starting points for the development of political communication systems in England and France, respectively.

The first stage in the development of a political communication system was provided by the special messengers, or *nuncii*, eventually sent by kings to different areas. This was a very old practice, which dated back to the very beginnings of the German kingdoms. In the Carolingian Empire, Charlemagne's messengers, the *missi dominici*, were sent to all provinces once a year. They fulfilled the functions of royal judges and inspectors, but also acted as spokesmen of the royal court to which they provided accurate information on the actual state of affairs in the different areas of the empire.[2] Throughout the Middle Ages, messengers went forth with letters, proclamations, inquiries, offers of peace, and declarations of war. Couriers carried letters to and from the papal curia. Towns employed their own messengers, who carried messages on municipal business. Travel expenses varied, and carriers frequently bargained for adequate or beneficial rates. In England, the first regulation concerning a systematic supply of horses for a letter post to Dover dates from the mid-sixteenth century. In France, a *poste royale* came into existence in the mid-fifteenth century, although the University of Paris had operated its own communication system since the thirteenth century. German commercial towns also supported a postal service since the thirteenth century, which subsequently extended throughout central Europe and as far south as Italy.[3] The relatively late development of a postal service throughout Europe thus forced the kings to develop alternative channels of communication in their dealings with their own subjects and, similarly, with foreign princes.

In thirteenth-century England, the royal household contained a varying number of permanent messengers, often amounting to a dozen or more. These officers were known as *nuncii regis*; they followed the king as he traveled and received a regular allowance of shoes and clothes every year. Toward the second half of the thirteenth century, they were complemented by additional messengers, distinguished from the former as *cokini* or *cursores*. These were actually messengers of an inferior status who did not use horses and were therefore cheaper to employ. As they were not regarded as regular servants, they did not at first receive privileges as members of the household. Their number varied according to the circumstances, increasing considerably in times of war, when the king had to obtain the latest information on political and military situations spread through different and distant areas. Between 1296 and 1297, Edward I employed forty-one extra messengers, apart from the twenty regular messengers at his service hitherto and in addition to the *nuncii regis*. Edward II maintained twelve messengers on a regular wage, who accompanied him on all his journeys. They received threepence a day when they were on the road, and four shillings and eight pence a year to buy shoes.[4] By the reign of Edward III, a man might enter the king's service as a groom or casual letter carrier, be accepted by the wardrobe as a cursor, and finally succeed in receiving a permanent post as messenger, which often bestowed on its holder the hope of a pension as the final reward for faithful service.[5] The use of messengers was widespread in the kingdom of France as well. Throughout the fourteenth century, the kings kept almost 100 messengers, while local lords contented themselves with two or three. A messenger riding a horse received some eighteen francs a day for a distance of about 34 miles, while a messenger on foot was rewarded with about nine francs for an average distance of 18 miles. For night journeys, they received double payment.[6]

Although the use of political messengers was not a trait unique to the Central Middle Ages, their increasing activity and the gradual institutionalization of the office turned them into an integral factor in foreign affairs. In the Early Middle Ages, kings were usually forced to lead most issues of foreign policy in person. They arranged meetings with foreign princes in neutral places such as the border region between the two countries, which allowed free deliberations without undesirable interference. The kings of France, for instance, customarily dealt with the dukes of Normandy in the border areas between Gisors and Trie. Diplomatic negotiations were sometimes carried out on a river, due to the security advantages of such a place and the secrecy it provided. In 921, the meeting between Henry the Fowler and Charles of France was arranged on a ship on the Rhine.[7] In 1476, Afonso V, king of Portugal, also proposed to Ferdinand the Catholic meeting on a ship anchored on the Castilian border.[8] Despite the inconveniences and delays implied in such a practice, personal meetings between princes remained quite customary throughout the Middle Ages.

The growing sophistication of the political system, the king's reluctance to leave his kingdom for prolonged periods, and the security problems still inherent in lengthy journeys increased the readiness of medieval kings to use the services of personal messengers. Royal messengers could carry their messages verbally or in writing, or in both forms at once. Sometimes, through the customary letter of identification, the sender informed the recipient that the carrier could supply further confidential information, "since we have opened our heart before him, about all those matters which we wished to explain to your majesty."[9] The lack of reliable channels left its mark on the complexity characteristic of the communication system. King John Lackland sent several copies of the same letter to Pope Innocent III in 1205 through various messengers, "because of the many perils of the ways."[10] Besides, it was almost impossible to honor any scheduled timetable. Although in most cases it was quite reasonable to expect that prelates would be found in a specific place, the accuracy of such an expectation decreased drastically with regard to secular rulers due to the peripatetic nature of their courts. The counts of Flanders, for example, had castles at Bruges, Ghent, Ypres, Courtrai, and Oudenaarde. Up to the thirteenth century, the kings of France also used to travel through the royal domain without any fixed itinerary. Messengers who had to convey messages to political leaders could not, therefore, rely on their chances of finding their addressees in a specific place, nor could they accurately fulfill any time schedule established in advance.

This fluid state of affairs forced kings to leave a maximal freedom of action to their messengers in order to facilitate their geographical mobility and further adjustment to the changing conditions of medieval diplomacy. Such messengers could be nominated for a certain term or for life, for a specific purpose or for more general goals. They could be sent to a certain country or be given full power to change their destination according to their own judgment and the development of their mission. This high flexibility resulted also from the inaccessibility to a regular communication between the sender and his messengers abroad. The wide authority conferred on messengers gave rise to their careful selection within the ranks of courtiers, the men closest to the king, whom he could reasonably trust. Such a position was enjoyed by the friars during the reign of St. Louis, who used their services for various purposes. In 1241, a group of friars was sent to the emperor of Constantinople in order to bring the relics of the Lord's Passion for which the Sainte-Chapelle was built. Edward III often appointed his messengers from among people of modest origins, such as clerks of the chancery, the wardrobe, the council, or the chamber. They usually started their public career in the lower ranks of the court and gradually made their way up, thanks to their personal qualifications and their unquestioning loyalty to the crown. This bestowed on them royal gratitude shown in their nomination to one of the richest ecclesiastical benefices, such as bishoprics or monasteries.[11] For especially important diplomatic missions, most prominent men were employed, laymen as well as prelates, whose personal prestige fitted the occasion.[12] The services of merchants were also welcomed when their skills were more in demand, as in the many businesses of the kings of England in Flanders.[13]

The main negotiator and other prominent men were sometimes given a safe-conduct in the form of a *letter patent*.[14] The king or his council determined in advance the exact remuneration that these envoys were to receive. The amount included the expenses of the principal officer and the delegation that accompanied him on most important missions. The high cost of the delegations and the qualifications required for the proper completion of diplomatic missions sometimes encouraged the use of the same messengers sent by foreign princes in order to return the expected answer. Edward I, for instance, employed the messengers sent by the German king to whom he supplied his own letters of credence. They carried Edward's reply to the king, the clergy, and the nobility of the Empire alike.[15] A messenger was sometimes given a number of missions among different people and destinations: a Venetian envoy to Ussono Cassano received an additional commission upon his return journey from Stephen Voivode of Moldavia to the pope. Stephen also requested from Venice a doctor to cure an ulcer on his leg, and the city's Council complied.[16] The economy implied in such a practice, however, compensated for the lack of expedience in using the services of foreigners who had no special interest to make the best of a mission bestowed on them by a third party. Though medieval kings did not disregard the possibility of using foreign messengers, they usually preferred to send their own envoys whose qualifications they knew and whose loyalty they could reasonably trust.

The more active diplomacy both in Europe and in distant countries encouraged the attempts to clarify the actual functioning of political messengers while creating a clearer juridical basis for the proper accomplishment of their mission. This tendency was mostly justified in light of the need to define the degree of authority conferred on messengers: hitherto they could often be authorized to lead a negotiation, but they lacked the power required in order to sign the resulting agreement, unless they were explicitly sent for this purpose. This state of affairs complicated the mission and affected its efficiency. The diffusion of Roman Law in the twelfth century provided a suitable solution for such inconveniences through the Roman office of *procurator*.[17] The *procuratio* turned the messenger into the full representative of the sender, bestowing on him all the authority of his principal.[18] A procurator became a sort of agent, useful primarily in business affairs but also for diplomatic purposes. Unlike the *nuncius*, the procurator when properly authorized could negotiate and conclude agreements between states, while obviating the need for the customary meetings of princes or for an interminable succession of messages sent back and forth between principals.[19] Procurators were appointed to negotiate diplomatic agreements since for minor business the employment of *nuncii* remained quite customary.

The widespread use of *procurationes* improved the efficiency of the communication system as it freed kings from their former need to conduct diplomatic negotiations in person. Philip the Fair, for instance, did not leave France during his reign except on one occasion in 1285, when he visited Catalonia. Procurators were subordinated to the king's council from which they received detailed instructions for their mission and to whom they had to report the results. The *procurationes* thus expanded the role of messengers beyond their former functions of conveying or receiving messages while turning them into more active participants in medieval diplomacy. In 1217, for instance, they were involved in the delicate negotiations between the English barons and Prince Louis of France, the future Louis VIII.[20] Edward II and his wife, Isabelle of France, took advantage of the same practice in order to advance their interests in England, France, and the papal court in Avignon.[21] Before his journey to Italy in 1311, the German King Henry of Luxembourg tried to advance his candidacy for the Empire through procurators sent to

Mantua and other Italian city-states.[22] The use of procurators became a fairly customary practice for improving the unstable position of the German kings in Italy. Lewis of Bavaria also used procurators fifteen years later in order to bring about a favorable reception during his journey to Rome.[23]

The effectiveness implied in the use of procurators, however, was conditional on the readiness of kings to stand behind the agreements signed by their representatives. Despite the fact that procurators often went beyond the authority they were conferred on their credentials, they enjoyed the unconditional confidence and support of their senders, thus allowing them to advance the interests of the political institution they represented. Furthermore, the instability characteristic of medieval diplomacy and the difficulty in maintaining regular contacts with their principals, brought about the use of *blanks*, which widened even more the procurators' scope to maneuver. Blanks were empty official documents signed by the political authorities, empowering their representatives to fulfill them according to their understanding and the varying requirements of the mission in hand. In 1257, Henry III employed blanks in his negotiations concerning the Kingdom of Sicily. The king's procurators, Simon de Monfort, Peter of Savoy, John Maunsell, and the Archbishop of Tarentaise, were commissioned to the papal court in order to request the services of a legate ad hoc to facilitate the negotiations with France and the amelioration of the papal demands in Sicily. Simon and Peter, the chief negotiators, received full authority and "twenty white and void schedules sealed with the seal of the king; and eight void schedules, sealed with the seal of the lord Edward; and ten void schedules sealed with the golden bull, under the name of the Lord Edmund, son of the king." If the two main negotiators would go to Rome, they should use these documents and blanks just as they deem expedient "for the honor of the king and the affairs of Sicily." Otherwise, the blank schedules should be canceled and returned to the king.[24]

Although the use of blanks widened the procurators' scope to maneuver, it could not, however, solve all kinds of difficulties arising from diplomatic negotiations, nor could it fulfill the need for more regular channels of communication in foreign affairs. The growing needs of medieval diplomacy ultimately brought about the emergence of more *permanent representations*, which aimed at providing more reliable and up-to-date information by frequent written reports. These developments first reached full crystallization in Italy, a fact that led Neale to regard the Italian city-states as "the nursery of European diplomacy."[25] The use of more permanent ambassadors and the regulations for reporting methods were already fixed in Italy toward the second half of the thirteenth century. In order to improve its messenger system and to define some of its more ambiguous procedures, the Venetian Maggior Conciglio initially extended the stay abroad of its representatives. Before returning home, Venetian messengers were requested to receive the explicit approval of the city's Council. They were also required to submit a written report on the results of their mission within a fortnight of their return and, in addition, to inform all helpful information about the country they had visited and its inhabitants.[26] These careful regulations indicate the increasing tendency to regulate the political communication channels with foreign countries, alongside the increasing power of the political institution that promoted them.

In 1350, Alberico de Rosate defined the *ambassiator* as "the ruler's messenger in important matters." This definition reflects the casual nature that still characterized the early ambassadors, even though their mission was more prolonged than those of former messengers. A longer sojourn in a particular destination led to two essential changes: the connection between the sender and his ambassador became even looser, as the latter could hardly obtain regular information about current affairs at home, nor had he access to regular reports on the changing priorities of his government. The irregular communication between the ambassador and his government, therefore, still required the further employment of messengers ad hoc. On the other hand, the ambassador became a most important source of information about current affairs in his foreign residence, where he also played an important representative role. This state of affairs shifted the emphasis of the ambassador's office, turning its focus on the reception of information about foreign countries while weakening its effectivity in a constant transmission of messages. Philippe de Commynes faithfully expressed, in retrospect, the ambivalent attitude toward the early ambassadors. While he admitted their contribution in the field of diplomacy, he advised rulers as to the most effective ways to safeguard their interests in their presence:

> If (ambassadors) come from true friends of whom there can be no suspicion, treat them with good cheer and grant them frequent audience, but dismiss them soon, for friendship among princes does not endure for ever. If from hostile courts, send honorable men to meet them, lodge them well, set safe and wise men about them to watch who visits them and keep malcontents away, give them audience at once and be rid of them. Even in time of war one must receive envoys, but see that a keen eye is kept on them, and for every one sent to you, do you in return send two, and take every opportunity of sending, for you can have no better spies, and it will be hard to keep a strict watch over two or three.[27]

What might be now defined as supplying information was then perceived as a simple act of spying, and the distinction between diplomatic representation and espionage remained somewhat ambiguous. The ambivalent attitude toward early ambassadors did not encourage the emergence of permanent embassies, but instead brought about a slow process of institutionalization. The thirteenth century remain still characterized by embassies ad hoc. Only toward the end of the fourteenth century did permanent embassies become a more common phenomenon. The first known resident ambassador was Messer Bartolino de Codelupi, who arrived in Milan in August, 1375, as the representative of Lodovico Gonzaga of Mantua. Before the end of the fifteenth century, permanent diplomacy had extended to the great powers of Europe, which gradually nominated permanent ambassadors. Resident embassies then sprang up in all the great Western capitals. In 1479, Venice sent an ambassador to reside at the court of France, and extended the practice to England in 1496. As pointed out by Mattingly, at this stage "the Middle Ages in diplomacy was over."[28]

Alongside the improvement in communication systems with foreign countries, the socioeconomic process of integration characteristic of the Central Middle Ages encouraged kings to reorganize their communication channels with their own subjects. This policy resulted also from the tendency to strengthen the king's position at home, while undermining the feudal barriers that still prevented the immediate contact between the king and the community of the realm. In the traditional framework of medieval society, moreover, the custom of a village or an estate, of a borough or a hundred, was required at every turn in the transaction of royal business and the administration of justice. This state of affairs encouraged the elaboration of a communication system less exposed to the feudal patterns of former procedures. The imperative to obtain maximal information and, in parallel, to influence the political audience, thus fostered the development of a two-way system of reception and transmission of information.

The conquest of England in 1066 emphasized the need of the Norman kings to obtain maximum information about their new kingdom and the income it could supply. This search for reliable information was behind the compilation of the *Domesday Book* (1086), one of the richest reports of the Middle Ages; it contained socioeconomic and political data as collected throughout the kingdom by royal commissioners especially appointed for this task. The inquiry of the king's commissioners focused on two main issues, namely, the liability to yield profit and the existing rateable value of each estate, and its past, current, and potential commercial value. The survey was designed to discover past tax evasion and to provide the economic data on which a revision of the assessments could be based. The Domesday Book also furnished the king with information about royal rights in the shires and boroughs, together with an analysis of the tenurial relationships in the kingdom.[29] The tendency to record a maximum of information about their kingdom, its inhabitants, and income left its mark on the policy of English monarchs in later periods as well. Henry II (1154–1189) and Edward I (1272–1307) promoted similar inquiries by which the essential character of royal administration was fixed. By means of inquiries and writs and the accumulation of written records, control from above was brought into closer touch with local conditions.

Despite the use of written reports, the political communication system as a whole remained crucially influenced by the personal contact between the communicator, that is, the king in person or his messenger, and the audience. The importance ascribed to personal contact brought about the visitation of kings in all areas of their realm, a very common practice from early times.[30] The peripatetic nature of medieval courts thus appeared as an efficient means for strengthening a direct bond between the king and his subjects, besides the economic advantages implied in the implementation of the royal rights of hospitality. The great mobility of the kings of England is clearly indicated by the itinerary of King John between November 26, 1200 and March 12, 1201, which records it in detail (see map 4).[31]

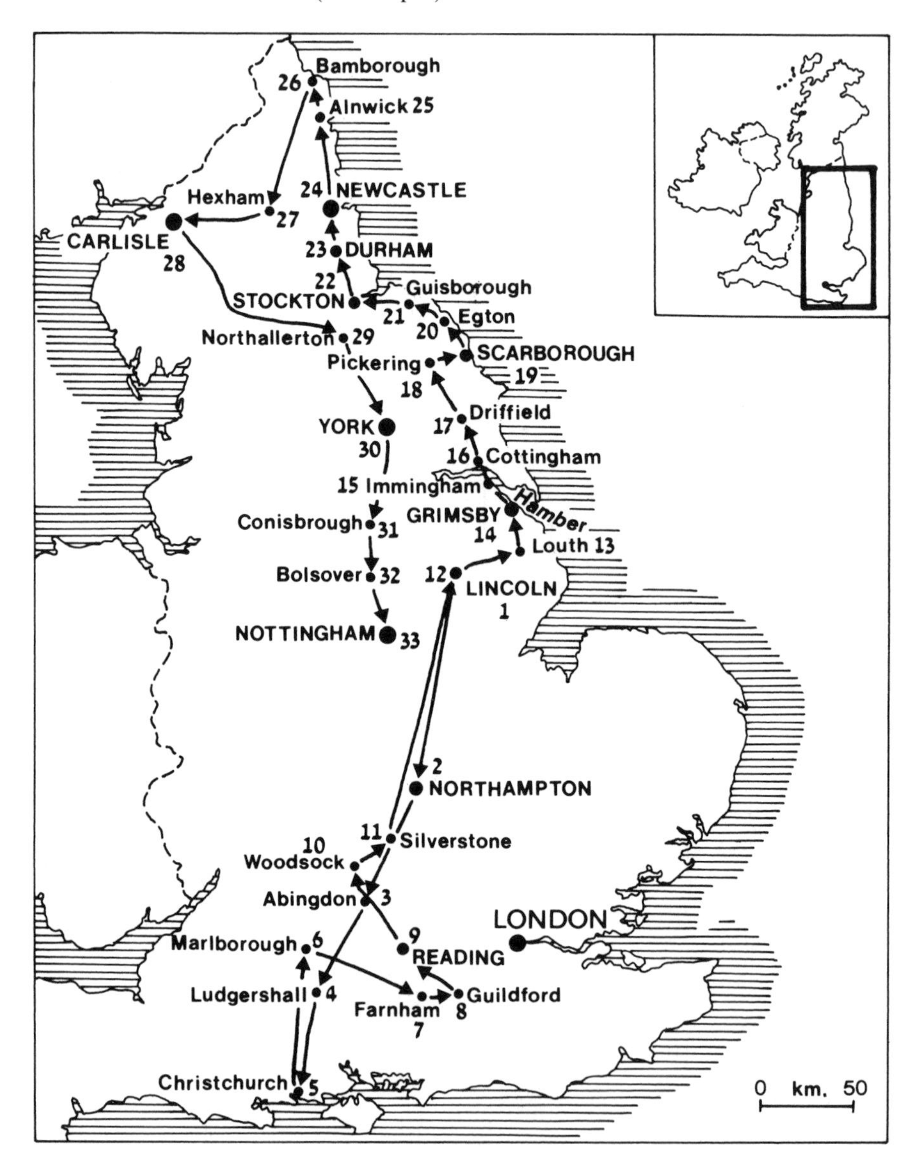

King John's itinerary.

In less than two months John managed to travel to almost every corner of his realm. He often used the same roads that served his contemporaries on their way to Westminster,[32] but sometimes deviated from the main roads in order to hunt, using one of the forest lodges. John probably utilized long carts, which, though slower than horsemen, could carry the many accessories of the royal entourage.[33] Contemporary commentators reported in a rather positive light the mobility of kings, which was regarded as reflecting their sense of responsibility to see that justice was done. According to Peter of Blois, Henry II, for instance, "does not remain in his palace as other kings do, but, going about the provinces, he investigates the deeds of all, judging most strictly those whom he had appointed judges of others."[34] Although the kings of France ruled over considerably larger territories, their itineraries were limited to the royal domain, a fact dictated by the political weakness of the early Capetians. Up to the thirteenth century, before the king settled more permanently in Paris, the Capetians often traveled between Paris, Soissons, Orléans, Compiègne, Senlis, Laon, Beauvais, Etampes, Bourges, Chartres, and Reims (see map 5).

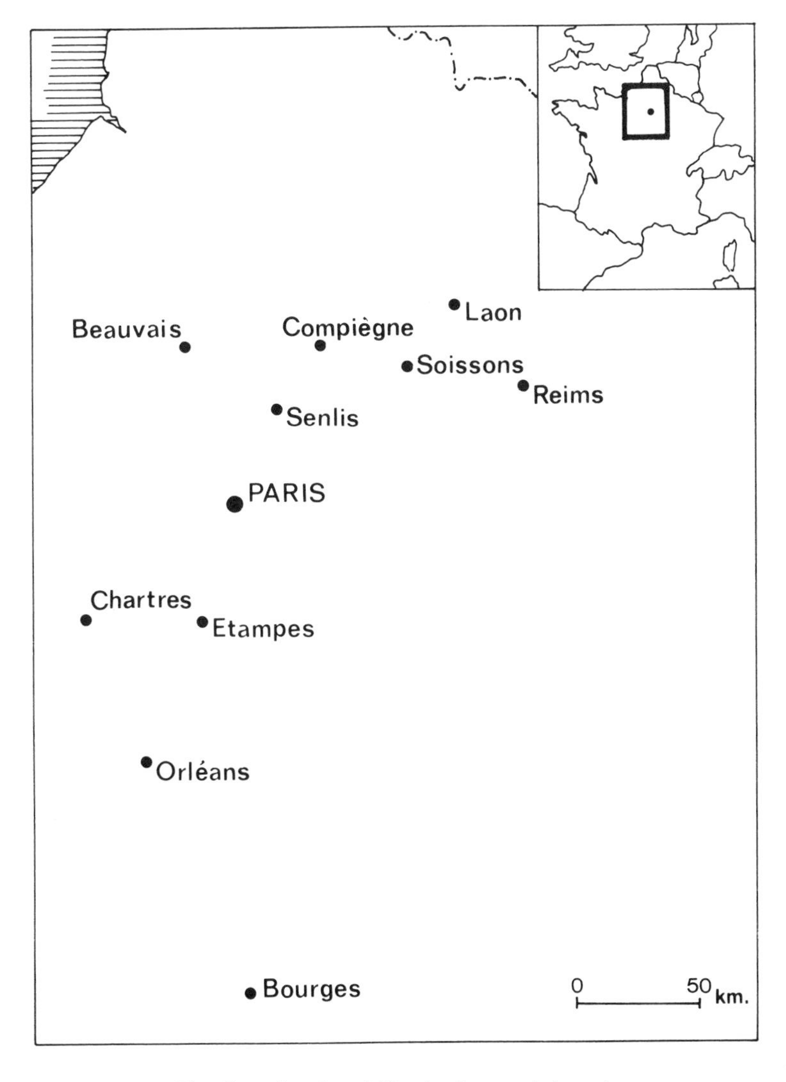

The Capetians' mobility in the royal domain.

Although royal wayfaring fostered the direct contact between the king and his subjects, it was basically intermittent and could hardly ensure the fluent communication of the royal court with the countryside. Moreover, in the eleventh and twelfth centuries, kings were not yet aware as to the importance of a communications system per se. Political communication systems emerged de facto, in order to solve fiscal or political needs in the short term, and could be generally categorized through three major systems. In France and the Crusader kingdom of Jerusalem, where feudal lords assumed the royal functions in the countryside, they were expected to represent the kings there. Though this system was quite effective and inexpensive, it led to a further disintegration of the royal rule into feudal categories. In Byzantium and the kingdom of Sicily, royal administration was operated by hired officers who received no financial or political prerogatives. Though most effective, it was a very expensive system that required a parallel method of supervision. Rulers could also rely on the popular institutions of the local communities. This trend developed in tenth-century England and in Germany, where royal authority was maximally assisted by local institutions. Despite its economic advantages, this system also reduced the monarch's ability to maneuver in his own kingdom.

The Norman conquest brought about the simultaneous implementation of the three methods in the kingdom of England. The Norman kings integrated prominent landholders in their communication system through the framework of the *Great Council*, a custom widespread in most European kingdoms at the time. The Norman kings also encouraged the further development of local frameworks, such as the *shire* or *county*, and its administrative subdivision known as the *hundred*. All these were subordinated to royal hired officers, the *sheriffs*. In the kingdom of France, where local institutions were less deeply rooted, paid royal officers were appointed regularly from the reign of Philip August onward. They were called *bailiffs* in the north and *seneschals* in the south, and supervised the former royal officers, the *prévôts*. The Great Council and the sheriffs or bailiffs supplied the main regular communication channels at the disposal of medieval monarchs within the borders of their kingdoms. The mobilization of the royal officers for communication purposes thus reflects the earliest stage in the development of the political communication system. It also indicates the lack of more specialized channels that could both supply information and transmit the king's message throughout the realm.

The earliest means of communication at the service of medieval kings was the *Great Council* to which the tenants-in-chief, the main vassals of the king, laity, and clergy alike, were expected to attend. They were called to the meetings through a personal summons, known as a *writ* in England and *edictum* in France, whether for purposes of consultation, propaganda, or legislation. These gatherings represented an expansion of the *king's council*, the advisory group of courtiers possessing extensive judicial and executive powers whose services were required in the administration of the realm.

From the king's perspective, mere attendance at the Assembly reflected his degree of power in actual practice, as it was strictly regulated by the feudal concept of *consilium*, which still systematized the links between the king and his tenants-in-chief. But from the perspective of the magnates the *Assembly* also acquired important communication perspectives as it provided them with a suitable forum for the interchange of information and deliberation on matters of general interest. The Carolingians used to summon prominent landholders to the Assembly, whose sessions they regulated by their needs or interests. This state of affairs actually cast a shadow over the constitutional weight to be attributed to the early Assemblies. In a curious episode concerning King Lothar in 966, he supposedly advised his son, the future Louis V, "to rule and direct the kingdom with the advice and help of the magnates, to regard them as relatives and friends, and never initiate any important maneuver without their consent."[35] If this report is indeed accurate, it suggests one of the first attempts to bestow consultative powers on the Assembly beyond its primary role in the fields of information and propaganda. If such a consultation took place in the Early Middle Ages, it was, however, essentially dependent upon the royal will, nor could it oblige the king, who enjoyed his exclusive prerogative of fulfilling or ignoring the advice of his magnates.

Furthermore, in most cases, the king listened only to the advice of his *curiales*, the courtiers close to him, while the other members of the Assembly were left with the theoretical prerogative of consent through acclamation. The Anglo-Saxon kings also used to summon the "wise men" of the kingdom to the *Witan* or *Wittena Gemot*, which deliberated once a year, at Christmas, Easter, or Whitsun. At the meeting of 931, for example, 100 people participated, namely, the archbishops of Canterbury and York, two French princes who were visiting the court at the time, seventeen bishops, five abbots, fifteen aldermen and fifty-nine *ministri*, that is, the permanent royal advisers who included the king's relatives, officers, and warriors. Yet this large number was rather exceptional, as some thirty people usually were present.

The Norman kings maintained the practice of holding old festival courts to which they summoned their tenants-in-chief three times a year, at Easter, Whitsun, and Christmas. This custom was formally established at the Provisions of Oxford in 1258 as a result of baronial pressure.[36] The early Capetians also used to summon their feudal lords fairly frequently, sometimes once a month, in order to counterbalance the weaknesses of their own administration system. The meeting place was established according to the king's residence at the time, a fact which dictated the peripatetic nature of the Assembly within the royal domain. The increasing power of kings and the resulting improvement of royal administration gradually encouraged a greater regularity of the Assembly in France as well. Before his departure for the Third Crusade (1190), Philip August formally established the regular summoning of the Assembly in Paris three times a year.[37] The institutionalization of periodical Assemblies in both France and England was closely connected with the process of sociopolitical integration characteristic of the times, which reduced the isolation inherent in the feudal system and encouraged a more regular contact between the king and his subjects throughout the realm. It affected not only the regularity of the meetings but also their nature.

The more regular meetings between the king, the nobility, the clergy, and the representatives of knights and townsmen provided an important channel of communication through which the king could raise important issues and reasonably expect their maximal diffusion in the realm. The use of the term *parliament* (from 1239) heralds this process of change, which gradually turned early Assemblies into the representative body of the community of the realm.[38] According to Humbert de Romans, parliaments gathered at certain times every year, so that "the more important public affairs might be resolved upon deliberation with the magnates, that account might be rendered by the king's ministers, and that legislation for the realm might be made."[39] Matters of peace and war, the relationship with the Church, departures on Crusades, and all questions concerning domestic and foreign policy were brought before the community of the realm through its representatives at the Assembly.[40] The concept of community of the realm, formerly reserved for the members of the aristocracy and clergy alone, now came to include knights and townsmen as well, who became more active participants in the political game of the times. Toward the end of the thirteenth century, parliaments became the composite of two Assemblies: the standing curial element still occasionally provided by magnates in large numbers and by the representatives of knights and townsmen, and the successive conventions of suitors and officials, which constituted a kind of continuous review of the provincial affairs of the realm.

Referring to the early development of parliaments, Marongiu distinguishes between three kinds of Assemblies that were motivated by the different but nonetheless complementary interests of the monarchy.[41] A first aim was intended for the "public relations" of medieval kings. These gatherings gave publicity and attached prestige to particular events, such as the announcement of royal weddings or the king's departure on Crusade. The attendance of the nobility and prelates from all over the realm assured maximal publicity for the royal deeds and encouraged further cooperation with the royal policy. Such was the case of Louis IX's Crusader vow in 1247: the king took advantage of the pomp and ceremony of the Assembly in order to weaken the reluctance of many of his nobles to further support the Crusade zeal of the saint-king.[42] As royal courts became more aware of the importance to be ascribed to supporting public opinion, they fostered the consultative functions of the Assemblies on issues of war, peace, and finance. In France, the relationship with Flanders and England was one of the main issues discussed in the Assembly toward the end of the thirteenth century, while the English parliament focused on the most suitable policy regarding Scotland and France.[43] The participants expressed their acceptance of the royal policy by acclamation, but they probably had not yet gained the right to protest or oppose the king's proposals. The development of both kinds of Assemblies toward the end of the thirteenth century led to the emergence of representative institutions known as parliaments.

At their origins, parliaments represented the organized sociopolitical groups of the realm, namely, the aristocracy and the clergy. The incorporation of representatives of knights and townsmen indicates a more advanced stage. King John encouraged the participation of knights in the parliament of 1213,[44] while the representatives of towns were formally incorporated by 1311. Knights and townsmen, however, had actually participated in parliaments at an earlier date. Already in 1177, contemporary sources reported the attendance of hundreds at the parliament summoned by Henry II in Winchester, namely, "counts, barons and knights of the kingdom."[45] The summons of Edward I for the parliament of 1274 provides one of the earliest evidences about the attendance of town representatives and mirrors the royal arrangements prior to the opening of parliament:

> Edward, by the grace of God King of England, lord of Ireland and duke of Aquitaine, to the sheriff of . . . ., greeting. As for certain reasons we have postponed to the morrow of the Sunday after Easter next (April 22, 1275) our general parliament which we proposed to hold with the prelates and other magnates of our realm at London a fortnight after the Purification of the Blessed Mary next (February 16, 1275), we order you to cause four of the knights of your county with knowledge of the law and also six or four citizens, burgesses or other good men from each city, borough and market town of your bailliwick, to come there on the aforesaid morrow of the Sunday after Easter to consider at the same time as the magnates of our realm the affairs of that realm. You shall also have our letters addressed to various persons in your bailliwick handed to or sent to them on our behalf without delay. And this you shall in no way omit, and you shall give us full information about the execution of this our command at the date aforesaid. Witness myself at Woodstock, on 26 December in the third year of our reign.[46]

In this particular case, sheriffs were given two months' notice in order to announce the postponement of the parliament and to deliver the new summons to the knights and townsmen by means of messengers carrying oral or written messages.

The "*Modus tenendi parliamentum*," a treatise written between 1316 and 1324, adds to our knowledge on the members to be summoned and the proceedings prior to the formal opening of parliament.[47] According to the *Modus*, the king had to give forty days' notice of his intention to open a parliament. The summons had to be sent to five main categories of people, namely,

> (1) archbishops, bishops, abbots, priors and other leading clergy who hold by earldom or barony by reason of such tenure, and no lower clergy unless their presence and attendance is required for reasons other than their tenure. . . ; (2) every one of the earls and barons, and their peers, that is, those who have

> lands and rents to the value of a whole earldom. . . . and no lower laity ought to be summoned and come to parliament by reason of their tenure. They should be summoned and come only if their presence is useful or necessary to parliament on other grounds. . . . ; (3) Also the king used to send his writs to the warden of the Cinque Ports to cause to be elected from each port by that port two suitable and experienced barons. . . . ; (4) Also the king used to send his writs to all the sheriffs of England for each to cause to be elected from his county by the county two suitable honourable and experienced knights. . . . in the same way word used to be sent to the mayor and sheriffs of London. . . . York and other cities for them to elect on behalf of the community of their city two suitable, honourable and experienced citizens. . . . (and) two suitable and experienced burgesses.

In addition to this detailed list, the *Modus* mentions a fifth category of personages who should participate in parliament on account of their office, without them being summoned by writs, namely:

> (5) the two principal clerks of parliament . . . and the other secondary clerks . . . and the chief crier of England with his under criers,[48] and the chief doorkeeper of England . . . the chancellor of England, the treasurer, the chamberlains and barons of the exchequer, the justices, and all clerks and knights of the king, together with sergeants of the king's pleas. . . .

With regard to the actual functioning of parliament, the *Modus* establishes that "on the first day proclamation should be made, first in the hall of the monastery or other public place where the parliament is being held, and afterwards publicly in the city or vill, that all those who wish to present petitions and plaints to parliament should deliver them not later than the fifth day following the first day of parliament." "Parliament ought not be dissolved so long as any petition remains undiscussed or, at least, the answer to it has not been decided."[49] Throughout the fourteenth century, parliaments were summoned every eleven months on average.[50] This relatively high frequency involved a heavy burden for all expected participants who became more and more reluctant to attend. Absenteeism from parliament resulted from the many expenses involved in such participation and the attempts to avoid the financial requests which had often brought about the summons. Moreover, the kings consistently reduced the participation of the clergy: while Edward I summoned some seventy abbots and prelates to the Great Parliament of 1295, by the reign of Edward III their number was reduced to nineteen, and throughout the fourteenth century their average number ran to about twenty-seven. For terms of comparison, the number of lay peers ran to between forty-five and sixty, although in times of peace there came about 100.[51] The social structure of parliaments thus reflects the changing balance of power and the royal tendency to take maximal advantage of them while advancing its own interests.

In France, too, the representative system emerged gradually from the customary meetings between the kings and the members of the nobility, laity, and ecclesiastics, who paid them homage. The ability of the Capetians to secure a reasonable attendance, however, was even less effective than in England. Absenteeism became a common phenomenon on the grounds of sickness, the difficulties of the journey or the lack of sufficient time between the royal summons and the gathering of the Assembly. Although the ecclesiastical hierarchy usually sustained the royal policy, one can follow conflicting attitudes with regard to the Assembly. The archbishops of Sens and Reims attended the meetings quite regularly, but the archbishops of Tours and Bourges were only rarely present. The archbishops of Bordeaux took advantage of the long distance from the court in order to evade regular participation, except during the fifteen years of Louis VII's rule in Aquitaine, and the archbishops of Lyon did not attend at all. Among the members of the secular nobility who participated most were those whose benefices were located near the royal court or whose interests were close to the Capetians, such as the counts of Flanders, Ponthieu, Vermandois, Soissons, Nevers, Blois, and Champagne. On the other hand, the dukes of Normandy, Burgundy, and Aquitaine, and the counts of Britany, Anjou, Toulouse, and Auvergne only rarely responded to the royal summons. From time to time, the Capetians summoned members of the low clergy and small vassals as well, a practice which raised the number of participants. Only the members of the high nobility and most important prelates received a personal summons, the *edictum*. Lacking an effective administration that could announce the forthcoming meetings in the provinces, the Capetians often required the services of prelates at this purpose. In times of crisis, all participants received a safe-conduct and, sometimes, they were also required to bring a suitable military force which had to escort them to the meeting place.

The kings' attempts to ensure maximal participation indicate the importance acquired by the Assembly or parliament and their resulting contribution in the emergence of a nationwide communication system. These supplied the royal court with a convenient stage where it built up a sympathetic public opinion; they further strengthened the king's supervision over the local administration, as the royal officers had to report to the parliament committees on their actions and income. Philip the Fair well understood the propaganda advantages implied in such practices when he turned the Assembly into a publicity arena to launch his campaign against Pope Boniface VIII and the Templars. Edward III also took maximal advantage of this forum to encourage wide support of his aggressive policy at the outbreak of the Hundred Years War. By the later Middle Ages, parliaments had five major functions: to determine civil and criminal causes; to deal with taxation and supply; to present petitions; to pass laws; and to discuss affairs of state. Thomas Bisson sums up the role of medieval parliaments as

> the periodic meetings of the king's counsellors who displayed especial competence and loyalty in handling the cases and affairs that came before the king in swelling volume after about 1250. Whatever the disposition of these matters, they had first to be discussed, to be "talked about," so that the term *parlamentum* was no less applicable to these sessions than to the less special-

ized assemblies which it continued to denote. The prelates, knights and clerks of parliaments were committees of the undifferentiated king's council.[52]

By 1311, parliaments emerged as a recognizable body of lords, judges, knights, and commoners, the latter sitting in a representative capacity with many functions varying from parliament to parliament. Yet the king could summon, postpone, or dismiss parliament as he wished; he could call whoever he wished even though the magnates claimed a hereditary right of summons through continual usage. The development of parliaments in the Central Middle Ages thus embodies an essential paradox. Although parliaments allowed the active participation of wide social strata summoned to discuss most important political affairs, they resulted from the centralizing policy of monarchs and their tendency to strengthen their links with all inhabitants of the realm. Although this state of affairs hints at the propagandistic role of parliament, it does not blur the contribution of parliaments in the emergence of a communication system. They provided the kings with a useful arena for propaganda as well as a means for information supply from all areas of the realm. They further contributed an effective channel for supervising the local administration: royal officers had to report before the parliament committees about their current activities and were also committed to bring about the implementation of the parliament's decisions in the province. Despite the fact that magnates, prelates, knights, and townsmen did not yet enjoy the prerogative of a veto, in the framework of parliament they became more active partners in the political interplay while profiting from the reception and exchange of information up-to-date.

The tendency to turn the parliament into a central communication channel was subordinated to the growing financial needs of the monarchy at the time. The development of royal administration and the involvement in expensive wars aggravated the economic situation of the kings who had not yet developed regular sources of income. This increased the tax burden on townsmen who became the main targets of royal taxation and, as such, its main opponents. Vincent de Beauvais voiced the prevailing opinion when he asserted that princes who extorted money from their subjects were like thieves. Yet they sinned even more gravely than thieves since their actions endangered public justice, which they had been instituted to uphold.[53] Large sectors of the nobility also joined the opposition to the royal fiscal policy due to its centralizing implications.[54] From the royal perspective, the need thus increased for an adequate system of propaganda that would mitigate the prevailing criticism and turn parliament into a suitable arena to obtain general approval. This state of affairs was further fostered by the Roman principle of "*Quod omnes tangit ab omnibus approbari debet*" (what touches all alike, must be approved by all), which formalized in juridical terms the duty of monarchs to obtain consent for their policies in general and for taxation in particular.[55] The growing financial needs of the monarchy and its tendency to foster favorable public opinion on the one hand, and the opposition to taxation and the growing consciousness of the juridical validity of consent on the other, turned the medieval parliament into a crucial communication channel for both supporters and opponents of the royal policy.

The degree of influence achieved by parliaments and the implementation of their legislation throughout the kingdom was still conditioned by the efficient organization of the royal administration in the local level. The current communication between the parliament and the local officers acquired maximal importance in times of war when the latter were responsible for the collection of taxes and the recruitment of soldiers. The *sheriffs* in the kingdom of England and the *bailiffs* or *seneschals* in France complemented the political communication system. They ensured the current communication between the monarchs and their subjects and were also expected to serve as the most faithful spokesmen of the parliament's decisions throughout the realm.

William I's policy to tighten his control throughout the realm while increasing his income turned the Norman conquest into a catalyst for the development of a communication system in the kingdom of England. In order to improve the functioning of local institutions and bring about their more effective subordination to the royal interests, the Norman monarchs divided the kingdom into thirty-nine *shires*. The *sheriff* became the main agent for the exercise of royal authority in the shire and the guardian of the royal estates. He acted as a middleman for all communication between the royal court and the inhabitants of the shire, a fact which implies his complete involvement within the Anglo-Saxon communal systems. Most laymen were indeed still subjected to the jurisdiction of the communal courts of the shire and the hundred. The sheriff presided over the meetings of the men of the shire at their regular sessions every four or six weeks, where the king's proclamations and orders were read and their resulting implications on the local level discussed. The hundred court met sometimes as often as once a fortnight; twice a year there was a special, full session presided over by the sheriff. In addition to the close supervision of the shire and the hundred courts, the sheriff had to bring about the actual implementation of their decisions and to collect the fines they imposed. He was also responsible for the maintenance of peace, the arrest of criminals, the cooperation with the ecclesiastical establishment, the collection of taxes, and the guarding of royal property in the shire. In order to accomplish these many duties, he was assisted by a regular staff, which mainly included clergymen.[56] The crucial role played by the sheriffs turned the appointing of suitable men to the office into a most relevant issue. Following the Norman conquest, most sheriffs were appointed from among the Norman nobility, a policy that threatened to turn the office into hereditary while weakening the king's control. Both Henry I (1100–1135) and Henry II bestowed on the office the nature of a hired bureaucracy while releasing it from feudal connotations. From the twelfth century onward, sheriffs were directly appointed by the king and always in areas other than the manors on whose revenues they depended. The *Dialogue of the Exchequer* explicitly established that "any knight or other discret man, may be appointed to a sheriffdom or other bailiffry by the king, even if he hold nothing of him in chief, but is the vassal of another. . . . It is within the king's power to require the services of anyone, if he have need of him."[57]

The ascension of the Angevin dynasty in 1154 underlined even more the centralizing policy of the English monarchs. Henry II encouraged the development of a supervisory system, which became quite necessary in view of his long absences from the realm. In order to strengthen royal control over the sheriffs and to avoid any mismanagement, the "*Justices of England*," "Chief Justices" or "Itinerant Justices," were appointed. They transferred the royal council to different areas of the kingdom while becoming the personal representatives of the king in the shires. At the time of their visitations, they suspended the regular functioning of the local officers and were engaged in all sorts of public business, mainly, pleas of the crown, which concerned the king and the efficient government of the realm. They also provided the royal court with a detailed, accurate report of the shires and the prevailing climate of opinion. The justices were fully authorized to receive appeals from all subjects during the king's travels to the Continent. They also acted as spokesmen of the royal policy, which they publicized throughout the kingdom. Henry II appointed to the high office personages close to the court who enjoyed his confidence, such as Richard de Luce, Thomas Becket, Glanville, the chronicler Roger de Hoveden and the storyteller Walter Map. From 1176 onward, the institution was fixed and there were quite regular circuits of chief justices almost every year. The kingdom was divided for this purpose into six regions, each of which was dealt with by a group of justices. Twelve men from every hundred in the shire and four from each village were then required to bring all available information before the justices for their scrutiny.

Beside the regular visits of the justices, the Angevins also appointed *special committees* from time to time. Such committees were mainly concerned with pleas against the king's officers, especially complaints that had risen while the kings were away. This state of affairs further underlines the identification between administration and communication in medieval society, while the very emergence of a communication system was completely subordinated to the changing needs of the kings. Following an absence of more than three years in Gascony (1286–1289), Edward I appointed a special committee to receive pleas from any subject of the realm. The goals of the royal committee were explicitly defined by the king's order to all sheriffs from October 13, 1289:

> To the sheriff of . . . . Notification that the king has appointed J. Bishop of Winchester, R. Bishop of Bath and Wells, Henry de Lacy Earl of Lincoln, John de Sancto Johanne, William le Latimer, Mr William de Luda keeper of the wardrobe, and William de Marchia, to hear any grievances and wrongs that have been committed during the king's absence from his realm by his ministers upon any persons of the realm, in order that they, after hearing the complaints and the answers of the ministers concerning them, may relate and explain them to the king in his next parliament to be duly corrected. The king therefore orders the sheriff to cause all and singular of his county who feel themselves to have been aggrieved during the king's absence by his

> ministers and who wish to complain of the same, to be distinctly and openly warned throughout the sheriff's bailiwick to come to Westminster on the morrow of Martinmas next before the king's subjects aforesaid to show and prosecute their grievances. The sheriff is charged to execute this order as he loves himself and his goods, so that he may not be found remiss or negligent and so incur punishment as a condemner of the king's orders.[58]

Though one might question the degree of efficiency of such public proclamations or their actual capacity to change customary ways of life, they reflect, however, the royal awareness as to the lack of more effective, regular channels of information. The rudimentary techniques of communication combined with the poor state of the roads left much room for mismanagement and abuse of royal authority. This ultimately encouraged the integration of wide social strata into the royal communication system in order to utilize the information they could provide for supervising the king's own representatives in the province.

The Capetians as well fostered the development of a political communication system along similar lines to those developed in England about 100 years earlier. Yet the French communication system suffered from the same inconveniences that affected the whole political organization, particularly, the lack of regular sources of income to secure its operation. This was the main factor behind the incapability of the Capetians to develop reliable communication channels until the end of the twelfth century. The royal representative in the province was the *prévôt*, who was assigned the management of the royal affairs.[59] The long list of ordinances dealing with the provosts' mismanagements, however, indicate the many disadvantages involved in the office. Based on feudal premises, it allowed the provosts to increase their power and income at the expense of royal prerogatives.[60] Philip August aimed at solving these problems through strengthening his control over the royal officers and providing his court with more reliable channels of information. Before his departure on the Third Crusade (1190), Philip established a new officer, the *bailiff* or *seneschal*, who was directly subordinated to the royal court. First appointed from the ranks of royal messengers, the exercise of his charge was directly established by the king who could remove or transfer him from one region to another at any time.[61] At the end of the twelfth century, the bailiff's office was still of a rather temporary nature; the bailiwick was not yet properly defined, and neither was the number of provosts subordinated to the bailiffs. During the thirteenth century, the functioning of the bailiwick was more properly structured, but the Capetians did not relinquish their prerogative to replace the bailiffs every two or three years. Between 1286 and 1289, almost every bailiff and seneschal was assigned to a new district, other than his original place of birth. Though these practices were rather inexpedient from an administrative angle, they actually prevented the further feudalization of the royal administration.

The bailiff or seneschal was ultimately responsible for every action that affected the interests of the king in his district, namely, keeping the peace and defending the borders; arresting malefactors and controlling the proper operation of law courts; collecting revenues and maintaining revenue-producing properties in proper condition; and enforcing royal ordinances and putting into effect the numerous royal writs with regard to transfers of land, establishment of rents, compromises over jurisdiction, and enforcement of parliamentary decisions. He was the highest judge and the final administrative authority, while appeals from his decisions could be presented only before parliament or the king's council. The due completion of the bailiff's office implied his continuous movement within the bailiwick and also out of it. He was expected to reach in person the different areas of the bailiwick, which were often larger than the French departments in modern times. He was also required to regularly attend the meetings of the Great Council or parliament and those of the regional assemblies. The bailiffs thus provided more regular and reliable channels of information either for the king or the inhabitants of the province, as they could transmit royal messages and also inform the king of the prevailing climate of opinion in the countryside. The considerable authority bestowed on bailiffs brought about their careful selection from among the ranks of expert and skillful men, who, at some stage of their career, were incorporated into one branch of royal administration. In a country as diverse as France, the bailiffs ensured, as far as was possible, uniformity in the implementation of royal policy.[62] The arrest of the Knights Templar is but one example of the high degree of efficiency achieved during the reign of Philip the Fair (1285–1314). The simultaneous arrest of all knights throughout the realm on October 13, 1307 required absolute secrecy and complete coordination between the different and sometimes distant bailiwicks.[63] The successful achievement of this mission indicates the advanced logistic system developed by the bailiffs and their unconditional loyalty to the crown, whose interests they advanced in actual practice.

The great distances from the court, however, and the still strong influence of feudal customs sometimes brought about the misapplication of the wide authority conferred on the office. The royal attempts to prevent abuses fostered new ordinances concerning the bailiffs' duty to remain at their posts for an additional period of forty days following the end of their charge. This extra period was meant to enable the local inhabitants to voice their complaints and to give the bailiffs the time required to make the necessary amendments. But these forty additional days could hardly bring about dramatic improvements, nor could they actually change customary ways of operation. The appointment of *enquêteurs* or *réformateurs*, who functioned in a way rather similar to that of the itinerant justices in England, aimed at ensuring the proper operation of local officers while providing the royal court with an additional, more direct channel of communication with the provinces.

According to Matthew Paris, St. Louis arranged the regular activity of *enquêteurs* before he departed on the Sixth Crusade:

> In that same autumn (1247) Louis, the most pious king of the French, sending friars Preacher and Minor through his whole kingdom to make diligent inquiry, caused an investigation to be made also through the bailiffs, and vocal

> announcements to be publicly proclaimed, that if any merchant or anyone else who had suffered any injustice by any forced transaction or extortion of money or food-stuffs, as is frequently done by the king's collectors, could produce a note or a record or a statement, or would swear or in any other way could prove (his claim), he was ready to restore all. And it was so done.[64]

In France as well as in England, the establishment of a supervisory system acknowledged the active participation of the king's subjects who were encouraged to provide additional information either by written or oral messages, but always under oath. In his advice to his heir, Philip III, St. Louis emphasized the importance he ascribed to the reception of maximal information, as the very implementation of royal justice was actually subordinated to it:

> If anyone has an action against thee, make full inquisition until thou knowest the truth. . . . Use diligence to have good provosts and bailiffs, and enquire often of them, and of those of thy household, how they conduct themselves, and if there be found in them any vice of inordinate covetousness, or falsehood, or trickery.[65]

The testament of Louis IX thus underlines the increasing awareness of medieval monarchs as to the importance to be ascribed to the reception of maximal information, a need which in the political language of the times was suitably justified by the royal genuine search of justice and righteousness.

The many difficulties plaguing the implementation of such goals and the king's interest in ensuring the most objective procedures brought about the careful selection of the *enquêteurs*. Most of them were appointed from the ranks of courtiers, the *palatines*, who enjoyed wide authority and whose services were required in embassies ad hoc. Of the thirty-one *enquêteurs* nominated during the reign of St.Louis, eight were Dominicans and seven Franciscans, a state of affairs which mirrors the high prestige the friars enjoyed at the time.[66] Others were appointed from the ranks of the secular clergy, or were laymen whose character and loyalty to the king were beyond doubt. They were sent in pairs, usually a layman and a clergyman, while the local bishops and bailiffs announced their coming to the area. The *enquêteurs* represented the king's person and received viceregal powers. They could dismiss or punish those royal officials whom they considered inefficient, repressive, or corrupt. Pierre Roche, judge of Minervois, was condemned and dismissed from his post on charges of mismanagement in 1309. Pardoned by Philip the Fair in 1314, and condemned again on basically the same charges, he was finally executed in 1318. Such investigations were performed in public while complaints against the king or his representatives could be delivered orally or in writing, but always under oath. After the testimony was given, the *enquêteurs* decided on the most suitable ways to correct the situation. They could not, however, reopen a case that had already been dealt with by former *enquêteurs*, nor deal with most difficult cases that were transferred to the royal court for its consideration. The testimonies of 508 sworn witnesses who appeared before the *enquêteurs* allowed Delisle to reconstruct the main issues under investigation, as follows:

> 1) How has N. acted in his bailiwick?
> 2) How has he acted in his position as bailiff in protecting the rights of the king, his possessions and the land?
> 3) Under N.'s administration, have the rights or possessions of the king been diminished at all?
> 4) How has he acted in handling cases and pleas?
> 5) Has he asked for, received or kept any loan or deposit?
> 6) Has he, his wife, his children, or anyone else for them, brought, sold or exchange any thing or, otherwise, made agreements with anyone else in which some advantage was taken?
> 7) Has he or anyone else for him asked or kept anything for making peace, for determining a settlement, or for doing justice?
> 8) Has he unjustly arrested, imprisoned or punished anyone through confiscation of goods or in person?[67]

This detailed list indicates the difficulties which plagued the emergence of a nationwide communication system. The centralizing tendencies of monarchs actually increased their need for more frequent and effective channels of communication. One of the mains goals of royal administration focused, therefore, on the supply of reliable information, which in many cases had fiscal connotations. Yet the primitive channels of transmission and the lack of a more efficient communication between the king and his officers in the province encouraged the development of alternative channels, which both provided additional information and controlled the way of operation of the royal officers. Lacking effective control over local officers and, consequently, the ability to check the accuracy of their reports, the monarchs were forced to rely on the reports of the same population whom their own representatives were expected to control.

As the number of royal officers increased as well as the cases brought before them, the need for more accurate channels of information became even more acute. Philip III (1270–1285) regulated the operations of the *enquêteurs* on a more permanent basis, and defined their supervision over bailiffs and seneschals.[68] During the reign of Philip the Fair, seventy-three *enquêteurs* operated in actual practice. Half of them were originally from Paris and had been attached to the parliament, by which they had been sent to critical areas that required exceptional action. The thirty-seven *enquêteurs*, originally from the provinces, were chosen from among the local nobility and knights, although there were also eight abbots and three prelates among them. The growing number of *enquêteurs* reflects the increasing awareness of the monarchy of the need to improve its communication system, as it appeared the most suitable means to protect royal rights while providing additional sources of income. At a time when the royal court was mostly in need of financial resources, the *enquêteurs* practical knowledge of the country and its inhabitants became an important source of information that could hardly be neglected. The rather high efficiency of the *enquêteurs* is satisfactorily proved by the readiness of countrymen to advance large sums of money to the royal treasure in order to exempt themselves from such an investigation. Continuous royal pressure on the matter increased the hostility towards the *enquêteurs*, voiced in the Assembly. Toward the mid-fourteenth century, the rising power of the Assembly led to royal concessions in this regard, and the fiscal activities of the *enquêteurs* were limited. The office thus returned to its original role of supervision and prevention of mismanagement by the royal officers.[69]

In his instructive study on the *enquêteurs*' activities throughout the reign of St. Louis, Wyse emphasizes their contribution to the development of the representative system in France later on: the sending of royal messengers to the provinces paved the way for a direct channel of communication between the king and his subjects. In time, this was reciprocally used in order to send the representatives of the local communities to the parliament.[70] Though this assumption may seem far-fetched, the *enquêteurs* did provide a more reliable channel for the transmission and reception of information, whether from the king to his subjects or from the provinces to the royal court. The development of the office and its successive functions further testify to the tendency of medieval monarchs to enlist their administrators for communication purposes, while those were eventually subordinated to fiscal and political needs. Yet royal manipulation encountered varying degrees of criticism, which was voiced at the Assembly. Despite the kings' initial success in presenting the *enquêteurs* as an additional expression of their genuine search of justice, the manipulation of the office for fiscal needs brought about much criticism, and the kings were ultimately forced to relinquish this beneficial but still unpopular practice. This state of affairs mirrors the dynamism inherent in the political communication process while large social sectors became more active in the political interplay. They were no longer only targets for royal propaganda but appeared also as the monarchy's active partners who, as such, did not restrain themselves to the theoretical approval of its policy.

The growing influence of the Assembly and its members' capability to stand against royal manipulation heralds one significant stage in the political communication process. It indicates the existence of a two-way system of communication utilized both by the king and by the enlarged political society of the times. This state of affairs suggests the dynamic nature of the political society and the resulting need to adapt the premises of research to long-range developments. In the first stages, eleventh-century monarchs were hardly capable of fostering a nationwide communication system, as they directed all forces for the very survival of monarchy in the feudal regime. The basic weakness of monarchs and the isolation inherent in the feudal regime had turned the focus of the communication system abroad, in the relationship with other princes and the papal court. The crucial role played by the magnates in the king's council, the personal essence of political communication, the changing interplay between the king's subjects and the king's messengers, the lack of a clear differentiation between administration and communication, and the subordination of the communication system to the fiscal and political needs of the monarchy, all mirror the many drawbacks intrinsic in political communication throughout the Middle Ages. The integration process in the socioeconomic level, however, encouraged the standardization of communication channels, first with foreign countries and, later, in the internal order as well. The primitive system of messengers ad hoc was gradually improved by the introduction of Roman patterns and the emergence of permanent embassies, which heralded a new stage of diplomatic intercourse. This process was concomitant with the gradual strengthening of monarchy at the interior plane. The increasing power of monarchs throughout the twelfth and thirteenth centuries reduced to some extent the former large prerogatives of the aristocracy, secular and clerical, and expanded the political audience to include the knights and the bourgeoisie. The representation of the community of the realm in parliaments, the use of paid, removable officers, the more widespread use of written reports, and the improvements in the reception process of information indicate the degree of development achieved by the political communication system in the Central Middle Ages. As such, it could both provide more accurate information and propagandize the message of the monarchy throughout the realm. The use of propaganda to legitimize the king's policy provides one of the earliest examples of its use in the long term, one which ultimately fostered the emergence of a national consciousness.

8

Propaganda and Manipulation of the Crusade Theme

In a review article in the *Encyclopedia of the Social Sciences,* Smith defined propaganda as "the relatively deliberate manipulation, by means of symbols (words, gestures, flags, images, monuments, music, etc.) of other people's thoughts or actions with respect to beliefs, values, and behaviors which these people ('reactors') regard as controversial" [18: 579]. The conceptual analysis of the propaganda phenomenon frequently emphasizes its novel, unprecedented character. Propaganda is viewed as a new social, cultural, and political force, sharply differentiated from the promotional-instrumental types of communication current in earlier periods. Propaganda in this specific, modern sense is said to be characterized by distinctive features such as detachment, concealment, and scientific sophistication.[1] The scientific analysis of psychology and sociology has also been depicted as a unique feature of modern propaganda. According to the prevailing views, it was absent from earlier types of promotional discourse.[2] Lasswell and Blumenstock contrast the attachment of former propagandists to their beliefs with the new breed of propagandist who "does not necessarily believe in what he says."[3] This approach has been largely corroborated by Ellul: "Propaganda no longer obeys an ideology. The propagandist is not and cannot be a believer."[4] Another trait ascribed to modern propaganda focuses on its devious character, as the originators of the message as well as the goals actually promoted are hidden from the audience.[5] Manipulation has been defined as the action of man trying to influence others to accomplish a specific program. The manipulator does not make use of force but rather of a system of signals, organized and expressed by representation. The means of manipulation cover a wide spectrum and can vary from the human voice or gesture to the use of written messages or images. The manipulator sometimes limits himself to the world of ideas in order to influence the beliefs of his audience, though more often he aims at bringing about some action. The main goals of manipulation had been systematically categorized through the fostering of fears, contempt, suspicion, or love.[6] Both terms, propaganda and manipulation, further took on a decidedly negative connotation. In the face of the increasing sophistication of propaganda techniques, nineteenth- and early-twentienth century-social critics such as Lippmann, Lasswell, and Mannheim connected their fears about the manipulation of popular opinion (and votes) by professional propagandists to the survival of democracy.[7]

This chapter deals with the use of the Crusade theme to legitimize the process of state-building in the late Middle Ages, an issue that, in our view, fits into the patterns established in the conceptual plan with regard to propaganda and manipulation. With the exception of scientific sophistication anachronic to the Middle Ages, it seems that most traits ascribed to "modern" propaganda developed 500 years ago, in the process of an emerging national identity looking for its legitimization. The manipulative use of the Crusade theme by royal communicators aimed both at fostering fears and love, contempt and suspicion: love for the king and identification with his policy; fears, contempt, and suspicion toward foreign forces, personified by the pope, the Templars, or a foreign king, all of whom represented the harmful forces of evil. To foster the reception of the royal message, the king's communicators made maximal use of all the communication channels at their disposal: oral appeals in front of a receptive Assembly, written reports, ostentatious celebrations, and poetry, as well as the use of well-known symbols such as those related to the chivalry ideal. As to the degree of identification of medieval communicators with their message, the very fact that the plans of the Crusade did not materialize led, at least, to the possibility of a gap between theory and actual practice. The use of the Crusade theme thus provides a suitable case study for the development of propaganda and manipulation in the process of state-building, the fall of Crusader Acre in 1291 serving as a catalyst for the manipulation of the Crusade ideology in a national context. Yet the criticism aroused by the royal policy also hints at the limitations of manipulation which, in fact, could not successfully change the prevailing views of sociopolitical elites. This state of affairs actually turned the townsmen into the main target of royal propaganda, thus revealing the changing audiences in the political society of the times.

The use of propaganda in the late Middle Ages was closely connected with the search for legitimization of the political establishment, a trend characteristic of societies in a process of consolidation. Legitimacy implies the recognition of a political order and appears, therefore, as a contestable validity claim, used mainly while the justification of an order may be disputed. The claim to legitimacy is related to the social-integrative preservation of a normatively determined social identity. Legitimations serve to justify the existence of a political body, namely, to show how and why existing or recommended institutions are fit to employ political power in such a way that the values of society will be realized. It is closely connected with the wealth of beliefs, symbols, and myths upon which the collective life of society is based. In early civilizations, the ruling families justified themselves with the help of myths of origin. The Roman emperors or the German kings represented themselves originally as gods. On this level, narrative grounds as mythological stories proved to be sufficient. The further development of political societies fostered the need for legitimization, not only of the ruler alone but of the political order as a whole. This goal could satisfactorily be achieved by cosmologically grounded ethics, higher religions, and philosophies that went back to the illustrious founders.[8]

In the Central Middle Ages, the demand for legitimization of the political body was oriented toward the religious *Umwelt* of the times through giving a new meaning to old concepts. The conservatism attributed to medieval society in this regard hints at its reliance upon accepted norms, a trait that is hardly unique to the Middle Ages. Moreover, the use of religious symbols did not imply a state of stagnation, as they were constantly adapted to meet changing needs during periods of transition. The manipulation of old concepts by the royal communicators, furthermore, reflects a more advanced stage in the integration process of medieval society, which, by the fourteenth century, became more communication-oriented. It also suggests the widening of the political audience whose support the royal communicators were seeking—as the former target of royal propaganda, the political and intellectual elites, were more consistent in the safeguard of their own interests and ideas and could therefore less easily be manipulated. The manipulation of abstract concepts, such as the Just War, the Holy Land, or the chosen people, quite dissociated from everyday reality, also corroborates the influence of former propagandistic campaigns and their success in enriching the wealth of symbols of medieval man. They eventually paved the way for more advanced stages in the communication process, as the effective use of manipulation was in fact conditional to the existence of common concepts to which a new significance could be ascribed. This state of affairs thus indicates the gradual progress of the communication system, which overcame the isolation inherent in feudal society and fostered the emergence of a more communication-oriented society.

The beginning of the fourteenth century heralds the emergence of national monarchies which in both France and England was concomitant with the royal tendency to strengthen the kings' power against the decentralizing tendencies of the feudal nobility. The last Capetians (1285–1327) faced an antinomian aristocracy that tried to benefit from the economic and dynastic crises that affected the kingdom; the secessionist tendencies of the Flemish added another factor to the unstable situation. Across the channel, the three Edwards (1307–1377) also faced a belligerent aristocracy that did not relinquish its customary claims with regard to the implementation of the Magna Carta; the separatist tendencies of the Scots on the northern border also became an important factor in the political crisis that England was undergoing at the time.[9] Although contemporary monarchs did not disdain the use of force to restrain antagonist tendencies at home, they were well aware to the fact that no political body could rely in the long run on the use of force alone. The appeal for supporting public opinion thus became a characteristic trend of royal policy, while the use of the Crusade theme provided a powerful tool in the hands of royal communicators.

The use of the Crusade theme to legitimize the emerging national state, however, seems rather paradoxical, the Crusades being a faithful expression of the corporate effort of Christendom as a whole and, as such, the antithesis of the national body. The pan-Christian essence of the Crusade was faithfully expressed by Fulcher of Chartres:

> Consider, I pray, and reflect, how in our times God has transferred the West into the East. For we who were Occidentals have now been made Orientals. He who was a Roman or a Frank, is now a Galilaean, or an inhabitant of Palestine. He who was once a citizen of Reims or of Chartres, now has been made a citizen of Tyre or of Antioch. We have already forgotten the places of our birth, they have already become unknown to many of us or, at least, are not mentioned. . . . Different languages, now made common, have become known to all races, and faith unites those whose forefathers were strangers. . . .[10]

The universal nature of the Crusades, moreover, resulted not only from the historical process that forced a common way of life upon people from different and, sometimes, inimical countries. It resulted rather from the wide identification of the Crusaders with the biblical heritage and the role of the chosen people, to be fulfilled anew by the Christian armies in the Holy Land. This facet had contributed the main axis of Crusade propaganda, which, in fact, ascribed only marginal importance to the historical process, namely, the Eastern Christians' call for help and the support of the Byzantine Empire. Crusade propaganda had thus focused on the wealth of biblical and evangelical symbols, which were extrinsic to the historical reality of eleventh-century Christendom.

The identification of the Crusaders with the concept of the chosen people provides a suitable link between the Crusades and the process of state-building in the late Middle Ages. The Crusades and the emerging national state were both faced with similar problems of ideological legitimization, and they both looked for this legitimization out of the context of actual reality. Furthermore, at the beginning of the fourteenth century, the process toward the emergence of national monarchies demanded a revision of priorities between Church and State, priesthood and monarchy, or between one's fidelity to the king and his policy on the one hand, and to the Church's ancestral immunity on the other. While the tendency to encourage a sense of political solidarity among the inhabitants of a defined territory created a gap with the pan-Christian essence of the Crusades, Western political leaders appealed to the Crusade theme in order to dispose of an old accepted symbol, well-known to their contemporaries. This was a natural result of the almost complete absence of "national patriotism,"[11] and the magic appeal inherent in the concept of Crusade. Moreover, the kings of England and France themselves, like their contemporaries, were still committed to the wealth of old accepted symbols integrated in the Christian faith, in which the Crusades still had a place of preference.[12] As the Crusades had formerly been legitimized by concepts extrinsic to the political reality of eleventh-century Christendom, by the fourteenth century they served to legitimize the process of state-building, notwithstanding the gap between the pan-Christian trend in the idea of Crusade and the national nature of the emerging state.

The integration of the Crusade theme to legitimize secular policies was closely connected with the development of the Crusades in the late Middle Ages. Although the idea of Crusade remained always connected with some kind of a "Just War," its political and economic implications were becoming more and more evident. The interests of the papacy in the West had encouraged the gradual disjunction of the Crusade idea from the Holy Land while turning it into a formidable weapon at the disposal of the papal monarchy. Crusades were promoted by the papacy against the Albigensians and Frederick II, both regarded as heretics, as well as against the Italian city-states which challenged the centralizing policy of the Vicar of Christ in the Papal State. Furthermore, on at least two occasions (1215–1217 and 1263–1265), the popes made use of the Crusade theme to support the policy of the kings of England, and all those enlisted in the royal enterprise were bestowed by papal sanction with the Crusade privileges.[13] According to Pope Innocent III, the insurgent barons of England were less amenable than the Saracens because of their open revolt against King John, who had committed himself to the Crusade and had formally taken the vow.[14] In his call to the nobles of Bourges to support King John, Innocent III claimed that they should join "the army of Jesus Christ, and bravely defend His soldier and indeed His enterprise, so that at last in heaven they may be deemed worthy of receiving the triumphal palm of a soldier of Christ and the victor's crown of celestial glory."[15] The distortion of the Crusade idea and its further manipulation for political aims in Europe found an additional expression in the fiscal plan. Large sums of money contributed in support of the Holy Land and the Crusader enterprise overseas covered in actual practice the growing expenses of the papal court. In 1309, for instance, Pope Clement V withheld the money collected to support the Christian campaign against Granada on the grounds of the forthcoming Crusade to the Holy Land which, in fact, never materialized.[16]

The advantages of such a policy in the economic and political planes encouraged the readiness of Western kings to share with the papacy the benefits implicit in the Crusade. Such was the case of the Christian kings of Spain, who, being in the forefront of the *reconquista*, were rewarded with the granting of indulgences and the right to collect the tithe.[17] In its origins, this practice seemed highly justified: as the collection of the tithe was substantiated on the grounds of a Just War against the Moslems, it appeared quite justified to entrust the Crusade income at the disposal of those kings who devoted themselves to the defense of Christendom along the Iberian frontier.[18] But the readiness to undertake the Crusader vow resulted in other Christian kings becoming beneficiaries of the tithe income, an arrangement that did not always depend on the fulfillment of the vow in actual practice.[19] Thus, one of the first steps toward the manipulation of the Crusades in the national context was made with a papal blessing for a crucial practical issue: the income accruing from the tithe. This practice was not interrupted by the fall of Crusader Acre in 1291. The popes continued this beneficial policy to the kings, by which their vow to take up the cross rewarded the royal treasury with the papal authorization to collect the tithe. Some data will illustrate the crucial implications of this development in the fiscal plane. In large areas of thirteenth-century Valencia, the income from tithes surpassed the amounts collected by the royal treasury from Jews and Moslems together.[20] Between 1301 and 1327, the papal authorization to collect the tithe in England brought to the Exchequer the significant sum of 230,000 pounds, which represented the bulk of the royal income at the time.[21] The kings of France were also active partners in the collection of the tithe, and the apostolic authorization to impose the tithe on behalf of the last Capetians was renewed practically year after year.[22] A half-yearly account of 1316 records that just under fifty percent of the royal income came from Church taxes.[23] Similar practices continued at the time of the Valois dynasty. After Philip VI of Valois had publicly announced his decision to lead a further Crusade (October, 1332), he was accordingly bestowed the collection of the tithe for six years. Contemporary data suggest that large sums of money reached the royal coffers in this way.[24] Although the Crusade plans never materialized, the money was not returned to the Church, nor did it bring about any noticeable help to the Eastern Christians. It served, rather, to subsidize the war efforts against England on the eve of the Hundred Years War. Philip VI corroborated this state of affairs before Pope Clement VI in 1344. Yet the king found it highly justified in light of the fact that "what was good for France would be eventually good for the Crusades as well."[25] The apostolic sanctions authorizing the collection of the tithe reflect the papal tendency of subordinating the Crusades to the fiscal needs of the emerging national monarchies. In the papal document redacted on behalf of Edward I to collect the tithe in 1306, Clement V (1305–1314) mentioned the needs of the Holy Land as making such taxation obligatory.[26] But as early as in 1307, the Crusade theme disappeared as though it had never been, to be replaced by the needs of the royal court, which, according to Clement, largely justified the further collection of the tithe.[27] Similarly, the papal authorization accorded to Philip the Fair to collect the tithe for three years in 1305, was justified by the Capetian war effort in Flanders and the recovery of the French economy.[28]

The papal readiness to legitimize the use of the tithe for secular purposes was soon exploited by contemporary kings. They found in the ecclesiastical order an important source of income, which could hardly have been neglected in view of the extraordinary expenses of the growing royal administration and the war efforts at the time. When Philip the Fair required a double tithe from the clergy of Tours in 1305, he justified his claim on the grounds of the clergy's ancestral commitment to the kingdom of France, namely, to its "spiritual and temporal aid, in order to preserve, defend and guard the unity of this realm. . . . a venerable part of the Holy Church of God." The king also used the occasion to teach the clergy some principles of Christian ethics. He enthusiastically encouraged them to follow the example of Christ, thus avoiding any preference of their particular interests above the welfare of the people, "since it is for this welfare that Jesus Christ . . . exposed himself to death." From the royal point of view, furthermore, the reluctance of the clergy to pay the tax could entail the violation of its "sacred ministry." Although the fiscal implications of such a claim were clear enough, royal propaganda succeeded to some degree in facilitating the reception of the king's message and weakening the clergy's opposition to royal taxation.[29] In 1294, the abbot of Cluny justified his grant to the royal treasure on the grounds of the king's status as "the leader. . . . of the cause of God and the Church and the fighter for all of Christendom."[30]

Brown has convincingly depicted the intrinsic animosity towards taxation in medieval society, with the prelates appearing in the forefront of the opposition to royal taxes.[31] The Crusades had undermined this attitude to some extent, paving the way for regular taxation. Yet the readiness of European society to pay the Crusade taxes was not an expression of a new economic concept, but resulted rather from the wide identification with the goals supposedly served by the papal tax, namely, the Crusade. No identification of this kind could be detected with regard to the royal fiscal policy at this stage. On the contrary, it was often a matter of controversy between the king and the representatives of the community of the realm in parliament. The use of the Crusade theme to justify taxes aimed to weaken this opposition to same extent, while the use of well-accepted norms encouraged a more favorable reception of the royal demands. The collection of the tithe thus provided the royal treasuries with a regular source of income and released the monarchs from the restrictions inherent in the customary feudal aids, limited hitherto to the payment of the king's ransom, the knighting of his eldest son, and the dowry of his eldest daughter. By strengthening the fiscal maneuvering of medieval kings, the tithe also fostered the transition process of feudal suzerains into national kings. Kantorowicz has pointed out the logic behind this development:

> What was good for the *regnum Christi regis*, Jerusalem and the Holy Land, was good for the *regnum regis Siciliae* or *Franciae*. If a special and extraordinary taxation was justifiable in the case of an emergency in the Kingdom of Jerusalem and for its defense, it seemed also justifiable, especially in the age of purely secularized Crusades. . . . to meet the emergencies of the Sicilian Kingdom or those of France in the same fashion. After all, emergency begins at home.[32]

By the late Middle Ages, royal propaganda succeeded in giving "home" a new meaning. For the former Crusaders, "home" had been represented by the Holy Sepulchre and the Holy Land but, by the fourteenth century, it had lost its former universal meaning to be replaced by the immediate territory, that in which the Christian was born and was later to be buried. The collection of the tithe in the service of the emerging monarchies thus played an important role in the floating identification of the individual as a Christian and as a subject of a national identity, giving more emphasis to the latter. The former readiness to die in the defense of the Holy Land was gradually replaced by the Roman motto of "*pro patria mori*," while the concept of *patria* weakened the hitherto universal essence of Christendom.

The popes' generosity toward the Western kings and their resulting support of the royal aggressive policy, however, aroused widespread criticism in both England and France, from prelates and laymen who directed their diatribes against popes and kings alike. In his poem *Sirventes contre Philippe de Valois*, Raimon de Cornet accused Philip VI of misusing the Crusade funds. He advised the king to use his forces against the "Turcs" and leave the English in peace, as they did not cause any blame to fall on France.[33] The criticism aroused by the continued collection of the tithe after the fall of Acre reflects one of the major problems faced by royal propaganda in the late Middle Ages. The growing national state lacked a fiscal basis which could ensure its regular function. The problem turned more acute as the growing expenses of the royal administration became a source of resentment, arousing an unwillingness to cooperate with the secular state, since it still lacked legitimization. This was the major reason for the manipulative use of the Crusade theme, used to justify taxes. But the climate of opinion of the times was still attached to the original principles of the Crusades and contemporaries were not too receptive to the manipulation of the Crusade for political and fiscal purposes. Despite the popularity enjoyed by medieval kings and their success in fostering a favorable public opinion at home, their communicators did not completely succeed in bringing about similar results in the Crusade-fiscal context.[34] This state of affairs, however, did not restrain the manipulation of the Crusade theme in the political sphere, the popes again serving as pioneers in the propaganda campaign.

The English policy of coercion in Scotland, and that of the French in Flanders, were presented by papal documentation as a "Just War," which assured its participants, especially the kings who led the armies, of all the benefits of the Crusades. Pope Clement V and John XXII (1316–1334) accused the Scottish and Flemish insurgents of being obstacles to the renewal of the Crusades, "*impeditores negotii terrae Sanctae*,"[35] their fighting being considered harmful to the Holy Land.[36] In his letter of 1316, John XXII admonished the count of Flanders and the townsmen as by their harmful insurrection against the Capetians, they were actually obstructing the implementation of a new Crusade. The papal document categorically identified the French court with the Crusades as, according to Pope John, the Capetian devotion to the faith, the Crusades, and the Apostolic See, not only justified but virtually commanded the papacy to support the French policy against the Flemish insurrection.[37]

The papal manipulation of the Crusade theme in the political sphere was soon imitated by the political establishment of the times. Those same Scots accused by the pope of obstructing the Crusades, reacted by blaming Pope John of treachery against both the Holy Land and the Crusades and denounced the papal complicity in the belligerent policy of the kings of England (Declaration of Arbroath—1320).[38] The use of Crusade terminology to legitimize secular aims gradually became a common practice in the political lexicon of the times.[39] The chancellery of Edward the First also exploited his Crusader's background in order to deny any legitimacy both to the Scots and to the no-less insurgent barons of England.[40] But if in the case of Edward the First the use of the Crusader halo was fairly justified in light of his participation in the Crusade,[41] quite different was the case of other kings whose theoretical readiness to take the vow entrusted them with all the Crusaders' advantages. This development found its peak in France during the reign of Philip the Fair.

The use of the Crusade theme by the communicators of Philip the Fair was integrated in their tendency to develop the Cult of Monarchy on religious premises to which they ascribed a new meaning. The mythical images of Charles Martel and Charlemagne were then incorporated in the propaganda campaign as they proved the age-old commitment of the French monarchs to the battle of survival of the Catholic faith against Islam. Expressions such as "*ad regni regimen a Deo positi*," we have been appointed by God to rule the kingdom[42] or "*Christus est nobis via, vita et veritas*," Christ is our road, our life, and (source of) truth,[43] quoted often by the royal chancellery, emphasized the tendency of the royal communicators to turn the religious essence of the monarchy into the source of legitimation for the royal policy. This trend took also advantage of the well-known devotion of the kings of France to the Crusades, personified by St. Louis.[44] Philip the Fair himself had five immediate predecessors who had all gone *Outremer*, three of whom had died during or returning from a Crusade. In the early fourteenth century, no French monarch could afford to ignore the legacy of Louis VII, Philip August, and Louis IX, who had turned the Crusade into the ideological platform of the French monarchy. The unquestionable commitment of the Capetians to the Christian enterprise in the Holy Land were indelibly publicized in the *écu*, the gold coin minted by St. Louis in 1269, which bore the legend of *Christus vincit, Christus regnat, Christus imperat*, together with the shield of the militant pilgrim (see plate 21). The three clauses became the symbol of the crown of France, while the king actually committed his rule to Christ the victorious, the royal, and the imperial.

Philip the Fair succeeded in regaining the leadership of the Crusade, which after St. Louis had been transferred to Edward I, king of England. Though Philip's vow to take up the cross never materialized, the royal spokesmen succeeded in perpetuating his Crusader halo for generations while the king's theoretical commitment to the Crusade was suitably complemented by the devotion of his ancestors to the Holy Land in actual practice.[45] The attempts to strengthen the links between Philip and his meritorious grandfather lay behind his efforts to bring about the canonization of Louis IX and the removal of the saint's body to Paris, with great pomp.[46] All these served as major means for royal propaganda to exhibit the policy of Philip the Fair as the natural outcome of that of St. Louis, who, within the prestigious Capetian lineage, achieved pride of place.[47] Yet not all contemporaries fell as easy prey to the king's manipulation of the Crusader prestige of his ancestors. In his epilogue of the *Life of St. Louis*, Sir Jean de Joinville exhorted Philip to follow in actual practice the Crusader zeal of his grandfather.[48] The unquestionable devotion of the Capetian dynasty to the Holy Land, nonetheless, supplied the royal chancellery with the legitimization required for the policies of Philip the Fair. The sentence, "following the precedents of our ancestors," became a common formula in royal documentation, used whether the king was aiming at reducing the prerogatives of ecclesiastical inquisitors, at controlling the loans offered by Jews, or at glorifying the royal commitment to welfare.[49]

The Crusader vow of Philip the Fair and his three sons was presented by royal propaganda as yet another result of the Capetian ancestral devotion to the Holy Land and the Crusades.[50] On this occasion, royal propaganda was not limited to the cultural elite, but by the intensive use of propagandistic tools such as processions, festive meetings, and celebrations, it ensured the widest participation in the Crusader zeal of the king. The appeal of royal propaganda to wide social sectors indicates a more advanced stage in the integration process and the gradual emergence of a more communication-oriented society. The sociopolitical elite lost its former privileged status while artisans, knights, merchants, and commoners in general expanded the audience of the royal message. In 1313, all Paris celebrated the knighthood ceremony of Philip's sons for one week, which was formally accompanied by the Crusader vow of the whole royal family. The chroniclers describing the event found it difficult to find some precedent which equaled the solemnity and festivity of the ceremony. They emphasized the attendance of three kings: Philip himself; his eldest son, Louis, king of Navarre; and his son-in-law, Edward II, king of England: "this was the most beautiful and solemn festivity we have had for many a long time. . . . and all for garlanding the king and his sons. . . . This festivity, indeed, has bestowed great honor on the king and his sons but also on all the inhabitants of Paris."[51] The ceremony of knighthood acquired the dimensions of a festival that preserved something of the meaning it had enjoyed in primitive societies. It appeared as the supreme expression of their culture, the highest mode of a collective enjoyment, and an assertion of solidarity. The more crushing the misery of daily life, the stronger the stimulants that were needed to produce that intoxication with beauty and delight without which life would be unbearable.[52] The Crusader vow of the royal family and the publicity given to the event further suggest the impact of the Crusade theme and its important contribution from the angle of royal propaganda: the image of a distant king surrounded by his counselors was then superseded by one of a more accessible monarch who personified the union between the dearest ideals of medieval society: knighthood and Crusade.

The gap between Crusade ceremonials and actual practice, however, was still to be bridged. Despite the fact that Philip and his three sons, the future Louis X (1314–1316), Philip V (1316–1322) and Charles IV (1322–1328), maintained close contacts with the Christian armies in Cyprus, Rhodes, Armenia, and Romania, and despite the outbursts of administrative and planning activity in 1308, 1316, 1318, 1321, and 1323, and the collection of ships in 1319 and 1323, the journey *Outremer* remained a desirable but nonetheless unrealizable mission. An anonymous French chronicler found it difficult to estimate the enormous amounts of money collected for the Crusade at the time but he found it necessary to lay down that "the pope (Clement V) guarded the money together with his cousin, the marquis; and the king and all those who had accepted the cross remained here; and the Saracens live in security across the sea, and, I believe, they may sleep on undisturbed."[53] The aforesaid view reflects one important trend in the climate of opinion of the times that eventually focused on the criticism of the Crusade policy of the papacy, considered as an outcome of greed, nepotism, and avarice. Although the Western kings remained the main beneficiaries of the papal fiscal demands, they were not proportionally affected by the waves of criticism and could therefore continue this desirable practice by which they got the money and the popes the hatred. This state of affairs reflects the success of royal propaganda in channeling the criticism to external factors, mainly to the papacy. The linking of the Crusade with knighthood, the axis around which medieval men tried to understand the motives and the course of history, excluded per se the pope while favoring the reception of royal propaganda. The knighthood ceremony of Philip's sons and the Crusader vow of the whole royal family appeared as another expression of the close identification between the Crusade and knighthood, one that was personified by the Capetians in actual practice.

To make a maximal profit of the concept of knighthood, it had to overcome a development similar to that undergone by the Crusade in order to serve as a source of legitimation in the process of state-building. The manipulation of chivalrous concepts to advance the centralizing policy of monarchs ultimately encouraged the decline of the feudal structures which had given knighthood its *raison d'être.* It provides, therefore, a suitable example of the long-range influence of manipulation, when concepts were exploited in order to undermine the historical reality that gave them birth. On the other hand, the success of royal communicators in manipulating chivalrous ideals reflects the advanced communication stage of fourteenth-century society, whose acquaintance with chivalry was actually dissociated from everyday experience but resulted rather from a communication process. It further indicates that large social sectors became more communication-oriented and, as such, more exposed to royal manipulation.

The manipulation of both the Crusade and the chivalrous ideals at the service of royal propaganda could hardly have succeeded without the widespread belief in the supremacy of France in both fields. Already in the mid-thirteenth century, an anonymous cleric depicted France as a "new Athens," which inherited from both Athens and Rome the excellency of their knights.[54] Fourteenth-century sources further presented the French knights as the personification of the superiority of France over all other nations. In *Le roman de Fauvel*, God had chosen France to sow there the flowers of peace and justice, of faith and freedom, and the flower, meaning the best, of knighthood. And Gervais du Bus thanked God for "this garden of France," asking for His protection against the forces of the devil.[55] An anonymous addendum even glorified the Crusader zeal of Philip the Fair, presenting the king of France as a divine response to the former prayer of Gervais du Bus:

> This was Philip, who has been once . . .
> Ha! Lord God, how he navigated
> By sea, with courage, and wandered on the earth
> To conquer the Holy Sepulchre![56]

Guillaume de Nangis, too, emphasized the predominance of French knighthood over other knights all over Christendom. He saw in the three petals of the *fleur de lys* the representation of faith, learning, and military power, France being outstanding in all three virtues, which flourished more abundantly there than in the other kingdoms of Christendom.[57] The anonymous author of the *Eloge de la cité de Paris* (1323) depicted "France shining above other kingdoms, having the triple superiority of the three components of the Holy Trinity which symbolized power, science and goodness . . . (France), therefore, is powerful because of the courage of its knights, clever because of the science of its clergy, and good because of the generosity and clemency which has always characterized its princes."[58] Several years later, when Henry de Ferrières tried to prove the special predilection of God for the kingdom of France, he, too, emphasized the three divine gifts given to the whole kingdom: the sacred relics, the excellent clergy, and also "the good knights, audacious, virtuous and honest men."[59] The preeminence of the French knights complemented by the religious excellency of the clergy thus turned France into the most suitable arena of the Crusade. This state of affairs substantiated the claim made by Philip the Fair himself, namely, that "the Crusade issue touches all (Christians) but mainly those from the kingdom of France who, as is widely known, had been especially chosen by the Lord's grace for the defense of the Catholic Faith."[60]

The Crusader zeal of the king thus appeared as the natural outcome of the knightly excellence of his subjects, both qualities being an expression of God's preference for the kingdom of France and its eminence over all other kingdoms of Christendom. These premises favored the reception of the concept of the Just War and its translation into the language of political practice. Already at the end of the twelfth century, the anonymous author of the *Chanson de Roland* had bestowed the halo of martyrdom on the warriors of Charlemagne, who had died on the battlefield.[61] True, they had supposedly fought against the Moslems of Spain and could therefore be regarded as equal to the Crusaders. Capetian propaganda, however, bestowed on the Moslems their former, more abstract definition as "heathens," while they were, in fact, replaced by the enemies of the crown as a whole. The poet Richier summed up this concept when he stated that all those killed while protecting the French crown from its enemies should be saved in the life after death, equaling, therefore, not only the Crusaders, but the saints and martyrs as well.[62]

The use of the Crusade theme appears also as a formidable weapon to be used against the enemies of the king. Pope Boniface VIII and the Knights Templar were both accused by Philip's spokesmen of being the major cause of the loss of the Holy Land, although Boniface VIII had ascended to the papacy five years after the fall of Acre, and contemporary sources attested to the heroism and self-sacrifice of the Knights Templar in defending the last Christian fortresses in the Holy Land. Both the arrest of the pope at Anagni on September 7, 1303,[63] and the arrest of the Knights Templar in France on October 13, 1307, moreover, had faced royal propaganda with the imperative to legitimize the unprecedented acts of the king, not only before his subjects but also before all Christendom. Furthermore, both events were connected with the new political equation, which required a reconsideration of accepted norms, mainly between one's ancestral fidelity to the Church's immunity and the unconditional loyalty to the king and his policy. The problematic nature of both the arrest of Boniface and the Templars thus faced royal propaganda with new challenges and favored further changes in the Crusade theme. It was gradually incorporated into one of the first propaganda efforts to spread the idea of sovereignty in Western society.

The propaganda campaign sustained by Philip's communicators against Pope Boniface VIII focused on the many dangers inherent in the papal policy with regard to the French ancestral sovereignty. As noted by the Dominican F.Renaud d'Albignac: "Let know that what the king did was done for the salvation of your souls. And, because the pope aimed to destroy the kingdom, we have all to ask the prelates, counts, and barons, and all those from the kingdom of France, that they will fight to maintain the status of the king and his kingdom."[64] This claim was complemented by serious accusations of heresy raised against Boniface VIII, which shadowed the very orthodoxy of the Vicar of God on earth. In 1302, Philip twice summoned the prelates, barons, and the representatives of the townsmen in Paris in order to inform them about the papal subversive policy against the sovereign status of the French kingdom. In 1308, the king turned anew to the Assembly in Tours in order to achieve wide support for his policy against the Knights Templar. On both occasions, the Crusade theme proved to be a useful weapon among the charges of heresy raised against Boniface VIII and the Templars. Both Boniface VIII and the Templars were accused of having harmed the Holy Land, be it through the misuse of the Crusader funds by the pope[65] or by the secret negotiations sustained by the Templars with the Saracens.[66] Both crimes had brought about the same, undesirable consequence: the loss of the Holy Land. Thus, to the long list of heretical practices ascribed to both Boniface and the Templars, the Crusade theme contributed an additional perspective, which was found most helpful for the purposes of royal propaganda. In his opening harangue against the Templars in 1308, William of Nogaret employed the legend of *Christus vincit, Christus regnat, Christus imperat*, which became synonymous with an invocation of St. Louis's guidance while appealing by its usage to the memory of and approval by the Crusader king.[67] The Assembly of the three estates provided the most suitable audience for royal propaganda, in which the king and his spokesmen found wide support. Philip's appeal to the almost 1,000 participants in the Assembly bestowed on the royal policy the legitimation of a "national" consent, which was achieved by the intensive use of all-available communication media. The oral appeal of the king's spokesmen, often delivered in the vernacular, was complemented by the royal written summons to the prelates, barons, and townsmen throughout the kingdom. This intensive propaganda campaign was largely justified. For the first time in medieval history, a king, blessed with the flattering appellation of the *Most Christian King*, tried to enlist a national-Christian front against the official representatives of the Church, either a pope or a Military Order, which enjoyed ecclesiastical immunity.

The national connotation of the Crusade theme was solemnly formulated during the Capetian campaign in Flanders in 1302. An anonymous preacher took the occasion to claim that any opposition to the king of France abetted heresy and was actually harmful to the Holy Land. Basing his sermon on I Maccabees 3:19–22, he depicted the French struggle against the Flemish within the cosmic dimensions of a battle between justice and injustice. The irrefutable justice inherent in the French policy resulted mainly from the purity of the royal blood, the ancestral devotion of the Capetians towards the Holy Church, the procreation of holy kings, and the many miracles accomplished by the kings of France in their lifetime. This resulted in the irrefutable conviction that "since the most noble kind of death is the agony for justice, there is no doubt but that those who die for the justice of the king and of the realm (of France) shall be crowned by God as martyrs." On the other hand, "he who carries war against the king of France, therefore, works against the whole Church, against the Catholic doctrine, against holiness and justice and against the Holy Land."[68] The Crusade theme thus corroborated the holiness of the king by the grace of God, conferring on his wars in any place and against any enemy the halo of the Just War.

The use of the Crusade theme further applied to all facets of political experience. In 1312, when Philip the Fair tried to encourage the candidacy of his brother, Charles of Valois, to the Holy Roman Empire, his messengers sustained the king's position before Pope Clement V on the grounds of the Crusade and the many advantages that such an election would bring to the Holy Land.[69] In 1317, Philip V issued a ban on tournaments and jousts on the grounds that they would hinder the preparations for the forthcoming Crusade, although his targets were, most likely, provincial leagues of nobles who under the guise of jousters could train and conspire against the king.[70] Several years later, the king also requested in the name of the Crusade he never accomplished, a series of reforms of the French monetary system, the unification of weights and measures, and the recovery of the royal domain, which had been alienated illegally.[71] With the outbreak of the Hundred Years War, the Crusade theme again appeared as an important weapon to be used by both French and English propagandists, each side accusing the other of having hindered the Crusade. Philippe de Mézières, for instance, blamed the English for thwarting Philip VI's Crusade, thereby depriving many souls of heaven.[72]

The damaging impact of such an accusation was clear enough to cause Edward III to make some effort to improve his poor Crusader's image. This became an issue of maximal importance at the outbreak of the Hundred Years War, due to the imperative to encourage favorable public opinion for the English policy throughout the Continent. Although the king of England had never made the slightest material effort to support the Crusade, in December, 1335, he promised to help the Christian armies in Armenia and discussed the Crusade issue at the parliament of 1336. In November, 1337, he went even further and made public his readiness to join the forthcoming Crusade with 1,000 knights.[73] The manipulative use of the Crusade was shared by Edward III's main opponent, Philip VI, who made maximal use of the political advantages inherent in his status as captain of the forthcoming journey. His willingness to head the Crusade provided him with a means of binding important neighbors, such as the counts of Flanders and Hainault, the duke of Brabant, the king of Bohemia, and the powerful lords of Languedoc and Gascony, as well as bankers, merchants, and shipbuilders, to French interests. The Crusade further confirmed Philip VI's position not only in France, but also in Europe, and, according to royal propaganda, even in paradise. As the distance from the Crusader kingdoms increased, the projects for the recovery of the Holy Land provided an additional feature of the use of the Crusade theme to legitimize the process of state-building. The French lawyer Pierre Dubois used the Crusade theme to develop the foundations of a modern state, in which the king of France would become a superior judge and keeper of European peace,[74] legal possessor of all the ecclesiastical property within the kingdom,[75] and the leading force of both internal politics and education.[76] True, the political philosophy of Pierre Dubois was sometimes too radical to be considered an authentic reflection of the state of mind in the early fourteenth century;[77] it reflects, however, the contribution of the Crusade in developing a national ideology within the class of lawyers who became the loyal servants of the crown and, as such, publicized its policy among the intellectual elite of the times.[78]

The manipulation of the Crusade in both the fiscal and the political planes reflects one of the most successful propaganda campaigns sustained in the Central Middle Ages. The Crusade theme reinforced the religious halo of national kings independent of papal tutelage and provided the royal treasury with a regular source of income. It also served as a propaganda weapon to be used against the enemies of the crown at any time and at any place. This process reinforced the sovereign status of the king by the grace of God and, in parallel, legitimized the emergence of the national monarchies. Those Crusades, which in the late eleventh century expressed the pan-Christian essence of European society, served as a source of legitimization 200 years later for the emerging national state, the antithesis of the Crusade and one of the factors behind its decline. The use of the Crusade theme for legitimizing the secular state, therefore, provides a new insight as to the continued evolution of the Crusade idea after 1291, despite the widespread apathy of Western Christendom toward the Crusades throughout the thirteenth century. The decline of the Crusade in the thirteenth century was both a cause and a result of the gradual emergence of the national state. When the Crusade was brought to the service of the Western kings, it became more synchronic to the prevailing climate of opinion and therefore recovered its former attraction. The approach to the use of the Crusade theme in the late Middle Ages in terms of "propaganda" and "manipulation" seems, therefore, largely justified in the technical limitations of the times. The idea of Crusade in the late Middle Ages was formulated in previously accepted terms such as the "Holy Land" and the "Just War," but these terms had lost their former universal meaning. During the reign of Philip the Fair, France became "the Holy Land" and her "Most Christian King" fought the "Just Wars" of the Church. As the Crusades had used biblical legitimization, the emerging national state similarly looked for its legitimization in the same biblical background but enriched it with the Crusade idea. In this context, the fall of Crusader Acre acquires new perspectives, appearing as a catalytic agent for the adoption of the Crusade ideology in the wealth of national symbols. According to Tyerman, "the ascent of the Capetian kings of France to a dominant position in Europe during the thirteenth century was, in part, both a cause and a function of their consistent enthusiasm for the defence and, after 1291, the recovery of the Crusader strongholds in the Holy Land."[79] Although one cannot ignore the political and economic implications of this "enthusiasm" which, in fact, did not materialize, the contribution of the Crusade theme still appears to have been crucial in the process of state-building. This is but one example of the dynamism inherent in religious symbols throughout the Middle Ages and their being manipulated for changing objectives.

9

Symbols and National Stereotypes in the Hundred Years War

Converging ideas and information by the use of certain signs or well-known parables is a very old practice. It corresponds with the need to give a more material expression to abstract concepts. For the purposes of this study, symbols, whether slogans or emblems, have been considered as those objects and expressions which have influenced social life through the emotional content they have acquired. They are the projected patterns of ideas and serve to focus emotional response. As such, symbols release powerful emotions, which establish favorable or unfavorable mental attitudes toward the different social objects. They become socially important through the habit-forming tendencies of the mind, for the mind undergoes experience and organizes itself with reference to indicators of the social world outside. These indicators are concepts or, more popularly, stereotypes to which the mind develops habitual responses.[1] For the sender, the use of symbolism implies the selection of symbols to represent and elicit meanings; for the receiver, it is the process of assigning meaning to symbols.

This chapter deals with the use of symbols for purposes of political propaganda during the first stages of the Hundred Years War, and their further development into national stereotypes in France and England. Although the awareness of the advantages inherent in the use of symbols was not unique to fourteenth-century Europe, the endless conflict between the two major powers of Christendom (1337–1453) encouraged an unprecedented use of symbols through adjusting them to the new national context. The uncompromising struggle between France and England created a propagandistic challenge for the royal communicators who could not, eventually, make indiscriminate use of the Just War concept to justify the aggressive policy of each Christian monarch against the other. This state of affairs dictated two major roles to be fulfilled by the national stereotypes: they had to contribute a suitable response for the need of self-definition and, in parallel, for the definition of the enemy, which ultimately provided a group of reference.

The use of symbols was quite widespread in medieval society and left its mark on all aspects of daily life. Through all the ranks of society, a definite hierarchy of material and color distinguished between classes and gave to each rank an unmistakable form, which preserved and enhanced the feelings of its own dignity. Green, for example, was the privilege of queens and princesses, whereas it had been white in preceding ages. Besides, metaphors were often used to express ideas through the personification of institutions and objects. Almost each house and bell had a distinctive name. Also, the coat of arms and the heraldic figures therein acquired lofty values as they expressed intricate mental contexts. Whole complexes of pride and ambition, of loyalty and devotion, were condensed in the images of lions, leopards, eagles, or lilies. The prevailing use of symbols involved the need to both properly understand and explain their meaning, a task which faced medieval communicators. St. Augustine referred to the most effective ways of understanding written symbols, especially those contained in Holy Scriptures. According to Augustine, the difficulties in understanding written concepts result from the use of unknown or ambiguous signs. When the reader encounters a word or a phrase with which he is unfamiliar, Augustine advised using "old" information to interpret the "new." He also encouraged acquaintance with the original languages of the Holy Scriptures, that is, Greek and Hebrew, to try and understand the whole context of the biblical passage and, whenever possible, to ask the advice of more educated men. In addition to a general education in Liberal Arts, especially grammar, he also recommended the study of dialectics, "for this art teaches the communicator to recognize valid and invalid reasoning and improves the art of arriving at and/or testing conclusions."[2]

These statements indicate the awareness common to the Fathers of the Church, as to the difficulties intrinsic in the use of symbols. Although St. Augustine appealed to the intellectual elite who could read his treatise, he did not assume its correct understanding of symbols, neither of those related to the Holy Scriptures. The common difficulties in understanding and explaining the meaning of symbols became more critical in the Early Middle Ages, due to the fragmentation of the political society and the illiteracy that characterized medieval society as a whole. This state of affairs did not prevent the widespread use of symbols by early missionaries. Symbols still appeared as the most useful channels through which the Christian message could be spread among populations, largely illiterate, whose ability to absorb abstract messages was practically nil. The annals of the Christian mission among the German peoples were characterized by the frequent use of symbols, which transmitted to the common man, in a fairly simple and familiar language, the principles of the monotheistic religion. They brought the Christian dogma closer to everyday reality and ensured its acceptance among large masses of nonbelievers. Alcuin was well aware of the power of symbols and explained their function in a metaphorical and rather convincing way: "As clothing was first devised to protect mankind from the cold, and in course of time became an ornament of his person and a symbol of his rank, so figurative language was first used to satisfy a need, and later was given new popularity by the pleasure it provided, and by the ornamental effect it produced."[3] In other words, when symbols become rooted in the mental climate of society, they often replace the abstract ideas that they had originally represented, giving to them a higher significance. In the process, the objects, words, and phrases that become social and national symbols often free themselves from the circumstances out of which they have evolved, and develop along lines of their own, which tend to follow the social process.[4] St. Augustine and Alcuin thus provide two different, but nevertheless, complementary expressions of the use of symbols. Their statements suggest the awareness of Christian communicators as to both the difficulties and the advantages implied in such a practice. To the arguments brought by Alcuin as to the widespread use of symbols in the Middle Ages, one can further add their crucial influence in the biblical and the German cultures. Despite its awareness of the inconveniences implied therein, the Church's readiness to utilize well-known symbols was therefore related with its general tendency to adapt former beliefs to the Christian doctrine while giving them a new meaning. This policy fostered the reception of the Church's message among traditional societies and paved the way for the predominance of Christianity in the long term.

Although biblical philosophy was founded on the absolute supremacy of man over nature, man being a creature of God and allowed by Him to command the fauna and flora to his designs (Gen. 1:28–9), biblical narratives served as an endless source of animal and floral symbols expressing political or religious messages. The apocalyptic vision of Daniel (Dan. 7:2–6) presented the old kingdoms through the images of lions, eagles, bears, and leopards, and the animal symbolism left its mark in the Gospels as well (Mark 1: 13; John 1:36). The lion was often regarded as a symbol of sovereignty, strength, and courage, while its fierceness and cruelty led to its further use in symbolizing a wicked enemy. As an emblem of power, it represented the tribe of Judah (Gen. 49:9). The powerful teeth of the lion are referred to as symbols of strength (Joel 1:6) but also as a symbol of the cruel and the mighty (Ps. 10:9, 35:17). The wolf denoted persons of a cruel or persecuting nature (Gen. 49:17) and symbolized bloodthirstiness and cruelty (Ezek. 22:27). The leopard suggested swiftness (Hab. 1:8), and Daniel used it to indicate the speed of the conquests of Alexander the Great (Dan. 7:6). The solicitude of the eagle for its young, the swiftness of its flight, its inaccessible nesting places, and its longevity brought about its use as a divine symbol (Ezek. 1:10). It also suggested the care of God for the children of Israel (Deut. 32:11). In Assyrian monuments, on the other hand, the eagle was frequently presented dominating the lion, while representing the superiority of mind over physical power.[5] Flowers also enriched the wealth of symbols of biblical literature. They have often been regarded as symbols of purity and beauty, but also of the intensity of human passions (Song 6:1–3). The rose symbolized the trust in God that the children of Israel should have in all afflictions, for its heart is directed upward even when it grows among thorns (Hos. 14:5). It was considered an expression of the uniqueness of the children of Israel, the chosen people (Song 2:1–3).

The incorporation of animal symbolism into the wealth of concepts of medieval man also suited the tendency of primitive German societies to anthropomorphism before they reached any clear distinction between man and beast. Animals had been admired as mysterious forces capable of assisting or injuring humanity by means of some secret occult influence. The widespread presence of animal images in medieval heraldry further indicates the longing of medieval men for the strength, boldness, stamina, or agility of either animals or birds. The use of such images in coats-of-arms was thought to bestow a sense of self-confidence on their bearers and, it was hoped, would arouse fears in their opponents. The Christianization process of former pagan symbols is reflected in the *bestiaries*, one of the most popular literary genres of the Middle Ages, which dealt with the supposed habits and appearances of various animals and birds. They contained illuminated miniatures of each animal and explained the moral lessons to be learned from its behavior (see plate 26). In the bestiary written by Philippe de Thaon, the term *lion* is interpreted as "king," the king of the animals. After describing its physical characteristics, Philippe emphasizes its symbolic meaning: the lion symbolizes both divine and human traits; it is St. Mary's son, the King of the World, Who, on the Day of Judgment, will know no mercy toward Jews. The lioness symbolizes St. Mary, and her cub is seen as the Christ-child. The roar of the male symbolizes the divine capacity by which Jesus Christ was resurrected from the kingdom of death. The panther, too, symbolizes Christ, as its name has an etymological logic which could hardly be questioned: the prefix *pan* means everything in Greek, and God is the one and single everything.[6]

PLATE 26. Lambert of St. Omer, *Liber Floridus,* Saint Bertin, c. 1120. Lion and Porcupine (120 × 200 mm).

Animal symbolism was also used to characterize large social entities, thus providing the first stage in the development of national stereotypes later on. In the *Livre des secrets aus philosophes*, Jehan Bonnet reported the main features of various animals on which he based his meticulous categorization of most European nations of the time. The German, the Flemish, and the English were depicted as phlegmatic, in contrast with the Lombards, Portuguese, Spanish, Catalonians and French, who were sanguine, a more desirable national character, in the opinion of Jehan. These basic trends of character encouraged these nations to show feelings of superiority over the inhabitants of areas such as Burgundy, Auvergne, Provence and Gascony. On the other hand, the melancholic nature characterizing the Bretons, Scots, Welsh and Irish is typical of animals, such as the hare and the fox, and of birds such as the heron and the goose, whose meat, according to our author, "is not tasty to man's palate."[7]

Both the biblical and the German traditions encouraged the tendency of medieval communicators to ascribe animal and floral symbolism to national entities. They also accorded to these symbols the role expectations of the subjects they were expected to describe. The unprecedented nature of the Hundred Years War in time and intensity, and the tendency of kings to ascribe national consent to their policies, brought about the spread of the national stereotypes beyond the narrow limits of the intellectual and political elites. In an instructive article, "The Promise of the Fourteenth Century," Strayer emphasized the positive message of the fourteenth century for generations. The endemic crises characteristic of the time could, in his opinion, provide a suitable example of the ability of Western society to overcome the crises, real or imaginary, while fostering a deeper awareness of the creative potential of mankind.[8] The emergence of national stereotypes provides another facet, not necessarily marginal, of the creative talent of medieval man and the mechanisms of defense he developed in face of the crisis, when the material foundations of its existence were collapsing and he had to find new resources to face the injuries of men and of nature. The diffusion of national stereotypes, however, was conditional on the emergence of a national consciousness which, per se, reflected the decline of the feudal system and, in parallel, the gradual disintegration between the different components of medieval life. The corporate essence of former kinds of solidarity in the framework of guilds, manors, villages, or universities, was now replaced by a new kind of *political solidarity* among members of a definite territory who spoke the same language and acknowledged the same ruler. This embryonic national consciousness no longer depicted real experiences but developed on a completely abstract level. It therefore required objects of identification while encouraging the use of old symbols after they were adapted to fulfill the new needs. The national stereotypes, with their biblical and German connotations, provided one of the most effective tools used by royal propaganda. Although attempts to a quantitative research in this regard are condemned to failure, the increasing number of stereotypes in medieval literature and the fact that stereotypes turned into proverbs used in daily practice indicate the wide diffusion they achieved among the literate and the illiterate.

Fulchignoni defines national stereotypes as "the total of opinions and beliefs which a people elaborates with regard to another people, including its entire characteristics or a part of them, without this concept presuming to reflect "objectively" the traits of the people on which the stereotype was elaborated."[9] The continuity of a stereotype, therefore, does not reflect its degree of reliability, but is rather conditioned by the sociopolitical conditions of the society that created it. Changing needs influence the nature of the stereotypes that a society elaborates on others.[10] Yet there is a lasting tendency to attribute positive traits to "loved" nations, while negative characteristics are applied to hostile or less friendly social groups.[11] The national stereotypes consequently provide a hierarchic picture of the whole social system, while the position of one's own nation is developed as a function of the position adjudicated to the other. The symbolic image of one's own nation acquires in the process important dimensions of security and insecurity. The security of one nation is crucially influenced by the insecurity ascribed to the other.[12] Elaborating the essence of stereotypes in the early 1920s, Lippmann laid down the methodological premises that have since been confirmed by more recent research. According to Lippmann,

> Stereotypes are an ordered, more or less consistent picture of the world, to which our habits, our tastes, our capacities, our comforts and our hopes have adjusted themselves. They may not be a complete picture of the world, but they are a picture of a possible world to which we are adapted. In that world people and things have their well-known places, and do certain expected things. We feel at home there. We fit in. We are members. . . . A pattern of stereotypes is not neutral. . . . A stereotype is the projection upon the world of our own sense of our own value, our own position and our own rights. The stereotypes are, therefore, highly charged with the feelings that are attached to them. They are the fortress of our tradition, and behind its defenses we can continue to feel ourselves safe in the position we occupy. . . . Its hallmark is that it precedes the use of reason. It is a form of perception, imposes a certain character on the data of our senses before the data reach the intelligence.[13]

The diffusion of national stereotypes in the first stages of the Hundred Years War was related to the increasing need to give a more concrete form to the emerging national consciousness, a need further aggravated by the intensity of the conflict. Yet the outbreak of the Hundred Years War could hardly be considered an unprecedented event in the annals of the kingdoms of France and England in the Middle Ages. Since the middle of the twelfth century, a state of continuous hostility characterized the relationship between both realms, further harmed by the feudal links between the kings. The Norman conquest had turned the duke of Normandy, one of the peers of France, into a suzerain in England, who developed a centralizing policy in his own kingdom, and at the same time aimed at expanding his feudal rule in France.[14] Nevertheless, against the feudal nature of former conflicts, the Hundred Years War created a new political code in international affairs. It obliterated the feudal patterns that hitherto had provided a conventional basis in settling foreign relations and, in parallel, served as a catalyst for the process of state-building. It also relegated to a marginal position the former mediatory role ascribed to the popes as the acknowledged leaders of all Christendom.[15]

Fitzralph of Armagh, bishop of London, was fully aware of the problematic nature inherent in the Hundred Years War, and laid down its implications in the political as well as the ethical level. In the procession for the welfare of the king and the princes (1345) he asked the people to pray that the king

> may obtain a just and happy result in his military campaigns. . . . Wherefore do men pray improvidently that the king may overcome his enemies, and also slay in battle. For those who pray thus, in their praying they offend God, and hinder their lord the king. They offend God in acting contrary to his command: "Thou shalt love thy neighbour as thyself" (Matt. 22:39). They hinder the king, withdrawing from him their spiritual petitions. . . . Such men who beseech God to pour out the blood of their adversaries, are violating the rule of prayer which insists that each shall seek and pray for all men that which they would desire to be done to them by others.

This rule of conduct, however, did not prevent Fitzralph from vigorously criticizing the "pacifists" of the time, who questioned the moral justification of the English invasion of France. The bishop replied on the grounds that "according to the judgment of our realm, the territories of England and France are properly one kingdom, the indivisible realm and dominion of the English king. By hereditary right the rule over both devolves upon him and his successors."[16]

Evangelical love as well as hereditary rights thus provided legitimacy for the policy of Edward III. Though expressed essentially in feudal terms, these were actually used to legitimize a new national equation that operated against the feudal system. Moreover, from 1337 onward the king of England had sustained an aggressive policy against France which had brought him the important victories of Crécy in 1346 and Poitiers in 1356. In the eyes of medieval man, those appeared as divine sanctions for the royal claims to the French crown, an argument deliberately exploited by English communicators.[17] The successes in the battlefield thus created a positive background for royal propaganda. Yet English communicators had still to deal with the critical attitudes regarding the ethical justification of a military campaign carried out against a Christian country. Moreover, if the justification for the war was exclusively based on the hereditary claims of Edward III,[18] the war might accordingly have become the personal concern of the king who could not on these grounds justify his increasing demands for manpower and money from the whole kingdom. Former feudal concepts could furthermore hardly provide a suitable legitimization for the king's innovative policy, which had in fact to be justified on new ideological grounds. Rather different was the perspective from the opposite side. Though the French could boast of the legitimacy of carrying out a Just War, the defensive nature of their struggle provided but little comfort, in view of the English successful offensive on all fronts. The continuous attacks on French soil, with large areas remaining under foreign rule, raised doubts and pressures on the royal court, which was forced to supply appropriate answers to the crisis. Besides, the first stages of the Hundred Years War were carried out against the setting of natural disasters and plagues on a scale unprecedented even for medieval man, already acquainted with suffering for generations. French communicators were thus challenged by the prevailing feelings of: "My God, my God, why hast thou forsaken me?" (Mark 15:34), a cry that faithfully expressed the anguish of many. In their attempts to explain convincingly the historical development, they had also to strengthen the reliability of the monarchy as a means to stabilize the shaky political solidarity among their countrymen.

Despite the gap between conqueror and conquered, victorious and defeated, royal propaganda in both France and England was faced with the same imperative of supplying convincing answers to the sociopolitical reality in a process of continuous change. Moreover, the many difficulties in everyday life and the more laborious struggle for survival fostered the illusion of the approaching doomsday. This state of affairs encouraged further searching for collective mechanisms of defense. It created a fertile ground for the elaboration and diffusion of national stereotypes, which became an integral part of war propaganda in both countries. They strengthened the feelings of solidarity and the personal involvement of the community of the realm with the lasting conflict. Expressed in a familiar language that provided the individual with a retreat from the crisis, the national stereotypes also supplied medieval people with the framework of a social group who shared the sense of crisis as well as the hope for a change.[19] The very need for new identification objects corroborates the decline of the feudal regime and the gradual emergence of a new kind of political solidarity, which fulfilled the role played hitherto by the corporate frameworks. Furthermore, the development of national stereotypes could not rely on the lack of information and the isolation that characterized the feudal regime. The reception of national stereotypes in medieval society was therefore conditional to the existence of common cultural denominators on which the stereotype could be eventually elaborated.

The historical process also left its mark in the stereotype, but it was nevertheless distorted, and actual reality was ultimately assimilated to the stereotype. This state of affairs becomes more evident in light of the stereotypes created by French and English communicators toward the other's country, notwithstanding the mutual knowledge and the many links between both sides of the Channel. From the thirteenth century onward, both England and France developed similar political institutions, defined themselves as Christian, and represented, in fact, the two major powers of Christendom. French was the common language of the intellectual elite of England; many Englishmen arrived in France as travelers, settlers, or merchants, and actually settled there before, during, and after the Hundred Years War. Many Frenchmen also reached England, and there was a high rate of exchange of teachers and students between the universities of Paris and Oxford. Thus, the national stereotypes had not only to overcome former feudal concepts but also the sociocultural links between both countries. From this perspective, one can further argue that the national stereotypes did not reflect concrete historical situations but, rather, served as promoters of change. The close relations between England and France did not prevent the diffusion of national stereotypes that distorted actual reality and relegated objective data to a marginal position. This state of affairs confirms the premise of Lippmann: "For when a system of stereotypes is well fixed, our attention is called to those facts which support it, and diverted from those which contradict. . . . We do not see what our eyes are not accustomed to take into account. Sometimes consciously, more often without knowing it, we are impressed by those facts which fit our philosophy."[20]

The crises of the fourteenth century and the outbreak of the Hundred Years War provided a fertile soil for the reception of national stereotypes, elaborated upon well-known symbols taken from the flora and fauna. Rooted in the wealth of concepts of medieval man, symbols provided a useful media for the spokesmen of the royal policy to explain the logic of the war and to foster a sense of political solidarity between members of a same nation. Incorporating animal symbolism facilitated the projection of positive characteristics to one's own nation, and of negative, even diabolical traits to the enemy. This state of affairs is faithfully reflected in some political poems written in France in the first stages of the Hundred Years War. In the "*Dit de la rebellion d'Engleterre et de Flandre*," the anonymous author presented Flanders through the image of a lion, while England was represented by a leopard. Neither should be allowed to graze on the lands of France,[21] since

> There is no Englishman who loves a Frenchman,
> Today they are at peace,
> By tomorrow they will be at war.[22]

Another poem of the times, "*Le roman de la fleur de lys*," emphasized the superiority of the French over the English through the symbol of the fleur de lys. It reflects the halo of holiness and heroism of the royal house of France and the special grace with which God had blessed this, His chosen kingdom. Through the symbolism of the fleur de lys, the distinctive prerogatives of the French were elaborated. They widened the gap between them and all other nations of Christendom, which were also represented by floral or animal symbolism:

> Some use emblems of leopards,
> Some of eagles, which from both sides
> Watch and are double headed.

The symbol of the kings of France, nonetheless, is such that

> Both eagle and gryphon will fear,
> While leopards will flee.[23]

The fleur de lys became the distinctive symbol of France and its king, representing the special divine grace toward the whole kingdom and suggesting its many virtues. The excellence of France in Christianity, peace and justice, love and chivalry, was condensed in the symbolism of the fleur de lys. According to Gervais du Bus, the many virtues implied in the fleur de lys amply justified the favorable attitude of God toward the kingdom of France.[24] To the letter of formulated political messages, essentially rigid and explicit, the flowering imagery of symbols provided some kind of musical accompaniment, which, by its perfect harmony, allowed the mind to transcend the shortcomings of the logical expression.

In the kingdom of England as well, the animal and floral symbols provided the basis for national stereotypes that aimed to express the superiority of the leopard over the fleur de lys (see plate 27). In "*An Invective against France*," England is described as "the queen of the world, the rose, the thornless flower," which was victorious in marine struggles. Consequently, all Englishmen won the complimentary description of "a humble people, graceful . . . which has won the French with joy and left them rolling in their blood."[25] While praising the English victory at Neville's Cross (1346), another poem of the times presented the king of France as a lion, *rex leo*, falling an easy prey into the hands of the eagle king, personified by Edward III.[26] The use of well-known concepts taken from the world of nature thus facilitated the acceptance of national stereotypes. The more familiar a phenomenon, the more quickly it was recognized. Less effort was needed for the medieval population to accept recognizable objects such as lions, leopards, or flowers than unusual ones. Moreover, as the background and culture of medieval people often served as a guide in the creation of perceptual predispositions, the use of well-known animal and floral images fostered the reception of the national message and left little room for questioning the very essence of this message. From the moment the stereotype of the national object had been determined, the actions of its protagonists, either English or French, received a new meaning, different in some way or another from the expectations that had existed prior to the definition of the stereotype.

Animal symbolism, however, was not always used for praising the royal policy or the many virtues of the king's faithful subjects, nor was it monopolized by the spokesmen of the royal court. Also, the opponents of the royal policy made use of the same symbols to express their bitterness against the king's overlooking of ancestral customs and privileges. "*The Battle of Lewes*" expressed the prevailing criticism against Edward I through the use of animal symbols, which were complemented by a rather original etymological commentary:

PLATE 27. Edward III, gold noble, before 1360. Edward III thought it derogatory to copy his rival's coins and so altered his type from that of the French real coins to one having for its type the leopard of England, in order to emphasize the fact that it was the king of England who claimed the throne of France. There were three distinct issues of the leopard before 1360 being equivalent to the half-noble.

> To whom shall the noble Edward be compared?
> Perhaps he will be rightly called a leopard.
> If we divide the name, it becomes a lion and a pard:
> A lion, for we have seen that he was not slow
> To meet the strongest, fearing the attack of none,
> Making a charge in the thick of the battle
> With the most unflinching bravery . . .
> He is a lion by his pride and by his ferocity;
> By his inconstancy and changeableness he is a pard,
> Not holding steadily his word or his promise,
> And excusing himself with fair words. . . .[27]

The use of animal symbols by both supporters and opponents of the royal policy indicates their wide diffusion in the political society of the times. Besides, while the images of lions and leopards were often positively associated with virtues such as speed or courage in the battlefield, they still left enough freedom of action in the hands of those authors who satirized the political situation behind the shield of animal or floral figures. This seemed an easier way than directing their barbs against the real personages against whom their criticism was, in fact, intended. In the prophecies of Jean de Bridlington, the king of England was symbolized by a bull as well as by a leopard, and the king of France by a rooster, while the king of Scottland received the rather problematic image of the crab.[28] According to Jean, the bull symbolized the physical power, while the leopard suggested the boldness of the king of England in face of his enemies.[29] And if the Flemish merited the collective epithet of "swine," according to our pseudoprophet, it resulted "from their filth and beastliness."[30] One is not surpised that the Flemish did not identify themselves with this description but chose rather to represent themselves through the lion's mouth. In a contemporary poem, the lion's mouth was warned to beware of the fleur de lys and not to trust it, a delicately phrased warning of the French coercive policy in Flanders at the time.[31]

The national stereotypes thus came to include not only the symbolic or personalized images of the political institutions but also the detailed images of role expectations. The widespread use of floral and animal symbolism representing national identities was dictated by the basic limitations of traditional societies, which could reach only a certain degree of complexity, a limitation hardly unique to medieval people. When the complexity of reality or of the political message became intolerable, medieval man retreated into the symbolic images, which offered him a more familiar and more ordered world. The reluctance toward accepting a multidimensional value order gave more impetus to the use of stereotypes and their acceptability in medieval society. They answered the limitations and needs of most contemporaries while defining all the inhabitants of a country in symbolic categories of a clear-cut nature, which, as such, left no room for fine nuances. The representation of nations by lions, leopards, or eagles, which hate, fight, avenge, or love, reduced the abstract dimension of the war, eventually presented as a conflict between opposing elements. The use of stereotypes, therefore, provided a useful medium to narrow the gap between the individual and the political entity, which lost its former abstract essence in the process. Stereotypes fostered the reception of the Roman concept of fatherland, which became more familiar and, as such, more beloved. At the same time, foreign countries and people became even stranger and, as such, more hated, as the stereotype sharpened the gap between one's own nation and all others.

Jervis has claimed that "hostility needs no special explanation."[32] Yet medieval men did feel the need both to understand and to explain the nature of the lasting conflict between Christians. Although the hostility toward foreigners and the threat inherent in their very existence were endemic phenomena in medieval society, they became more acute in times of crisis. When trying to understand the historical development, the explanations provided strengthened the stereotype of one's own nation and, in parallel, of the others. The importance ascribed to this defense mechanism increased in times of military defeats when the need arose to found some escape from the threatening reality. Pierre Langtoft found comfort when he invoked providence against all those who challenged the rule of England. Somewhat pathetically, he voiced his hopes that:

> May Wales be accursed of God and of St. Simon!
> For it has always been full of treason.
> May Scotland be accursed of the mother of God!
> And may Wales be sunk down deep to the devil![33]

This prayer was a direct outcome of the misfortunes of England in the 1310s. The continuous victories of the Scots in the battlefield were further regarded as extraordinary events that contradicted the rules of the universe:

> . . . New prodigies are now performed,
> When the daughter takes upon her
> To lord it over the mother.
> England, the matron of many regions,
> To whom tributary gifts were given,
> Is now, alas!, constrained too much
> To be prostrated to the daughter,
> By whom the maternal crown is injured.[34]

In face of the changing order of things and the inability to renew past schemes in actual practice, the national stereotypes provided collective mechanisms of defense, fostering the feelings of justice and security of the "right" albeit defeated side. The individual was no longer left alone against the abstract conflict as God, the eternal justice, and the reward promised to the pious, stood at his side. Those who collaborated with the Scots were consequently named Pharisees and regarded as suitable heirs of Judas Iscariot.[35] The "*traytours of Scotlond*" themselves were eventually depicted as such "who love falseness, and will never leave it."[36]

In the first stages of the Hundred Years War, the struggle with France brought about a displacement of the national stereotype from the Scots to the French. The epithet of "Pharisee" was then applied to the entire kingdom of France, which also received the profile of "lynx, fox, she-wolf and siren." The wide-ranging use of such animal symbolism was intended to point out the many demerits of the French, especially their brutality and vanity. The king of France himself served as a source of inspiration for the national stereotype of his subjects. *Philipo Valeys* was depicted by English communicators as a wolf, a rather realistic image in their opinion, because of his thirst for Christian blood.[37] Against such an unpleasant background, it was not surprising that the many virtues of the kingdom of England were portrayed, such as its excellence in raising beautiful flowers, a symbolic reference to its brave knights.[38]

The Bible also enriched the national stereotypes with the images of its legendary heroes: it bestowed on the participants in the Hundred Years War all the vices and virtues of their mythical forefathers. The anonymous author of "*On the Battle of Neville's Cross*" did not hide his mockery from the Scots "who regard themselves as equals to Samson, the Maccabees and Gideon, and actually run away from the battlefield in a way more suitable to rats or ostriches."[39] They had to be more properly called "Leviathan" (Is. 27:1; Job 3:8), as their rebellion against England is as faulty as the former insurgence of Korah, Dathan, and Abiram against Moses (Num. 16:27–35). And if those epithets were not enough to reveal the impious character of the Scots, our author summarizes his diatribe with the astonishing news that the Scots belong to the same despised race of Gebal, Amnon, and the Hagrites, of whom it was said "they have taken crafty counsel against thy people" (Ps. 83: 3–8).[40] Biblical symbolism thus served as an important source of inspiration for stereotyping the enemy: it deepened the gap between the adversary and its many faults, and the virtues of one's own side. It facilitated the self-definition of medieval political societies, expressed in well-known symbols and enriched by the legendary aura of the mythical past. The concept of the chosen people provided a most helpful point of identification when facing the many sufferings encountered in actual reality. In the case of the French, it was expected to work against the harmful implications of the war in both the self-confidence of contemporary society and the reliability of the Valois dynasty. Still, the image of the chosen people, which had been formerly a guarded privilege of the French,[41] gradually came to serve as the distinctive status of the English as well. At the end of the thirteenth century, the "*Song on the Scottish Wars*" had already grouped all nations challenging the rule of England under the pejorative name of "*infidorum*," the faithless. This "fact" gave consistency to the resulting expectation of being rewarded with divine grace as had been the case with the children of Israel in the distant past:

> May the Governor of the universe Whom we address as God,
> Who protected the Hebrew people through many difficulties,
> Give the English victory over their enemies!

This prayer was further justified in light of the unquestioning devotion of the King of England who:

> Labours to devote himself entirely to Christ.
> Edward, our king, is entirely devoted to Christ.[42]

The ruling dynasty thus remained the major source of inspiration for stereotyping the political society. Although the kings of England could not claim for themselves the mythical halo of Charles Martel, Charlemagne, or St. Louis, the royal communicators did not fall behind the French in their attempts to glorify the royal house. While Philip VI received the name of Saul "who is always late to go into war," Edward III was presented as a new King David.[43] The call for God's help, quite widespread in the sources of the period, was further justified by the links of Edward III with the House of Jesse, which eventually overshadowed the glorious aura of the Valois' ancestors.[44] The royal house of England could therefore pride itself on the most magnificent biblical dynasty, one which embraced the Messiah and, as such, strengthened the sacred halo of the Anointed King. Most of these concepts were spontaneously spread by contemporary poets and chroniclers for whom the royal house provided the inspiration for developing the national stereotype. Edward III himself exploited every occasion to emphasize the ancestral loyalty of his forefathers to God, the Catholic Faith, and the Holy Church. In his letter of 1339 to the Cardinals, the king of England presented his ancestors as "admirable champions of Christ, heroes of the True Faith, zealous lovers of our Holy Mother, the Roman Church, and loyal servants of its commands."[45] The selective attitude of providence toward the kingdom of England, together with the devotion of the royal dynasty to God, whether as a cause or a result of God's predilection for the realm, provided the two central motives developed throughout the national stereotype. At the beginning of the Hundred Years War, the whole English nation was identified with the "people chosen by Christ."[46] The special attitude of God toward the whole nation was mostly manifested with regard to its leaders. The English barons were consequently depicted as the actual personification of "Noah, Job, Daniel, Lot, Samson and Solomon," who excel in their modesty and love. To these meritorious virtues, English communicators added the nobles' unquestioned fidelity to their own wives,[47] a comment that hints at the moral traits ascribed to the French already in those days. Henry Percy, one of the heroes in the English offensive against Scotland, was accorded the description of "Maccabee," a title which gradually became the collective attribute of all English warriors. The identification of the English armies with the legendary heroes of Antiquity justify the further expectation of being properly rewarded with special divine grace in the battlefield, the English fight being connected with all the mythical attributes of the sons of Mattithias.[48] Through the image of the chosen people, the national stereotype thus melded together the king and the barons, both perceived as the representatives of a wider entity, the community of the realm. This provided a useful basis for a proper definition of the national body while supplied divine sanction a priori for its deeds. The further diffusion of the national stereotype consequently fostered wide support for the policy of the king and the barons, whose actions were seen in accordance with the expected roles inherent in the stereotype. On the other hand, those actions that contradicted the stereotype were often simply neglected. Hence the double role ascribed to the stereotype both in the individual and in the collective level of perception: it strengthened the self-image of the person or the social group on which the stereotype was elaborated while, by creating expected roles, it also influenced the prevailing attitudes to their policy in actual practice.

This process was hardly unique to the Middle Ages. According to Lasswell,

> Identification with any particular symbol by any person at any phase of his career line initiates a complex process of symbol elaboration. At the earlier,

> loves tend to be reactivated in relation to the new symbol. The individual who late in life experiences "conversion" and becomes an "American" or a "Communist" or a "Catholic," reads into this symbol the loves and hopes of his entire personality. His elaboration upon the symbol will depend upon the forms of expression with which his personality has been equipped through aptitude and training. If he belongs to those who require large emotional responses from the environment, and if he has a facile technique for the oral or written production of language, he may fill the auditoriums of his vicinity with rhetoric and the printing presses with poetry and prose.[49]

The tendency to use stereotypes in order to influence the immediate environment was, quite naturally, more emphasized among the royal communicators who could and, indeed, did make use of them to encourage public support for the royal policy. The important victories in the battlefield facilitated the reception of their message as they provided "proof" of God's sanction of the English policy.

The military failures in Crécy and Poitiers brought about a rather opposite, depressive atmosphere in the kingdom of France. In the early stages of the Hundred Years War, it acquired a melancholic nature, full of longings for the glory that had enhanced the times of Charles Martel or Charlemagne. The prevailing tendency to seek escape in the splendor of a mythical past was expressed in the words of Jean le Bel after the French defeat at Crécy: "How painful and pitiful is the situation to which the kingdom of France has deteriorated because of the bad advice of its leaders. This is France whose abundance in honor, reason, clergy, knighthood, merchandise and all kinds of goods was admired by the whole world. Today she is so miserably punished and dragged into disgrace by her enemies."[50] Other French chroniclers regarded the Hundred Years War either as an outcome of the influence of Satan, who always instigated bloodshed among Christians, or as a divine punishment for the many sins of the contemporaries, *propter peccatorum*,[51] a fairly common cliché of the times. The spirit of defeat surrounding the French sources hints at the difficulties in securing the acceptance of a national propaganda in times of war when the political message was not suitably backed by victories on the battlefield. In the second half of the fourteenth century, however, when France began to recover from its military reverses, the biblical myth spread with a new impetus, strengthening the positive aspects of the national stereotype. Eustache Dechamps added the name of Bertrand Duguesclin (1320–1380), the French hero in the French recovery, to the selective list of the Nine Worthies which hitherto had been composed of Hector, Julius Caesar, Alexander the Great, Joshua, King David, Judas Maccabeus, King Arthur, Charlemagne, and Godfrey de Bouillon.

The concept of the Nine Worthies, developed during the Hundred Years War, reveals an additional source of inspiration of the national stereotype. The biblical heritage was complemented by a wealth of mythological symbols while further fostering the reception of the national stereotype. Among the many myths widespread at the time, the Trojan myth achieved pride of place.

It strengthened the glorious pedigree of medieval kingdoms, since it provided the most illustrious national ancestry, that of Troy.[52] English, French, Scots, and Flemish communicators glorified their people in their sole origins in Troy, the noblest of nations, and pushed into the background their rather obscure origins somewhere in Scandinavia. The use of the same symbols for different and often opposing purposes reveals an additional feature in the national stereotypes. Although they aimed at sharpening the distinctive essence of each kingdom and its superiority over the others, national stereotypes of different and sometimes inimical countries were expressed in similar terms and symbols, a trait which is hardly unique to medieval society.[53] The annals of Troy also served as a kind of warning with regard to all those who persisted in carrying out "unjust wars," indicating the miserable fate awaiting the Scots, should they continue their hopeless conflict against England:

> Troy is ravaged by war, the land near about being full of camps,
> It is with its boundaries become the property of another.
> The son of Achilles rules over the cities and towns;
> Pyrrhus lays them waste, Aeneas has wept for Priam.
> Merlin writes that the proud crowd shall perish;
> The barking dog shall depart, and the ox shall go into exile.
> Then shall the Eutherian grove be stripped of its feathered branches;
> And the Albanian race will see their kingdom perish.
> Wretched Scot, lament, thy hour of weeping is now come;
> For the kingdom of thy forefathers ceases to be thine.
> Thou art deprived of a prince,
> And art so trodden down in the field,
> That by thy ill merits thou wilt always be an ass.
> A voice from the bottom of the Cambine waters calls thee,
> To be punished with such slaughter
> As the race of Adam has not yet seen.
> Hasten thither, and become the companion of the devil!
> Andrew will no longer be their leader.[54]

This poem faithfuly reflects the harmonious amalgam of the Trojan myth, the biblical narrative, and the Catholic scale of compensations. The punishment awaiting Scotland and its inhabitants had not been known to mankind since the first man had been expelled from paradise. Even St. Andrew, the saint patron of Scotland, would be unable to rescue the Scots from the just English offensive. One might reasonably question, however, the ability of most contemporaries to understand the symbolic roles attributed to Achilles, Paris, or Merlin, and consequently reach the expected conclusions. From a communication perspective, nonetheless, this poem actually achieved its expected goal: it strengthened the confidence of the English regarding the forthcoming divine justice. It also reveals the process undergone by symbols and words that had been dissociated from the circumstances that had given them birth, following the sociopolitical process. Beginning as ideals or yearnings, programs thus become phrases, slogans, or symbols of aspiration for desired changes, and undergo a certain augmentation of influence until the struggle is precipitated. Now, becoming slogans of conflict, they touch off tremendous charges of emotion until, used again and again, symbol and emotion become almost indistingishable from each other.

Quite different was the stereotype of the Scots developed by Froissart, who put aside mythical sources, preferring to present his own view. Somewhat influenced by the sympathizing French attitude toward their uncompromising struggle against England, Froissart presented the Scots as men who were

> right hardy and sore travailing in harness and in wars . . . They take with them no purveyance of bread nor wine, for their usage and sobriety is such in time of war, that they will pass in the journey a great long time with flesh half-sodden, without bread, and drink of the river-water without wine.[55]

This description reflects the "kernel of truth" of the stereotype, which, despite its essentially subjective approach, contains some basic characteristics of the social group it claims to depict.[56] The "objective" approach that characterizes Froissart's attitude toward the Scots, however, completely disappears when he comes to portray the English. In one of the later versions of his chronicle, he depicted in a rather negative light the utilitarian attitudes of the English toward their monarchs.

> They do not love their kings, neither respect them, unless they can pride themselves on victories and initiate long wars against their neighbors, especially the strongest and richest among them . . . The English have surprising qualities, as they are mostly restless and bitter, they are easily enraged, calmed with difficulty and completely ignorant of the rules of courtesy. They find special pleasure and satisfaction throughout wars and killing. They are most jealous of the property of others and cannot easily adjust themselves to conditions of peace or harmony between nations. They are vain and boastful.

After such a description, it is not surprising that Froissart summed up his opinion by expressing doubts whether "there could be found other dangerous and evil people such as the people of England in the whole universe."[57] The *Chronique des quatre premiers Valois*, on the other hand, finds it satisfactory to focus its criticism on the arbitrariness characteristic of the English "who want all their wishes to be turned into actions at once."[58]

French hostility toward the English, sharpened by the many defeats they suffered in the early stages of the Hundred Years War, encouraged the tendency of French communicators to emphasize the moral superiority of their own countrymen. This tendency which over the years acquired the characteristics of a collective defense mechanism, was expressed in a series of proverbs. The maxim: "the loyalty of the English is not worth a pound from Poitou," then acquired the prestige of an unquestionable truth. Similarly, there were popular sayings related to "the generosity of the French" or "the loyalty of the English" as most amazing phenomena, the mere existence of which would cause wonder.[59] The author of the "*Dit de la rebellion d'Engleterre*" linked the treachery considered characteristic of the English with their physical features. Although he conceded that the English had "honest faces and often spread promises," he rhetorically asked "who will find anyone ready to fulfill them in actual practice?!"[60] Stereotyped attitudes toward the English were not exceptional in the popular culture of the fourteenth century. On the contrary, one can hardly find any nation to which were not attributed distinctive features, which in time became its identifying attributes. The tendency to give each particular case the character of a moral sentence or an example, the crystallization of thought, finds its most general and natural expression in the proverb. In the mentality of the Middle Ages, proverbs performed a very active function. There were hundreds in current use in every nation.[61] "The treachery of the English" was but one facet of this mental attitude that tried to categorize peoples and nations into clearly polarized notions of good and bad, virtues and vices. "The ostentation and slyness of the French," "the unrestricted rage of the Germans,"[62] "the plotting spirit of the Italians" became common expressions in everyday language, thus proving the wide reception of the national stereotype and its legitimization by actual experience. They deepened the role expectations from the subject of the stereotype, while strengthening even more the links between individuals and their fellow countrymen and widening the abyss between them and all foreigners.

When one comes to sum up the main characteristics of national stereotypes during the first stages of the Hundred Years War, it is rather difficult to dissociate those elements unique to medieval society without being influenced by the stereotypes and symbols that are part and parcel of our own everyday reality. The eagle displayed as the symbol of liberty and hope for Western democracies; the tiger, which you are kindly requested to put into your tank; the lion, roaring at us from the cinema screen; the names of many cars; Walt Disney's cartoons of delightful animals that can speak, love, or hate . . . not one has lost its appeal in the Atomic Era, notwithstanding the fact that the gap between man and beast has become well-defined. And, actually, why should symbolism lose its attraction? At this stage the historian has to leave the stage for the psychologist and, apart from emphasizing the longevity of the phenomenon, has to follow the German physiologist Du Bois-Reymond and repeat after him, *Ignorabimus!*[63]

III

Heresy

10

The Challenge of Heresy

The communication systems developed in the ecclesiastical order and the Western monarchies had emerged in the framework of conventional norms and, as such, enjoyed legitimacy and wide support. Prelates and kings were accordingly blessed with divine grace and presented as those who voiced the *Vox Dei* in actual reality. The last section of this study deals with sects that were refused such legitimacy, while medieval society as a whole condemned them to continual scorn and, ultimately, to their physical extinction. Lacking socioreligious legitimization, the so-called heretical movements were forced to rely almost exclusively on the efficiency of their communication systems to swell their ranks with new sympathizers, while clandestinely maintaining the current contacts among their members. The emergence and spread of heresies provide, therefore, an excellent case study on the capability of nonestablishment groups in voicing their message, while challenging the monopoly of the recognized representatives of God in Christendom, either the Church or the secular rulers. Moreover, the very emergence of heresy reflects a more advanced stage in the development of communication. Essentially a challenge to traditional norms, heresy could develop in a society that was no longer isolated nor autarkic. It indicates the gradual decline of the traditional society, which integrated the ideology of its members into one wholeness that prevented the emergence of divergent currents of thought. The development of heresy depended upon the emergence of a more communication-oriented society, in which people were receptive to different messages and were further able to discern one from the other.

In the early seventh century, Isidore of Seville defined heresy and heretics as follows:

> *Haeresis* is called in Greek from choice, because each one chooses that which seems to him to be the best, as in the case of the Peripatetic (Aristotelian) philosophers, the Academics (Platonists) and the Epicureans and Stoics or, as others do, who contemplating their perverse dogma, recede from the Church of their own will. And so heresy is named from the Greek, from the meaning of choice, since each (heretic) decides according to his own will whatever he

> wants to teach or believe. But it is not permitted to us to believe anything on the basis of our own will, nor to choose to believe what someone else has believed of his own will. We have the authority of the Apostles, who did not choose anything out of their own will to believe, but faithfully transmitted to the nations the teaching they received from Christ. Even if an angel from heaven should teach otherwise, it would be called anathema.[1]

The argumentation of Isidore was given juridical force in Canon Law, which stressed the individualistic approach of all heretics and, foremost, their refusal to acknowledge the magistracy of the Roman Church, to which every Christian was bound by the sacrament of baptism. This resulted in the obstinate character of heretics who did not abandon the error of their ways, after being properly warned by the Church on two or more occasions.[2]

Heresy appeared, therefore, as the voluntary rupture with the community of the faithful and the prevailing order of society, the most precious pillar of social well-being. This dissociation from traditional society fostered a distinctive scale of values within the sect, characterized by extreme norms of behavior that emphasized the growing contradiction between its own credo and that of the Church. As time went by, this contradiction acquired the never-ending antagonism between orthodoxy and heresy, the status of the latter being established by the Church from which it had emerged and whose norms it denied. Heresy was consequently defined by reference to orthodoxy. A doctrine, a sect, or an individual became heretical when condemned as such by the Church. As a phenomenon rather than a transitory event, heresy was neither exclusively doctrinal, nor exclusively social: there can be no heresy without belief, and no sect without adherents. The first gives it identity, the second significance. Accordingly, the formation of a heresy must be sought in the coalescence of doctrinal and historical factors.[3] The polarity between heresy and orthodoxy, however, does not indicate seeds of disbelief from the side of those declared heretics, but rather the opposite. Heresy often reflected a deep religious belief that could not easily be channeled into the frameworks of the traditional society, thus bringing about a growing dissatisfaction with the existing order. This became a common denominator of all sects, often accompanied by a strong intellectual curiosity and a fervent willingness to study in detail the principles of the true faith.

From a sociological perspective, heresy usually developed around a charismatic leader who was educated and possessed rhetorical skills. He succeeded in creating a group of followers whose norms of conduct he dictated, and which was consequently named after him. He became the "prophet" of the social group and, as such, provided a convenient target to project all expectations for a radical change in actual reality. This state of affairs indicates the influence of the intellectual elite in the emergence of sectarian groups, without turning heresy into an unique outcome of this class.[4] In its earliest stages, heresy appeared as an isolated phenomenon, geographically discontinuous and doctrinally idiosyncratic. It first spread in limited areas of Sardinia, Italy, and Spain about the year 970. Later outbreaks of heresy were reported in

Châlons-sur-Marne (c. 1000, 1046–1048), Liège (1010, 1024–1025), Aquitaine (1018), Toulouse (c. 1022), Orléans (1022), Arras (1025), Monforte (c. 1028), Csanad (1030, 1046), Ravenna, Venice, and Verona (1030, 1046), Upper Lorraine (1051), Sisteron (1060), and Nevers (1075). The synods of Charroux (1027–1028) and Reims (1049) also denounced the spread of heresy in the surrounding areas. Albeit fragmentary, these data reflect the location of early heresies, limited to large areas of Provence and Lombardy, where they reached all social classes: aristocrats and clergy at Orléans, aristocrats at Monforte, townsmen at Arras and Liège, and peasants and craftsmen as well. Just as the Church was a mirror of society, reflecting its extremes of rank, wealth, and intellectual opportunities, so were the movements that evolved from it; they embraced whole regions rather than specific socioeconomic groups.

The spread of heresy among different social strata prevents its categorization in terms of a class struggle, or an uprising against the feudal system. Yet the socioeconomic process of integration left its mark on the timing and the geographical development of heretical movements: they spread in areas characterized by rapid social changes and economic expansion, such as the Languedoc, Flanders, and Lombardy, in which trade and industry were developing alongside demographic growth. The ideological rupture from society could not improve the economic situation of those who were forced to emigrate, nor was it intended to. It provided, rather, a collective defense mechanism through the identification of sectarians with the image of the chosen people, which facilitated their sublimation of material distress. The conviction of being divinely appointed to carry out a prodigious task bestowed on the disoriented and the frustrated new bearings and new hopes. It provided them with not merely a place in the world, but with a unique and splendid opportunity that justified their dissociation from traditional society. It also gave to their resulting persecution by the establishment the meaning of having been elected and, as such, persecuted. Confusion and frustration, however, were not the sole elements of heresy, nor were they limited to the poor. When medieval man tried to voice his insecurity in the face of the process of change, quite naturally he expressed himself in the only terminology he knew, which had a predominantly religious essence. Being a faithful mirror of the values and hopes of medieval man, the Christian terminology also became a depository of his frustrations and confusions that went beyond class differentiations. On the other hand, the difficulty of many to adjust themselves to the socioeconomic changes encouraged their flight from the world, the pursuit of purity, and the rejection of all material things—mainly, for human flesh and its uncontrollable desires. The prevailing reluctance to accept any kind of materialism encouraged a sharp criticism of ecclesiastical norms, the Church being a key element of the existing order and, as such, regarded as one of its main beneficiaries. From the perspective of sectarians, the Church's aberration from the simplicity of Christ and the Apostles harmed the moral basis of priesthood, especially the right of immoral priests to perform the sacraments. They further questioned the validity of some sacraments such as baptism and penance, and denied the worship of the cross and of the images of Jesus Christ and the saints. The more rapid the process of change, the stronger the pressures exerted on the Church as the representative of God on earth and, as such, one expected to offer appropriate answers to all expectations, even the most unrealistic and anachronistic ones.

Despite the gap between the evangelical Christian community in first-century Jerusalem and the urban-monetary economy of the Central Middle Ages, the so-called heretics turned to the Gospels, hoping to find appropriate models on which to project their expectations. And the Gospels, indeed, provided a very clear message as to the egalitarian foundations of an ideal society based on the common disposal of property and the voluntary distribution of all goods to the poor. In His Sermon on the Mount, Jesus laid down that: "Blessed are the poor in spirit, for theirs is the kingdom of heaven" (Matt. 5:3). Beyond the divine blessing bestowed on the poor in spirit, St. Paul indicated the actual implications of Christ's commitment to absolute poverty: "For ye know the grace of our Lord Jesus Christ, that, though he was rich, yet for your sakes he became poor, that ye through his poverty might be rich" (II Cor. 8:9). Despite the fact that the Lord owned all the property in the world, He chose to become poor in order to teach mankind the right path to spiritual bliss. All those who voluntarily relinquished their property and shared it with the poor (Luke 12:33; Matt. 19:21), consequently become members of the heavenly kingdom while sharing in a mystical way the eternal rule of Christ.[5] The validity ascribed to the voluntary renunciation of property was further emphasized by the Fathers of the Church,[6] and encouraged the call of St. Jerome to follow naked in the footsteps of the naked Christ, *nudus nudum Christum sequere*. This concept was eventually shared by Hilarion, Ambrose, Clement of Alexandria, Gregory of Nissa, Basil, John Chrysostom, and Augustine.[7]

In the socioeconomic reality of the Early Middle Ages, however, it was rather easy to relegate the ideal of evangelical poverty to the dimensions of a desirable aim but still practicable only by a few, namely, monks or priests. In the background of the feudal regime, poverty, as much as could have been found, had not yet become a mass phenomenon and could satisfactorily be dealt with in the traditional social frameworks. Neither was a clear tendency to turn the ideal of poverty into the ideological axis of early heresies. Although as a social ideal, the call to poverty helped to the dissociation of sectarians from Christian society, it was not yet regarded per se an expression of heresy. The call to embrace evangelical poverty of Robert d'Arbrissel, a renowned priest and preacher in Anjou, though encountered disapproval, did not bring about his excommunication but to the foundation of a new Monastic Order (Fontevrault, 1096). He was forced, however, to restrain his criticism of the Church and to submit to the authority of the clergy while limiting his wanderings. On the other hand, Henri de Lausanne, a monk from Cluny who also preached evangelical poverty in Le Mans, was formally excommunicated by the Council of Pisa in 1135. Even within the ranks of those declared heretics, the call to embrace poverty was only a part of their creed and not always the most important one. For the heretics of Arras, their commitment to poverty was part of a broad program, which included not only manual labor and complete seclusion from this world, but also chastity, nonviolence, and brotherly love. In the heresy of Monforte as well, poverty was seen as secondary, being part of a complex of strict rules of conduct, which ascribed to absolute chastity a more fundamental role. Also, the Patarenes of Milan placed the concept of poverty on a secondary level, as an option for a way of living. They concerned themselves more with condemning the immoderate desires of the clergy for money and power, and demanding priests to imitate the poverty of Christ while adopting more spiritual attitudes founded in humility. Eleventh-century heresies, therefore, had not yet developed clear attitudes as to their own adherence to evangelical poverty, nor had they turned poverty into the essence of their creed. Nonetheless, there was a clear tendency to criticize the clergy's "greed and avarice," two terms that indicate the prevailing dissatisfaction with the ecclesiastical practices.[8]

The socioeconomic process characteristic of the Central Middle Ages turned poverty into an integral part of daily life, which, as such, acquired a crucial importance in the dissociation of heretics from the traditional society. Although until the beginning of the fourteenth century the masses of the poor had not yet become a common phenomenon,[9] demographic growth and the resulting emigration process increased the number of the disoriented and unemployed. It involved hundreds and sometimes thousands of peasants who were now faced with a new market system whose rules they did not know, and with socioeconomic gaps they were unable to bridge. In addition to poverty, as great as that suffered by any peasant, the masses of journeymen and casual laborers underwent disorientations such as could scarcely occur under the manorial regime. There was no immemorial body of custom that they could invoke in their defense; there was no shortage of labor to lend weight to their claims. Above all, they were not supported by any network of family and social relationships comparable to those that might sustain a peasant. Journeymen and unskilled workers, peasants without land or with too little land to support them, beggars and vagabonds, the unemployed and those threatened with unemployment, the many who for one reason or another could find no assured and recognized place—such people, living in a state of chronic frustration and anxiety, formed the most impulsive and unstable elements in medieval society and offered a most fruitful arena for developing sectarian feelings.[10] In addition to the accelerated process of change, between the tenth and twelfth centuries there was an unprecedented chain of natural disasters in France, Germany, Italy, and Catalonia. Between 975 and 1040, there were forty-eight years of famine in France and thirty-three more between 1059 and 1176.[11] Hundreds and sometimes over 1,000 men gathered at the gates of monasteries in order to get some food: 1,500 poor were registered each day in Val-Saint-Pierre in 1197, and their number reached 1,600 in Cluny.[12] The distress of daily life and the problems of survival encouraged the readiness of medieval people to listen to the message of evangelical poverty. As has been pointed out by Christine Thouzellier, "*la faim tenaille les estomacs comme le doute mord les esprits.*"[13] When the effects of hunger affects the stomachs, increasing doubts affect the spirits as well.

Poverty thus became not only an integral part of daily life but was at the very heart of existence. The awareness of this new state of affairs, however, was not the monopoly of the laity nor of the heretics. Between the tenth and twelfth centuries, a deeper commitment to poverty and to extreme norms of conduct became the common challenge of the new Monastic Orders of Cluny, Hirsau, Chartreux, and Citeaux, which succeeded in presenting before the faithful a more harmonious implementation of evangelical poverty.[14] These aims were ultimately pursued also by the Reformer papacy, which succeeded in bringing about a temporary decline of heresy from 1075 onward. The Church approach to evangelical poverty, however, was conditioned by its institutionalization process and, as such, could hardly give satisfactory answers to the large masses of the poor. In essence, it distinguished between individual poverty, which was divinely sanctioned in the Gospels, and the right bestowed by St. Peter and St. Paul on the pope and the whole ecclesiastical order through him, to keep the Church property in common. This convenient distinction was justified on the grounds of apostolic practice as perpetuated in the Holy Scriptures:

> And the multitude of them that believed were of one heart and of one soul: neither said any of them that ought of the things which he possessed was his own; but they had all things common. . . . Neither was there any among them that lacked: for as many as were possessors of lands or houses sold them, and brought the prices of the things that were sold, And laid them down at the apostles' feet: and distribution was made unto every man according as he had need (Acts 4: 32–35).

Emphasis, therefore, was laid on the voluntary poverty of individuals, which did not prevent the right of the ecclesiastical community as a whole to keep property, a right on which the medieval Church was not ready to compromise.[15]

Furthermore, the Gregorian Reform was intended as a long-range program not only for the amendation of the clergy's practices and the implementation of individual poverty, but also for the emergence of the papal monarchy. It did not bring about a sudden termination of all ecclesiastical abuses but rather encouraged the further institutionalization of the Church and the papal monarchy at its head. This process created well-defined borders between those who acknowledged papal supremacy and those who denied it and, eventually, increased the number of those who were rejected from the ranks of orthodoxy. The existence of much popular zeal reacting against a conservative Church still much in need of reform, a kind of primitive rationalism no longer satisfied with the fortress Church of the barbarian age, a new understanding of certain texts of Scripture that had entered the popular consciousness and could not be brought into accord with the ecclesiastical practices—all these actually undermined the ability of the clergy to safeguard its own leadership and facilitated the spread of heresy throughout Christendom.

This state of affairs reflects the paradox in which the Church was trapped in the Central Middle Ages. Against the sociopolitical process of integration, it had responded with the institutionalization of its own frameworks, thus further reducing its capability to adapt to new situations. Although the Gregorian Reform had rehabilitated to some extent the validity of evangelical poverty in ecclesiastical practices and improved the care of the poor through the new Monastic Orders, it was unable to raise the material status of the ordinary poor and was hardly interested in changing the social premises of feudal society. Moreover, the Church's theoretical adherence to the principle of evangelical poverty on the one hand, and the Church's position as one of the richest institutions on the other, provided one of the main factors adduced by sectarians as to their own right for existence, since the Church had betrayed the evangelical creed. This paradox was faithfully reflected in medieval sermons. For example, Jacques de Vitry refers to a hermit who, having relinquished all his possessions, was asked who had robbed him. Pointing at the Gospels he answered: "This copy of the Gospel, which teaches us to give all (our) things to the poor." He was asked: "How have you given all to the poor and still have this?" Straightaway he sold the Holy Book and contributed the money to the poor.[16] Medieval preachers further emphasized the didactic value of Christ's poverty for the salvation of all mankind: "*Qwhan that thys kyng that is kynge of all kynges and lorde off all lordys wylffully forsoke worschip to be made pore ffor our sake.*" Still, the numerous possessions of the Church throughout Europe remained as a fact that could hardly be justified on the grounds of the convenient distinction between individual poverty and the right of the ecclesiastical community to own property. It neither prevented the Church communicators from further encouraging the poor to submit to their distress, considered a reflection of the will of God: "*Heere men may see, who-so biholdeth wel, gret poverte in the aray at this lordes birthe. And bothe pore and riche moun lerne heere a lessoun, the pore to be glad in her poverte and bere mekely hire a-staat, seynge hire lord and hir maker wylfully to zeve hem suche ensaumple.*"[17]

The idealization of voluntary poverty by Christian communicators did not bridge the gap between those poor in spirit who were rewarded with eternal glory, and the common poor, who were often subjected to scorn and mockery.[18] Moreover, a distinction was made between holy voluntary poverty and idle parasitism that could encourage strong temptations to theft and perjury.[19] Huguccio (c. 1188) divided the poor into three categories, namely, those who were born poor but willingly endured their poverty for the love of God; others who joined the poor by renouncing all their possessions to follow Christ. Against both categories of voluntary poverty he placed a third sort of necessary or involuntary poverty, of those who were filled only with "the voracity of cupidity." It was well understood, therefore, that the experience of poverty, like the experience of pain, might bring spiritual enrichment to a man who was capable of accepting it voluntarily, but also that, in itself, poverty was an unpleasant affliction that might produce quite undesirable effects. Medieval Canonists, however, refrained from any pejorative categorization of the ordinary poor. Johannes Teutonicus (c. 1216) claimed that "poverty is not among the number of things evil," that is, things criminal or morally reprehensible. This opinion was shared a century later by Johannes Andreae, who argued that "poverty is not a kind of crime."[20]

Although poverty was not regarded as "a kind of crime," it was certainly not regarded as a kind of virtue, and the increasing numbers of poor could hardly find the moral support they were looking for in the ecclesiastical order. Moreover, the quicker the process of change, the stronger the criticism raised by the Church's practices, as the gap between the desirable evangelical path and the clergy's behavior became more noticeable.[21] This gap between the existing and the desirable ultimately provided the *raison d'être* of medieval heresies, which nurtured from old Christian concepts to which they gave a new meaning. Hugues de Rouen argued that the beliefs of sectarians "are not new but rather old and well-known," a claim that suggests the awareness as to the heretics' use of traditional concepts.[22] In fact, both renovation and reformation combined together in the development of heretical movements. They may be evaluated as progressive, following the changes they encouraged in the existing order. But the model of these desirable changes contained nothing new, as it was still being nurtured from the concepts of the Christian apostolic community in Jerusalem.

The *Waldensians* turned evangelical poverty into the pillar upon which all the ideological foundations of their sect were based. According to Bernard Gui, a fourteenth-century inquisitor from the Toulouse area:

> After they have been received into this society, which they call a fraternity, and have promised obedience to their superior and that they will observe evangelical poverty, from that time they should observe chastity and should not own property, but should sell all that they possess and give the price to the common fund, and live on alms which are given to them by their believers and those who sympathize with them. And the superior distributes these among them, and gives to each one according to his needs. . . . They commonly call themselves brothers, and they say that they are the poor of Christ or the poor of Lyons.[23]

Bernard Gui also reports the main stages in the development of the sect, which eventually brought about its gradual dissociation from the safe wings of orthodoxy:

> The sect of the Waldensians began in about the year 1170. Its founder was a certain citizen of Lyons, named Waldes or Waldo, after whom his followers were named. He was a rich man who, after having given up all his wealth, determined to observe poverty and evangelical perfection, in imitation of the Apostles.[24]

The personal example of the founder was emulated by his adherents, who saw themselves as the faithful followers of the evangelical creed spread by Christ and the Apostles. Their confidence in implementing the will of God gradually brought the Waldensians to question the magistracy of the Catholic Church while sustaining the right of each moral being, including women, to preach, thus strengthening the scrutiny of the individual with regard to the clergy and the dogma as a whole. Largely in rebellion against the corrupt and bureaucratic Church of their times, they also rejected the role of the priesthood in the sacraments of baptism and confession. They also opposed praying to saints and martyrs, the collection of tithes, the elaborate and costly construction of ecclesiastical edifices, and the Church's monopoly of preaching and interpreting the teachings of Christ. They totally denied the offering or reception of oaths or any judgment in which a person claims the right to inflict physical punishment or a death sentence on another. Beyond their primary and unquestionable allegiance to poverty, the Waldensians developed a peculiar code of evangelical Christianity, which has since become universally familiar.

Gerard Segarelly of Parma and his heir, Dolcino, laid down similar principles. They eventually brought about the emergence of the so-called *Sect of the Pseudo-Apostles*, which developed in Italy between the thirteenth and fourteenth centuries. As their preaching was also based on the Gospels, they liberated themselves from obedience to any human institution. Following the Apostles, they claimed their right to obey Christ alone, Whose ideals and lifestyle they emulated in actual practice. This "fact" bestowed on them the privileged status formerly enjoyed by the Roman Church, which it had eventually lost following the sins of priesthood. Also, the keys of heaven would be returned to the popes only if they returned to the purity and humility of St. Peter. The popes were thus called to avoid vain persecutions while *allowing each man to live in accordance with his own understanding*, a far-reaching claim in the climate of opinion of the times. Until such desirable changes materialized, the sect of the Apostles alone could guarantee true salvation. In order to properly fulfill their meritorious task, they committed themselves to evangelical poverty in a most extreme fashion. They lived from the believers' contributions, and never kept any goods with them, not even remnants of food, in order to fulfill the evangelical command to follow naked in the footsteps of Christ. They performed their ceremonies in public—in squares and in urban streets—and called upon everyone to repent while advancing their chances of salvation through joining the ranks of the true inheritors of the Apostles.[25]

The call to repentance also became the hallmark of the *Flagellants*, a movement launched in 1260 by a hermit of Perugia, which by 1261–1262 spread to the towns of South Germany and the Rhine. They saw themselves as the pioneers of a new era, charged with a redemptive mission for all humanity.[26] Usually led by priests, scores of men, youths, and boys marched day and night, with banners and burning candles, from town to town. When they came to a town, they arranged themselves in groups in front of the church and beat themselves for hours, thus emulating the agony of Christ on the cross. A chronicler reported that during the Flagellant processions, people behaved as though they feared that God was about to destroy them all by earthquake and by fire, as a punishment for their sins. They called out ceaselessly: "Holy Virgin, take pity on us! Beg Jesus Christ to spare us! Mercy, mercy, peace, peace!"[27]

The conviction that "You have chosen us from among all peoples" to carry out a prodigious mission was not unique to the Flagellants but, rather, became an outstanding feature of all groups declared by the Church as heretical. Also, *millenarist movements* shared the feelings of having been elected by providence to hasten the redemption of humanity. Their unconditional submission to a leader, whom they believed enjoyed the aura of the biblical prophets, allowed them to join his holy mission and even to share his supernatural powers as a reward. In the civilization of the Central Middle Ages, such fantasies spread through the whole population, but at times of general disorientation they became even more obsessive and compelling. Fulk of Neuilly, an ascetic thaumaturge active at the time of the Fourth Crusade,[28] and Guiard of Cressonessart, the fourteenth-century Angel of Philadelphia (Rev. 3: 7–13),[29] led such groups. They saw themselves as the "fugitive remnants," and as such, pioneers of the new era that heralds the final victory of Christ over the Antichrist.

The identification with the chosen people offered both an escape valve from actual reality and a solidarity group that shared with the individual his frustrations and hopes in the face of the crisis. Another kind of defense mechanism was provided by the dualistic philosophies, which turned the endless conflict between the existing and the desirable into the very essence of life in its more universal dimensions. Scattered among the intellectual elite of the Eastern Empire, they reached Italy first and later spread westward to France and throughout Christendom. Based on Manichean principles, they approached the whole existence as an endless battle between the good and the evil gods, between Christ and the Antichrist, indicating the basic opposition between all components of daily life, such as light and darkness, the body and the soul. The whole material world was perceived as a manifestation of the forces of evil, while the realm of God was consequently limited to the spiritual sphere. The lack of recognition in God's omnipotence, on the other hand, elevated the self-estimation of sectarians, as they believed to play an active role in the cosmic battle between Christ and the Antichrist, the former being incapable of overcoming His adversary alone. Their assisting role at the side of providence further encouraged the identification of dualistic sectarians with the well-known image of the chosen people while reducing even more their already low readiness to acknowledge the magistracy of the Church or to submit to its commands. Dualistic philosophy reached its zenith among the *Cathars*, a sect whose development fitted the patterns of a schism rather than those of heresy. The Cathars dissociated themselves not only from the Catholic Church, but also from the Christian faith as a whole, since they did not acknowledge the basic belief in one almighty God. Rainier Sacconi, a thirteenth-century inquisitor who had been a member of the sect before joining the Dominicans, alleged that:

> The common opinions of all the Cathars are these, namely, that the devil made the world, and all things in it. Also, that all the sacraments of the Church, namely, the sacrament of baptism of material water, and the other sacraments, are not profitable to salvation, and that they are not the true sacraments of Christ and of His Church, but misleading and diabolical, and of the church of the malignants. Also, it is an opinion common to all the Cathars that carnal marriage is always a mortal sin, and that the future punishment of adultery and incest will not be greater than that of lawful matrimony, nor would any among them be more severely punished. Also, all the Cathars deny that there will be a resurrection of the flesh. Also, they believe, that it is a mortal sin to eat flesh, or eggs, or cheese, even in case of urgent necessity. Also, that the secular powers sin mortally in punishing malefactors or heretics. Also, that no one can be saved but by them. . . . Also, they all deny purgatory. Also, it is an opinion common to all the Cathars that whosoever kills a bird, from the smallest to the greatest, or quadrupeds, from the weasel to the elephant, commits a great sin; but they do not extend this to other animals.

Criticizing as "false, vain and delusive" their interpretation of penance, Rainier also maintained that the Cathars believed that:

> eternal glory is not diminished for any sin. . . . that the punishment of hell is not increased to the impenitent. . . . that there is no purgatory for anybody;. . . . Nor do they ever implore the patronage of angels or saints, or of the blessed Virgin Mary, nor fortify themselves with the sign of the cross. . . . They frequently pray. . . . as they consider it absolutely necessary when they take food or drink.[30]

To their reluctance to accept the most basic tenets of the Catholic Faith, the Cathars further added their own ritual with distinctive sacraments and a hierarchical order. The four sacraments prescribed by the Cathars included the imposition of hands or *consolamentum*, which bestowed the Holy Spirit and the remission of sins, the benediction of bread, penance, and orders. They also laid down the foundations of their own hierarchy, which acknowledged the existence of bishops, elder sons, younger sons, and deacons. Each of the three higher grades had a deacon or an assistant to replace him. They all were considered "perfects," as they had been entrusted with the *consolamentum*. From the very moment they had received the *consolamentum*, they dissociated themselves from this world, their family, and property, and contributed all their possessions to the whole community. They often fasted for prolonged periods and contented themselves for the rest of their lives with very little in order to restrain the desires of the flesh. Although there was no clear division of labor between them, the sons were mostly employed in visiting the members of the sect. The respect and reverence they enjoyed was pejoratively described by inquisitorial protocols as *adoratio*, worship in the strict sense of the word. The believers asked for their blessing and were answered, "God is blessing you," or in the Provençal tongue, *Diaus vos benesiga*.[31] The mutual appreciation between the perfect and his followers encouraged the unconditional commitment of the leader to the community and, conversely, the readiness of all members to sustain his needs and provide him moral support in most critical situations.[32] The elder son was elected by the congregation and promotions were made to the episcopate as vacancies occurred. It was generally, though not universally, held that those lower in grade could not consecrate the higher, and therefore in many cities there were habitually two bishops, so that in case of death, consecration should not be sought at the hands of an elder son.

Having developed their own ritual and hierarchy, the Cathars were therefore in the forefront of the opposition to the Church. They did not acknowledge its mediatory role, nor that conferred by Catholic dogma to the saints whether through purgatory, penitence, or prayer. Bernard Gui provided additional features of their peculiar attitude to actual life and their self-identification with the chosen people:

> They claim and repeat frequently to be good Christians who do not swear, lie or curse anybody, do not kill, neither man nor beast nor any living creature. They also believe that they live in accordance with the true religion of Christ and the Gospels, as the Lord Himself taught the Apostles whose place they claim to inherit. This is the main reason why the Church ministers, namely, the prelates, priests, monks and mainly inquisitors, persecuted them in the same manner as the Pharisees persecuted Christ and His Apostles, and called them heretics, although they are honest men and good Christians.[33]

The willingness to idealize their suffering as an integral part of their being elected by God created a common denominator between all sectarians, and emphasized their rupture from normative society with the Church at its head. The voluntary pursuit of pain in the case of the Flagellants, or the persecutions forced on the Waldensians, the Cathars, and the Pseudo-Apostles by the Church further widened the gap between them and the ecclesiastical establishment from which they had divorced themselves and whose norms they denied. This state of affairs indicates the basic nonconformism underlying the very existence of the so-called heretical movements: they denied the Church mediation, the sacraments, and the existence of purgatory. Their basic individualistic approach with regard to the Catholic dogma and the Church attempts to enforce it was, however, significantly undermined by the spirit of association of all sects, which actually forced a new dogma on all its members. The individualistic approach, which had originally provided the *raison d'être* of heresy, was therefore replaced by a process of institutionalization, while the battle for survival forced all sects to develop a whole system of rank and dogma quite similar to that developed by the "Synagogue of Satan." Beside the gap between majority and minority, between persecutor and persecuted, one might discover at this point a fairly wide range of similarities between orthodoxy and heresy: they had both developed from the same sociocultural background. Heretics and Catholics both acknowledged but one truth, that of Christ and the Apostles, while they disagreed as to its interpretation. Furthermore, their claim for a better understanding of the will of God encouraged both to bestow upon themselves the role of the chosen people, elected by God to carry out His commandments. Neither the Church nor the heretics were able to think in terms of tolerance with regard to each other, but their actions, either of persecution by the Church, or martyrdom by the heretics, were further justified by the will of God. From this perspective, the Inquisitor and his victims could both be seen as martyrs of the same faith, the only one that would tolerate no other.

As a logical outcome of their evangelical inheritance, the heretics believed themselves to have been blessed with an absolute monopoly of the true faith and the unique valuable understanding of the will of God. They saw themselves as the sole and true followers of Christ and the Apostles, as they had freed themselves from the bonds of materialism, the realm of Satan, and, accordingly, were entrusted with the path to eternal salvation. The essence of the heretical message could therefore be summarized by the imperative of all the faithful to join the sects, thus starting on earth the way toward the Kingdom of God. This attitude eventually prevented any possibility of finding a compromise with the Church. The conflict between heresy and orthodoxy could not thus be explained on the grounds of intolerance between two parallel streams of beliefs acknowledging each other, but rather as an expression of the insoluble contradiction between the false and the true. The anecdote reported by the Anonymous of Passau faithfully reflects the excluding attitude of heresy vis-à-vis orthodoxy, and conversely: When a certain heresiarch named Hainricus, a glovemaker from Thewin, was led to execution, he stated before all, "It is right that you should condemn us in this matter, for, if we were not a minority among you, the sentence of death which you exercise against us in this manner we would exercise against your clergy and religious and lay people."[34] The excluding attitudes toward each other encouraged the sectarians' fanatical hatred of the Church institution as a whole, complemented by anticlerical attitudes to its ministers.[35] In conversations recorded in Montaillou, some Cathars voiced their hopes that "all priests would die, including my own son, who is a priest."[36] Others argued the unquestionable fact that "Catholic preachers teach the people but a little. Not even one half of the population attends their sermons and there are even fewer among them who understand the ecclesiastical message."[37] Belibaste, the leader of the Montaillou Cathars, claimed that "the pope sucks the blood and sweat of the common people, as do bishops and priests. They are rich and always in pursuit of honor, in open contradiction to the example of St. Peter who left his wife, his children, his lands, his vineyards and all his possessions, in order to follow Christ."[38] Hence the antagonism between the Catholic Church, one which oppresses and extorts, and the Cathar, which confers forgiveness and grace on all its members without waiting for material compensation.[39] The Catholic Church was further portrayed as the harlot of which we read in the Apocalypse, the Temple, and Synagogue of Satan, the courtesan, the mother of all fornications. Nevertheless, as Le Roy Ladurie has clearly pointed out, the Cathars did not criticize the existence of the Church as such, but rather its failure to be the Church of the poor as well as its preference for material gain.[40]

Although the Church was not ready to become the Church of the poor, the spread of heresy faced it with a propaganda challenge that it could hardly disregard. Church's communicators emphasized the harmful influence of heresy, portraying it as a threat to the very survival of orthodoxy. Pope Innocent III claimed that "it is necessary to regard as manifest heretics those who preach or publicly profess ideas contrary to the Catholic faith and defend their error."[41] This view was shared by Robert Grosseteste, bishop of Lincoln, who underlined the stubborn character of heretics who hold "an opinion chosen by human perception contrary to Holy Scripture, publicly avowed and obstinately defended."[42] Such definitions reflect the Church's perspective of heresy, which did not focus on the beliefs or rites that had brought about its condemnation as such but, rather, on the heretics' tendency to *publicize* their faith in open contradiction with the principles of the Catholic faith. They indicate, therefore, the essential rivalry between the ecclesiastical order and the heretical movements, as they both appealed and sought the approval of the same audiences, used the same symbols, and based their argumentation on the only authoritative source acknowledged by Christian society as a whole: the Holy Scriptures.

One of the first stages in the elaboration of a supportive public opinion against heresy focused on the stereotyped image projected on the heretics. It was expected to strengthen the self-confidence and group solidarity between all members of Christendom and, consequently, the foreignness and the threat implied in heresy. The stereotype underlined the heretics' pride, for they had set themselves against ecclesiastical authority; their superficial appearance of piety, which could not be real, since they were, in fact, the enemies of the faith; and the secrecy of their rituals, which were suspected of immorality and sorcery. Heretics were also depicted as unlettered, since they lacked the skills of the orthodox faithful and were accused of counterfeiting piety while secretly indulging in libertinism. All the fears of medieval society vis-à-vis sectarians were summed up by the perception of heresy as a plague or leprosy whose carriers spread its deadly germs, or as a poison that actually turned the heretics into snakes. The Council of Toulouse (1163) compared the spread of heresy to a cancer growing in all directions should it not be properly removed in time.[43] Eckbert of Schonau used the whole range of images and depicted the Cathars as those who "have multiplied in every land, and the Church is now greatly endangered by the foul poison that flows against it from every side. Their message spreads like a cancer, runs far and wide like leprosy, infecting the limbs of Christ as it goes."[44] The terminology of disease puzzled medieval audiences and yet transmitted quite clearly the alarming nature of heresy. The Church message thus became clear enough in order to create a perfect, acceptable equation between heresy and disease or, as succinctly expressed by Moore, "heresy was to the soul what leprosy was to the body."[45] (see plate 28).

Against the threat implied in the very existence of heresy, the Church's communicators placed the wholeness of orthodoxy, described by Vincent of Lerins (AD 436) as "that which has been believed everywhere, always and by everyone."[46] The universal nature of orthodoxy dictated a normative attitude

PLATE 28. Andrea da Firenze, detail of the *Triumph of the Church,* Florence, Spanish chapel, between 1366 and 1368. The Dominican Order is depicted as the most important organ of the Church for purposes of spiritual guidance and the struggle against the heretics. While the Dominicans dispute with the infidels, white and black spotted dogs, the *domini canes,* hurl themselves in a vividly symbolic attack upon the wolves that have penetrated the fold of the faithful.

to all facets of daily life and resulted in the complete refusal to justify the very existence of parallel streams of thought. Heresy was depicted as a threat to the harmonious existence of Christendom, which appeared as a wholeness whose unity had not to be harmed. The importance ascribed by the Church to the stability of the social order was expressed through the Latin term *ordo*, which indicated the divine sanction of the status quo by which Christian society became a true reflection of the City of God. This resulted in the closer links between God and man who, having submitted his own will to the will of God, moves within a perfectly ordered society toward His kingdom, well-defined by Duby as the kingdom of the inmutable.[47]

Against the totality supported by the Church's communicators, heresy presented an alternative view to actual reality, but this was refused its very right of existence. The propagandistic efforts of the Church to counterbalance its harmful influence, however, corroborates the impact of heresy in medieval society and the relatively advanced communication stage it had reached from the eleventh century onward. Lacking socioreligious legitimization, the spread and development of all sects was conditioned by the efficiency of their communication systems. In other words, "heresy" meant "propaganda," and the survival of heresy despite the Church persecution reflects the existence of a more communication-oriented society with more sophisticated communication channels at its disposal. In spite of the fact that heretical propaganda developed clandestinely and was often involved in a struggle to survive, it succeeded in becoming a "contagious disease," which reached all social strata in Christendom. This success was largely due to the fact that the heretical leaders did not present themselves as the promoters of change, but rather as the representatives of an ideal, mythical past, which the Church had actually betrayed. The unquestionable authority of Christ, the Apostles, and the Gospels as a whole eventually served as a shield behind which individualistic, nonconformist attitudes had been voiced. The conservative tone of most heretical messages could not blur the essential innovation inherent in the right claimed by individuals to properly voice the *Vox Dei* while requiring, as did the sect of the Pseudo-Apostles, the Church's recognition of each one's right to live according to his own understanding. To sum up, the very emergence of heresy, as a challenge to the accepted norms, indicates the gradual decline of the traditional society in which religious individualism was barely practicable. The location of heresy in the nascent urban centers reflects a new state of affairs in which the integration process in the economic and the social realms encouraged the elaboration of more sophisticated channels of communication and the more intensive use of propaganda. Heresy thus appears both as a cause and an effect of more advanced communication stages, largely in contrast with the isolation and the autarky characteristic of the feudal-corporate system. Yet in the Central Middle Ages, individualism as a social phenomenon was still of an ephemeral nature and the heretical sects were eventually forced to develop corporate structures in order to ensure their survival. By the sixteenth century, the collapse of the traditional structures strengthened these tendencies, while bringing about the emergence of Protestantism.

11

The Transmission of Heresy

The use of communication by kings and prelates developed within the framework of the traditional society. As such, it was of limited scope and opposed the emergence of parallel communication channels. The heretical sects, on the other hand, developed within the framework of urban societies in a process of integration without the support of the establishment and despite its complete antagonism. This particular background encouraged the so-called heretics to utilize to the maximum all the communication channels available at the time. Lacking the support of the establishment, the heretical sects were not influenced by the limitations imposed by the traditional society but could pave the way for a "total communication," which aimed to reach all those ready to listen to their message. The transmission of heresy, therefore, provides the most faithful reflection of the nonestablishment, voluntary channels of communication in the Central Middle Ages. Although these channels were not peculiar to the sects, they reached, however, a maximal scope among them. Both their voluntary nature and their intensiveness caused the communication channels of heresy to approach the patterns of modern society, thus emphasizing their contribution to the systems with which we are familiar today.

Despite its image of a contagious disease, heresy did not spread by itself but needed suitable channels of communication in order to propagate its creed. Moreover, the very survival of heresy was dependent upon the continuous flow of communication between its members, away from the watchful eye of the Inquisition. Borst indicates three major categories which, in his opinion, reflect the communication channels in the different stages of the evolution of medieval heresy: 1) interpersonal communication between people and groups; 2) diffusion of heretical traditions between groups of different periods; and 3) oral transmission of the heretical message within the same sect from one generation to another.[1] Within these large categories, additional differentiations had to be made between the geographical foci of heresy, in the countryside or the urban centers, and its different communicators, either the heretical message was transmitted by an itinerant population or by a settled one. An additional distinction focuses on the degree of spontaneity or institutionalization undergone by the communication process, the former inherent in daily practices, the latter resulting from the deliberate effort of the sectarian leaders.

The first seeds of heresy had emerged in the country areas of Italy and France at the end of the tenth and the beginning of the eleventh century. From the countryside, it spread toward urban centers such as Milan, Mantova, Cremona, Genoa, Pavia, Béziers, Carcassonne, and Toulouse, or to chateaux, such as Monforte and Montségur. Heresy thus gradually became an "urban phenomenon" enclosed in the suburban areas of cities in a process of demographic growth and economic expansion.[2] The transmission of heresy from the countryside to towns was but one aspect of the migratory process characteristic of European society at the time. Yet migratory movements, as much as they might have contributed to the transmission of heresy in space, lacked continuity in time and could hardly ensure fluent communication between heretics. Peripatetic groups played a cardinal role in the transmission of heresy, their geographical mobility resulting from economic occupation and social status, or from having experienced heresy, a fact that forced on them a wandering way of life. Geographical mobility thus appeared as a keynote of many involved in heresy. International merchants who were constantly on the roads, traveling from one country to another, became the occasional communicators of heresy over large areas. Heresy was also transmitted by pedlars and craftsmen who offered their merchandise from house to house, while developing a more immediate contact between heretics. Weavers and cloth and leather workers who embraced heresy joined these large categories of footloose people. They became not only the main communicators of heresy, but also one of the major socioeconomic groups within its ranks. Nelli tries to explain this particular collocation of a large number of heretics, and of the Cathars in particular, on the grounds of their essential opposition to the land, being perceived as an outcome of the evil god and, as such, a symbol of materialism. Without disapproving of the labor of peasants, most Cathars were more willing to qualify themselves for craftmanship and trade, both of which imply much creativity without exploiting the manpower of others.[3]

Yet geographical mobility did not always respond to ideological considerations, nor was it limited to the ranks of craftsmen and merchants. The aristocracy also enjoyed a rather high mobility due to its privileged status. For instance, Italian nobles were constantly on the move between their estates in the countryside and their urban mansions, which, from the twelfth century onward, became their preferred place of residence. In addition to the socioeconomic advantages implied in the ownership of large country estates, they could also, if the need arose, provide suitable shelter from Inquisitorial persecution. The significant number of clergymen, hermits, wandering monks,[4] popular preachers, lecturers, and students who joined the sects eventually strengthened even more the itinerant character of their members. An eminent teacher such as Odon de Tournai gathered around him a considerable number of students who came from the Ile de France, Flanders, Normandy, and faraway places in Saxony and Italy, to listen to his words. Although this unusual movement aroused admiration, it was nevertheless accompanied by varying degrees of criticism. Such an international congregation roused the fears of those who felt concerned with the safeguard of orthodoxy. A contemporary chronicler complained that all the inhabitants of Tournai had become philosophers, a "fact" of which he vehemently disapproved.[5] William of Malmesbury shared the prevailing critical attitude toward wandering students suspected of contributing to the spread of heresy. He reported that the "heretical" ideas of Berenger had been spread in France by poor students who wandered through villages and towns, exploiting the innocence of the crowds and encouraging them to follow the path of heresy.[6]

The itinerant character of most sectarians became both a cause and a result of the growth of heresy. The substantial number of footloose people facilitated the flow of communication between heretics and the spread of their ideas among additional social groups. Moreover, wandering became a distinctive trait of heretics vis-à-vis the sedentary character of medieval society as a whole, especially the Church, which had formally laid down the clergy's duty of permanent residence. It also sharpened the dissociation of heretics from the frameworks of the traditional society whose norms of conduct they challenged. The high geographical mobility of a significant number of heretics was undoubtedly fostered by the socioeconomic process and its resulting migratory movements; but it was also conditioned by the experience of heresy itself and the state of continuous persecution forced upon its members. The primary need to flee from the grip of the Inquisition dictated a state of continuous movement to most sectarians, especially their leaders. As the Inquisition persecution increased, many of them were forced to leave for exile, divorcing themselves from the heretical community. This state of affairs, characteristic of the Languedoc after the Albigensian Crusade, did not terminate the very existence of heresy in the area, but complicated the means of communication between sectarians who were deprived of their leaders. More and more heretics were forced to move from the Languedoc to Lombardy in order to fulfill the ritual of their sects. They usually made their way along secondary roads by which they could evade the *exploratores* more easily. This was a kind of secret police employed by the Inquisition, who followed all those suspected of heresy, eavesdropping on their conversations, and sometimes making use of dogs in order to trace their hiding places. Dangers and persecution could also arise from among the ranks of the heretics themselves, namely, from those who had returned to the orthodox fold and shared the commitment of the clergy to the "Holy War" against the "contagious disease." Sicar de Figueiras, a former Cathar, declared before the Inquisition his readiness to bring about "the arrest of all heretics" by recruiting all his squires, "who are well acquainted with their hiding places, the caves and woods as well as the holes they dig in the ground to hide their goods."

This state of affairs suggests the many risks inherent in the development of communication between heretics. Their struggle for survival was affected by the very fact that they lived within a society that did not share their creeds but rather regarded them as a threat to its own existence. The bitter enmity of prelates and kings toward heretics, and their stereotyping as a contagious disease or poison, indicate the unbridgeable opposition between orthodoxy and heresy and also dictated extreme caution in creating the basis of a communication system between sectarians. On the other hand, the voluntary or forced exile of many leaders pressed for the elaboration of communication channels between the different centers of heresy, mainly between those located in southern France and northern Italy. Bonet de Sanche, a Cathar of Castelnaudary, for example, wanted to travel to Lombardy to be granted the *consolamentum* by some of the perfects in exile. He asked Guillaume Raffard to join him, but Guillaume refused on the grounds that he would have to sell his cattle first in order to be able to pay the expenses of the journey. Twenty years later, the same Guillaume Raffard left the Languedoc on his way to Lombardy, his cows running before him. He had managed to sell them in Montpelier and hired the services of a guide, Pierre Maurel, who led him across the Alps. In Pavia they stayed for about three months in the house of another Cathar, Pierre de Montagut. They reached Sermione, an important focus of heresy at the time, and the *consolamentum* was granted to Guillaume Raffard, as well as to Guillaume Bonet and Pons Olive, other Cathars from Mirepoix who had also come for the same purpose.[7]

The imperative to ensure unbroken communication between the believers and their leaders in exile encouraged the gradual improvement of traveling conditions. Sectarians who could afford the expenses hired experienced guides, French and Italians, who were familiar with the local languages and knew secondary routes and hidden places away from the watchful eye of the Inquisition. Traveling sectarians could also rest from the hardships of the road while enjoying the hospitality of former emigrants who had developed a chain of friendly inns alongside the main routes. The existence of a more organized system of hospices or lodging houses achieved maximal importance, as among the roving people were a considerable number of single women. They traveled alone either because they had been forced to leave their families after espousing heresy or for reasons of safety, as entire families on the road would more easily rouse the suspicions of the Inquisition. The gradual improvement of traveling conditions, such as the availability of reliable guides and lodging places, resulted, however, in a parallel improvement in the Inquisition's methods of persecution, thus creating a vicious circle in which both the persecuted and the persecutors were trapped. With the high degree of sociability common to medieval men, however, their persecution by the Church strengthened the esprit de corps among sectarians and their resulting willingness to support one another in face of their common fate. The tradition of hospitality developed by all heretical groups was a direct result of this solidarity despite the fact that, since 1229, any assistance given to heretics would give rise to suspicion of heresy.

The sectarians' high degree of geographical mobility and the peculiar problems it caused are clearly illustrated by another group of Cathars led by Pierre Maurel, who moved from the Languedoc to Lombardy between 1271 and 1272. The meeting place was fixed at Saint-Martin-la-Lande. They waited there for three women and a child in the house of a Cathar woman on the boundaries of the village. Four more Cathars joined the group in Béziers.

Through Beaucaire, they reached Acqui in Lombardy; there they stayed at the house of another woman who had emigrated from the Languedoc. At this stage, Pierre Maurel changed his name to Pierre Gailhard. In Asti they met Bernard, Pierre's brother. Afterward, they moved to Pavia and lodged at the house of Raymond Galterio, but they failed to escape the Inquisition's searches and a member of the group, a weaver, was arrested there. The others moved to Mantua where they met two other Cathars from Limoux. Passing through Cremona, they finally reached Milan, where the *consolamentum* was bestowed upon them. Via Coni, they returned to France and, safe and sound, they reached their final destination of Castelnaudary (see map 6).

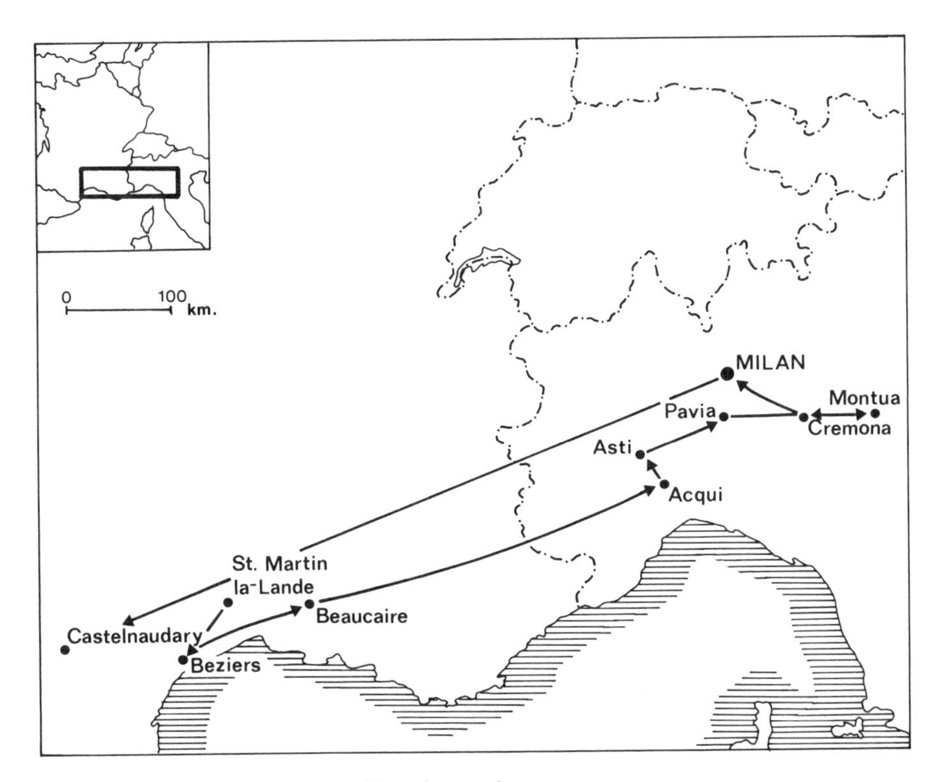

Heretics on the move.

Sometimes, when the heretics could not travel in person, those who could afford it hired the services of a third party, who performed the journey on their behalf. The Albigensian merchant, Bertrand de Montegut, paid a boatman, Marescot or Mascoti, to travel to Lombardy in order to bring back the perfect, Raimon Andrieu, expected to confer the *consolamentum* upon him. If Raimon could not be located, Marescot was allowed to bring whatever spiritual leader he could find. Marescot returned to the Languedoc with an Italian perfect, Guillaume Pagano, who performed for Bertrand the religious ceremony. Two other Albigensian merchants, Bertrand and Guiraud Golfier, also asked Bernard Fabre, a tailor who had emigrated from Albi to Genua, to bring them a leader for the same purpose. In this case, however, there is no information available as to the results of his mission.[8] Such attempts were not isolated phenomena. They reveal the existence of an effective communication system between sectarians from various and often faraway places. The common support for heresy created bonds of solidarity among people who would otherwise be complete strangers.

If geographical mobility was a key element in the life of heretics as a whole, it was even more significant among their leaders. Their basic commitment to their communities, their missionary efforts to encourage new sympathizers, and the even stronger imperative to keep one step ahead of the Inquisition, all gave rise to endless journeys between distant and widely scattered areas. In order to satisfy the peculiar needs that the heretical creed had created, the leaders had to spend long periods on the roads. They often moved in pairs, since the presence of a *socius* or partner was usually needed for performing the rituals. The visitation of the leaders provided the heretical congregation with the living message of the sect, their commitment to evangelical poverty being translated into actual practice by the simplicity of the leaders' dress and the modesty of their customs. Although most heretical leaders had messengers at their service, referred to as *nuncii hereticorum* in Inquisitorial sources, these messengers announced the approaching arrival of the leaders to some particular area but were unable to perform by themselves the liturgical ceremonies or the religious services required by the believers. According to Bernard Gui:

> They travel through the country, visiting and confirming their disciples in error. Their disciples and believers supply them with necessities. Wherever the leaders go, the believers spread the news of their arrival, and many come to the house where they are admitted, to see and hear them. All sorts of good things to eat and drink are brought to them, and their preaching is heard in assemblies which gather chiefly at night, when others are sleeping or resting.[9]

This state of affairs was well-known to the clergy and even resulted in some sort of admiration for those "masters of error." The Anonymous of Passau claimed that he had heard "from the mouth of a believer in their doctrine, that a certain heretic—whom I knew—for this purpose that he might turn him away from our faith and pervert him to his own, swam to him at night in winter across the River Ibbs." And he mourned the fact that "the negligence of the doctors of the faith shames us who do not show so much zeal for the truth of the Catholic faith as a perfidious Lyonist shows for the error of infidelity."[10] The basic rivalry between orthodoxy and heresy was, therefore, further sharpened by the sectarians' greater commitment to the diffusion of their creed, a fact confirmed by the testimony of the opposing side. The persecution to which they had been exposed by the Catholic establishment left its mark in the voluntary spirit of the communication system among heretics. Based on the personal contact of the leaders with their congregations, it also led to the spontaneous propagation of news and the gathering of assemblies zealously reserved for the members of the sect.

The geographical mobility enjoyed by a considerable number of sectarians, especially their leaders, also allowed the settled populations to obtain up-to-date information and the religious services they needed. The bonds between the wandering and the settled populations were further facilitated by the trades and occupations of many sectarians. In his confession before the Lombard Inquisition in 1387, Frère Antonino reported that out of 110 people, 25 members of his sect owned taverns, those being followed in number by millers. Both taverners and millers served in daily practices as the most suitable intermediaries between the itinerant and the settled populations, since they owned the most popular meeting places of the day. This state of affairs is confirmed by the confessions made in Pamiers before the inquisitor Jacques Fournier: six or even seven sectarians met in taverns where they could exchange information of common interest, such as the Inquisition activities in the area.[11] In mills as well, sectarians, mostly women, spread all kind of rumors and received the latest news.[12] Taverns, workshops, and mills thus became casual meeting places for the supporters of heresy. At Modena, certain mills were known in the late twelfth century as "the mills of the Patarenes," a term used to depict the Cathars. The narrow paths of medieval villages also encouraged the spontaneous meetings of sectarians and facilitated the exchange of information between them.[13] One inhabitant of Pamiers suspected of heresy, Arnaud de Savinhan, confessed before the Inquisition in 1318 that he had encountered Jean de Beubre, Pierre de Mayshelac, and Pierre Mercier on the bridge leading to Tarascon. They asked him for the latest news of Pamiers, and Arnaud took advantage of the occasion to tell them of the rumors he had heard from the Hospitallers, according to which the Antichrist had been born and doomsday was coming.[14] A chance encounter with people on the road, therefore, turned into a communication event that was remembered as such and considered important enough to be reported some years later. Moreover, our amateur communicator was aware of the importance attributed to his source of information and took care to report it, the prestige of the Hospitallers being expected to give further credence to the incredible news he spread.

One can argue that meeting places such as taverns, mills, and the village streets were not reserved for heretics only, but provided channels of communication for the entire medieval population. Yet the peculiar situation of heretics conferred great importance on such places, due to the secrecy required and the selective character of the communication between heretics. In the confessions made before Jacques Fournier, sectarians reported on meetings that took place in kitchens, usually in front of the family hearth, which provided some warmth on cold winter nights. The nobilities were able to provide better facilities, and from time to time their houses became meeting places for such gatherings.[15] Hospitality led to contacts. At Puylaurens, the mother of Sicard de Puylaurens lived with two of her sisters and another noble lady, all perfects, and men and women came and ate fruit from their hands.[16] In Montaillou and its vicinity, many sons of the large family coexisted under the same roof and were often joined by distant relatives, friends, or other sectarians who happened to be passing through.[17] The biography of Huguette, a

Cathar woman arrested by the Inquisition in Pamiers, clearly illustrates the importance ascribed to family links in facilitating communication among heretics dispersed over wide areas: Huguette was the daughter of a butcher from Vienne. After her father died, she was sent by her mother to Arles, where she remained for four years. She spent another year in Tarascon de Provence and returned to Arles, to be married to Jean de Vienne. Accompanied by the husband's sister, Petrone, the couple moved to Pamiers in order to improve their income, since Jean hoped to find a better market for his wine barrels in the markets of the growing city. Their move was also meant to assist Huguette's uncle, Raimond de la Cote, himself suspected of heresy, who suffered from a lingering disease. As to the fact that all the three stayed at his house, Huguette justified this on the grounds that "we speak the same language," a statement that suggests more than a merely linguistic understanding common to all members of the enlarged family.[18]

The biography of Huguette indicates a geographical mobility that until the fourteenth century was rather uncommon. It undoubtedly reflects the peculiar situation of heretics whose frequent moves were aimed at avoiding discovery and arrest at the hand of the Inquisition. But geographical mobility as a social phenomenon could hardly be considered the sole prerogative of heretics. The growing towns, the gradual transition into a monetary market with changing rules, dictated a continuous movement toward new economic opportunities, from the countryside to the towns and from one town to another. In the peculiar situation of medieval sectarians, however, their mobility acquired special significance due to their deliberate efforts to maintain continuous communication among their ranks. The obtaining of accurate information sometimes acquired the relevance of a dividing line between life and death. Family links provided sectarians with an important channel of communication that acted as a shelter, both in the moral and in the material spheres. The influence of heretics on children also helped to make heresy a part of their background. Bernard Mir remembered as a child going into a house of a perfect at Saint-Martin-la-Lande and being given nuts to eat and taught to bend his knees to ask for a blessing. Poor girls were often left with women perfects in order to be supported and educated in the ways of the sect. The Dominican, Jordan of Saxony, complained of the way in which parents put daughters into the houses of perfects and so, inevitably, to be drawn into heresy.[19]

Heresy thus became a part of daily experience, its communication closely related with this experience. The never-ending persecution and the limitations posed by the technical means of the times forced heretics to learn how to take the greatest advantage of all the alternatives at their disposal. Beyond the spontaneous meetings that resulted from chance encounters, there were also more formal contacts in which sectarians usually met under the guidance of a leader. Even church buildings served as meeting places. Jacques Authier preached in 1303 in the Church of the Frères de la Sainte Croix in Toulouse before sectarians, mainly Beguins and Cathars, who listened carefully to his sermon. Yet the more intimate and safe surroundings of the family house were chosen by Belibaste, the local leader of the Cathars in Montaillou, to explain the essential paradox inherent in the Catholic dogma of indulgences:

> Imagine a priest who comes to the pope and, in exchange for ten or twenty pounds, receives a formal bull, sealed by the pope, which states that anyone who gives him a donation, would be rewarded with an indulgence of forty days. He roams around the world with this charter, exploiting and deceiving the humble. . . . If such was the case, and if the days of indulgence could be compared to precious stones, a man who contributes a considerable amount of money would not be able to use all the indulgences he would be awarded.

And, in order to illustrate his point of view more forcefully, he mockingly begged with his audience: "In the name of God, give me donations so that I give you a thousand indulgences!"[20] Such meetings became an inseparable part of the lifestyle of sectarians and allowed an immediate contact between their leader and the believers. They strengthened their feelings of solidarity as a persecuted group, and contributed to a better understanding of the heretical doctrine. Other groups committed to voluntary poverty, the Beguins and the Beghards, gathered in similar meetings quite frequently and discussed most important issues related to their creed. Bernard Gui reported that they lived together in villages and towns, in small houses that they called "the houses of the poor," where all sympathizers met on Sundays and holidays; there they read or heard the summaries of the teachings of St. Francis and Olivi, pronounced in the vernacular. They were also in the habit of reading aloud the Ten Commandments, the Principles of the Faith, and the Lives of the Saints.[21]

Meetings of this kind suited the practices of the fraternities and were common to large sectors of medieval society. On the other hand, the secrecy inherent in the meetings among sectarians became an integral feature of their stereotyped image, which came to include magical practices, allegations of sexual orgies, and even narcotic addictions. Twelfth-century sources claimed that the Cathars met at night by candlelight to recite a demonic litany until they saw the devil appearing among them in the guise of some wild beast. Its appearance gave the sign for the beginning of diabolical orgies, covens, and magical journeys. The mention of a powder, lotion, or ointment said to have the power of attracting whoever took it to the cult, suggests, according to Moore, the symptoms of narcotic addictions, such as to the plants of the nightshade family, well-known in medieval Europe.[22] Stories of this kind were largely acceptable to medieval society, as they provided the ordinary man with a suitable target on which to project his own fears. But there were also some who rejected such possibilities and expressed their views more objectively. Jean de Capelli, a thirteenth-century Franciscan, provides a convincing explanation as to the heretics, mainly the Cathars, needing to maintain the secrecy of their meetings, since

> They are exposed to many defamations widespread by the common folk, according to which they gather in order to perform shameful and indecent

> deeds, in spite of the fact that they are innocent of all these charges. Furthermore, they are proud of their persecution, which proves, in their opinion, what has been written about them in the Gospels: "Blessed are they which are persecuted for righteousness' sake: for theirs is the kingdom of heaven. Blessed are ye when men shall revile you, and persecute you, and shall say all manner of evil against you falsely, for my sake" (Matt. 5:10–11).[23]

The diabolic image ascribed to the meetings of the sectarians did not prevent any further gatherings in the future but rather strengthened their identification with the image of the chosen people who had been persecuted by the Pharisees of their own days. Another result was greater caution to prevent any possibility of their meetings being noticed by the clergy. Bernard Gui reports in detail on the arrangements made by the Waldensians:

> Each year they hold or celebrate one or two general chapters in some important town, as secretly as possible, as if they were merchants, in a house hired long before by one or more of the believers. And in those chapters the leader orders and disposes of matters concerning the priests and deacons, and concerning all those sent to different parts and regions to their believers and friends, to hear confessions and to collect alms. He also receives the account of receipts and expenses.[24]

Despite the state of continuous persecution, heretical groups thus succeeded in laying down the basis of a communication system that operated in a manner quite similar to that of the Church. Beside the existence of local leaders, a more or less regular interchange of messengers was stipulated, while all dignataries had to report their actions and income to the general assembly once or twice a year. Furthermore, both the tendency to appear as merchants and to perform their meetings in the framework of the developing towns reflected the greater scope for maneuvering that the anonymity of medieval towns offered to the sectarians. They also indicated the familiarity of sectarians with commerce and urban life into which they could more easily integrate while evading the searches and possible arrests by the Inquisition.

In addition to the regular meetings once or twice a year, preaching undoubtedly provided the driving force and the main channel of communication of all heresies. Even their critics acknowledged the heretical leaders' rhetorical skills, which did not always find a suitable equivalent in the ecclesiastical order.[25] As the Anonymous of Passau testified: "The fifth cause (of heresy) is the insufficient learning of some people who sometimes preach what is frivolous and sometimes what is false." Against the decline of rhetorical skills in the ecclesiastical order, he emphasized the heretics' devotion to the Holy Scriptures, which reached its climax among the Waldensians:

> Men and women, great and lesser, day and night, do not cease learning and teaching; the workman who labors all day teaches or learns at night. They pray little, on account of their studies. They teach and learn without books. They even teach in the houses of lepers. . . . When someone has been a

> student (of theirs for as little as) seven days, he seeks someone else to teach, as one curtain draws another. Whoever excuses himself, saying that he is not capable of learning, they say to him, "Learn but one word each day, and after a year you will know three hundred, and you will progress. . . . they have translated the Old and New Testaments into the vulgar tongue, and thus teach and learn them. I have seen and heard a certain unlearned, illiterate rustic, who could recite the Book of Job word for word, and many others, who knew the entire New Testament perfectly.[26]

The success achieved by these revolutionary methods was quite naturally opposed by the Church communicators, who shielded themselves behind the ancestral customs of the traditional society. Pope Innocent III opposed the association of translations with unauthorized preaching and "secret conventicles," stating that "the secret mysteries of the faith ought not. . . . to be explained to all men in all places. . . . for such is the depth of divine Scripture, that not only the simple and unlettered, but even the prudent and learned, are not fully qualified to try to understand it."[27] Against the authoritarian-elitist perspective of the Catholic Church, the sectarians offered a more populistic approach, one which answered the increasing needs of an urban population in a process of demographic growth and economic expansion. Bernard Gui reported in detail the efforts of the Waldensians to close the gap between the rigid teachings of the Church and the desire for learning of the common folk whom they approached in their own language:

> He (Peter Waldo) caused to be translated into the French tongue, for his use, the Gospels, and some other books of the Bible, and also some authoritative sayings of Saints Augustine, Jerome, Ambrose and Gregory, arranged under titles, which he and his followers called "sentences." They read these very often and hardly understood them, since they were quite unlettered but, infatuated with their own interpretation, they usurped the office of the Apostles, and presumed to preach the Gospel in the streets and public places. And the said Waldes or Waldo converted many people, both men and women, to a like presumption, and sent them out to preach as his disciples. Since these people were ignorant and illiterate, they, both men and women, ran about through the towns, and entered the houses. Preaching in public places and also in the churches, they, especially the men, spread many errors about them.[28]

Although he portrayed the basic errors and the fatuity of the Waldensians in their presumption to preach, Bernard Gui still acknowledged their significant success among their audiences, which they turned into propagandists for their own creed. Medieval sources provide additional data as to the Waldensian didactic system, which succeeded in spreading the message of Holy Scriptures in a clearer and simpler way, more suitable for the ordinary, unlettered men:

> They also give titles to Psalms: *Eructavit* (Psalm 44), they call the "Revenge Psalm;" *De profundis* (Psalm 129) they call "The Calling Psalm," and like-

> wise with others. They teach and learn at secret times and places, nor do they admit anyone who is not a believer of theirs. When they assemble in a place, they first say, "Beware, lest there be a curved stick among us," that is, lest there be a stranger present. They order their teaching to be concealed from the clergy, so that some of them speak by signs which no one knows but themselves, and thus they transform words which no one knows but themselves: they call a church a "stonehouse," clerics "scribes," the religious "Pharisees," and the same with many other things. They never answer directly.

The experience of heresy contributed to the elaboration of a distinctive code of behavior and a lexicon peculiar to the members of each sect, another reflection of the advanced communication stage achieved by the heretical groups. Armanno Pungilupo of Ferrara reported that special signs were also used at the gates of houses, so that sectarians could easily recognize their fellows, thus further strengthening the esprit de corps within the sects.[29]

Their persecution by the Church, therefore, did not bring about the annihilation of heresy, but rather the improvement of the communication channels among its followers. Even the major critics of heretics were forced to aknowledge the wide popularity they enjoyed in large sectors of contemporary society. According to the Passau Anonymous:

> In all the cities of Lombardy and the province of Provence, and in other kingdoms and lands, there are more schools of the heretics than of theologians, and they have more hearers; they debated publicly, and convoked the people to solemn disputations in fields and forums, and they preached in houses, nor was there anyone who dared to stop them, on account of the power and number of their sympathizers. I myself have frequently been present at the inquisition and examination of heretics, and there are calculated to be in the diocese of Passau forty churches that have been infected with heresy. And in the parish of Kemenaten alone there are ten schools of heretics, and the priest of this parish was killed by heretics, and no judgment (against them) followed.[30]

This despicable situation was confirmed by Bernard Gui, who mourned that "Thus, multiplied upon the earth, they dispersed themselves through that province, and through the neighboring regions and into Lombardy."[31] Despite the subjective approach intrinsic in the ecclesiastical sources whose tendency was to maximize the danger inherent in the very existence of heresy, they still substantiate the considerable spread of heresy, which, as such, appears as an admission of their own failure. The development of heresy in medieval Europe was somewhat further influenced by movements such as the Crusades, which facilitated the transmission of heretical ideas among groups spread far and wide, in Europe and Asia. The Second Crusade fulfilled this intermediary role between the Bogomils' dualistic philosophy, rooted in Eastern Europe and the Cathar centers in Lombardy and the Languedoc.[32] Besides, there was a constant exchange of information between different heretical centers with regard to all matters concerning dogma and ritual without this bringing about much significant cooperation between them. Bernard Gui also reported on the widespread custom among the Beguins and Beghards, in which the older members transmitted the sectarian traditions to the younger generation, thus ensuring the preservation of their creed.[33]

Research on the communication channels of medieval heresies hints at their success in assuring their survival despite their lack of legitimacy both in the religious and in the sociopolitical sphere. Survival, however, could hardly be considered a propaganda success. Referring to the development and, eventually, the failure of medieval heresy to become a mass phenomenon, Le Roy Ladurie claims that heresy was condemned to failure in the era that preceded print as "Luther was victorious because of Gutenberg."[34] This claim actually turns print from a *technical* factor, which had undoubtedly fostered the rapid transmission of the Reformation, into a *causal* factor, largely responsible for the emergence of the Reformation and, ultimately, for its success. It seems to us that the "discovery" of print was closely related to the socioeconomic and cultural climate of the mid-fifteenth century, which asked for the improvement of the media without turning the movable metal type of Johann Gensfleisch (Gutenberg) into the cause of this need. The approach to printing as a major factor in the success of nonconformist ideas also disregards the predominance of the spoken word in the Middle Ages, when most people were still illiterate and print could therefore play only a marginal role in the transmission of heresy. Furthermore, the degree of success achieved by medieval heresies, as that achieved by the ecclesiastical order itself, should be considered in the long term not only according to technical media, but mainly with regard to the conservative mental structures of medieval men. From a long-term perspective, one can come to the conclusion that forcing sectarians to develop clandestinely, the Catholic Church's compulsory policy—and not Gutenberg—appears to have been one of the major catalysts for the success of Luther later, when the sociopolitical process of integration, including printing, came into maturity.

12

Heresy and Propaganda: The Franciscan Crisis

The development of heresy was conditioned by the reception of its message, when its basic coalescence with the socioeconomic process fostered its spread despite the Church's opposition. This chapter deals with the failure of the Franciscan Order to achieve wide support after Pope John XXII (1316–1334) condemned some of the Franciscan tenets as "heretical." The Franciscans' failure becomes rather problematic in light of the high prestige formerly bestowed on the Order and the crucial role it accorded to the implementation of evangelical poverty in actual practice. Moreover, the rhetorical skills of the friars had actually turned them into most successful communicators whose success was not limited to Christian society but reached distant areas in which they spread the evangelical message. The "Franciscan Crisis" during the pontificate of John XXII also reflects the ambiguous line dividing orthodoxy and heresy, the friars' readiness to submit to the papal dictates turning into the touchstone regarding the orthodoxy or the heresy attributed to the Order. Yet the minister general Michael of Cesena, William Ockham, and Bonagratia of Bergamo, not satisfied with their eloquent defense of the Franciscan creed, put forward a fresh interpretation of the papal plenitude of power that advanced European society toward a new era. Quite paradoxically, the individualistic approach of the Franciscan leaders seems to have been one major cause of their propaganda failure, as most people and, foremost, the members of the Church were not yet ready to change ancestral schemes of thought. Despite the fact that the Franciscan leaders used well-known concepts, the new meaning they were ascribed made it rather difficult to achieve a favorable public opinion among the clergy, and even within the ranks of the Order itself. This propaganda failure hints at the many difficulties that plagued the way of minority groups in spreading their message in the traditional frameworks of medieval society in general, and of the Church establishment in particular.

From its very beginnings, the Franciscan Order gave concrete expression to two fundamental Christian ideals, namely, the mystic tradition, which led to direct contact with God, and absolute poverty, as an essential element in the imitation of Christ. Mysticism and poverty had been integral to the Christian faith and received papal sanction within the monastic tradition, which nurtured the Franciscan Order as well. The biography of Francis of Assisi (1182–1226) reflects these basic features, which gave the Franciscan Order its *raison d'être*. Born to a rich cloth merchant of Assisi, Francis underwent a sudden conversion after an illness, renounced the pleasant life reserved for young Tuscan men of his rank, gave up his possessions, renounced his inheritance, and lived a life of prayer and poverty on the outskirts of his native town. As time went by, he was blessed with "divine strokes" or visitations, in which God revealed to him His will. His immediate contact with God was further confirmed by the stigmata or wound prints of the passion of Christ, which were believed to have appeared on his hands, feet, and side a short time before he died. Besides, Francis's celebrated marriage to "Lady Poverty" indicated his unquestioning commitment to mendicancy and poverty, both further emphasized later in his testament.

Francis and his first eleven companions brought about a new kind of monastic life which, though embracing poverty and celibacy, did not encourage an escape from the world. On the contrary, having no settled home, "the whole world became their monastery." They shared the shelters of the dearest creatures of God: the poor, the sick, and the lepers, the care for whom they regarded as the most suitable expression of their commitment to God. They worked on the harvest or whatever casual labor came to hand, approaching all life with joy, an attitude that in time brought them the nickname of "God's Jesters." They were known as *Fratres Minores* or Minorites, a name that indicated the humility of their life; they wore dark gray habits and went barefoot. They found all nature friendly and beautiful, as suggested by Francis's sermon to his "little sisters, the birds." Brother Leo, the "little sheep of God," and Francis' close friend and secretary, faithfully described the Franciscans' way of life at the very beginning:

> The most holy father was unwilling that his friars should be desirous of knowledge and books, but he willed and preached to them that they should desire to be founded on holy humility, and to imitate pure simplicity, holy prayer, and our Lady Poverty, on which the saints and first friars did build. And this, he used to say, was the only safe way to one's own salvation and the edification of others, since Christ, to whose imitation we are called, showed and taught us this alone by word and example alike.[1]

Humility, simplicity, poverty, and prayer thus became the four basic elements of the Franciscan creed, which brought the friars widespread support in thirteenth-century Christendom (see plate 29).

In a letter from Italy in 1216, Jacques de Vitry gave testimony to the unprecedented impact of the friars:

> I found but one consolation in those parts, and that was that many people of both sexes, both rich and poor, leaving all for Christ, have fled from the

PLATE 29. Bonaventura Berlinghieri. Altarpiece in the Franciscan Church in Pescia, 1235. The figure of St. Francis that fills the center of the panel is strange and forbidding, a sombre ascetic with set features whose proportions transcend any human scale. The position of the hands with stigmata and book corresponds to the Byzantine pictures of Christ in Benediction, and the small scale of the legendary scenes on the sides intensifies the isolation and the unworldly sublimity of the main figure.

> world and are now called Friars Minor. These people are held in great esteem by the pope and cardinals. They occupy themselves in no worldly pursuits, but work day by day with the utmost zeal and care to draw souls away from the vanities of this wicked world. And already, by the grace of God, they have had great success and have saved many. . . . They live after the manner of the primitive Church, of which it is written: "The multitude of them that believed were of one heart and soul." By day they go into the towns and villages in order to win others by setting them an example. At night they retire to some hermitage or lonely place and give themselves up to meditation. . . . I believe that, to the disgrace of the bishops who are like dumb dogs that will not bark, the Lord intends, before the end of the world, to save many souls by means of these poor and humble men.[2]

Being himself bishop of Acre, later to be appointed cardinal, the testimony of Jacques de Vitry indicates the awareness of the secular clergy as to the important social role of the friars. Establishing themselves in the growing towns, they filled the vacuum created by the incompetence of many prelates; they became most active in all fields of sociocultural and religious life, to which they gave a new impetus. The papal authorization to preach at any time in their own churches and on their journeys placed them in the very center of medieval communication. In contrast to the rather dreary and monotonous Sunday services that the parish churches offered, the friars provided something that was almost an entertainment. Their preaching was full of racy stories and the personal reminiscences of men who had traveled, offering an attraction that both the country folk and the town people found irresistible. Here was no monologue, but a display of pulpit fireworks, as they approached medieval audiences in their own language with humor and an endless number of anecdotes based on daily experiences. They also contributed fine manuals of devotion written in the vernacular and notable religious poetry, such as Francis's *Canticle of the Sun*, Thomas of Celano's *Dies Irae*, and Jacopone da Todi's *Stabat Mater Dolorosa*. In the field of education as well, the Franciscans provided one of the most vigorous streams of medieval society: they established a school for young friars in each house, and a school of liberal arts and theology in each custody. Their leading role in medieval universities is indicated by the long list of prominent scholars from their ranks, such as Adam Marsh, Bonaventure, Alexander of Hales, Nicholas of Lyra, Roger Bacon, Archbishop John Peckham, Duns Scotus, and William Ockham.

The rapid diffusion of the Franciscan creed in thirteenth-century England is symptomatic of their spreading throughout Christendom. Led by brother Agnellus of Pisa, nine friars came to England in 1224, four of whom were clerks and five laymen. Their friaries spread quickly, and by 1230 they had already established houses at Canterbury, London, Oxford, Northampton, Norwich, Worcester, Hereford, Salisbury, Nottingham, King's Lynn, Leicester, Lincoln, Cambridge, Stanford, Bristol, and Gloucester. In the next ten years, twenty more Franciscan friaries were founded.[3] Nevertheless, sent to foreign countries whose language they were not always acquainted with, the

Franciscans' commitment was often put to the test. Brother Jourdain de Giano wrote a vivid report of his own experiences in Germany as a member of the first Franciscan mission there (1219). Led by brother Jean de Penna, about sixty friars reached Germany without knowing the language. When asked if they needed to be housed and fed, they answered *Ja* and were welcomed. As they noted the good treatment they received in this way, they decided to answer in a similar manner to all questions. It happened that they were also asked if they had come from the ranks of those heretics who had already perverted Italy with their errors; the friars vehemently answered with the only German word they knew: *Ja*. Their affirmation involved them in a most bizarre situation, as many were arrested and ridiculed in front of the crowd. Finally, they escaped from Germany, which from then on was considered a most cruel country, one that became the target only for those brothers looking for martyrdom.[4]

Troubles of this kind reflect the communication problems inherent in the spread of the Franciscan message beyond the intellectual elite, when the friars appealed to people dispersed over large areas who spoke different languages and held different customs. Yet the spread of friaries throughout Christendom hints at the Franciscans' ability to overcome most obstacles while laying the foundation of an international Order. In 1213 the first Franciscan friary was established in Italy, on the mountain top of La Verna, and from there the Order spread out to Bologna, Rome, Florence, Milan, Trento, Vicenza, and Naples during the next decade. The General Chapter of 1217 envisaged two provinces in France where they were most active in areas touched by heresy, such as Mirepoix, Arles, Aix-en-Provence, Montpellier, and Perigueux. By the end of the century, they had established about 200 houses in almost every important town throughout the kingdom. In 1221, a new Franciscan mission was sent to Germany, and thirty-one friars attended the first provincial chapter later that year. After a preliminary mission between 1209 and 1210, more than 100 friars reached the Iberian peninsula between 1217 and 1219 and established the foundations of the Order there. The Franciscans also invested much of their resources in missionary efforts and encouraged the study of oriental languages, especially Greek, Hebrew, and Arabic. Franciscan missionaries worked in Morocco, Libya, Tunis, and Algiers and, from 1220 onward, built convents and hospitals for pilgrims in the Holy Land and Egypt.[5] Franciscan missionaries were also active in Lithuania, Poland, Prussia, Armenia, Bulgaria, and around the Black and the Caspian Seas. Through trade routes, they reached Persia, India, Sumatra, Java, Borneo, and China, and, by the special mandate of Pope Innocent IV, spread the Christian faith among the Tartars as well. In 1291, two Franciscans sailed around Africa with Genoese merchants, and by the fifteenth century they were preaching in Cape Verde, Guinea, and the Congo. This brief survey indicates the leading communication role of the Franciscans both in Christendom and out of it, while they became the most successful communicators of the Christian message. This success was undoubtedly influenced by the socioeconomic background of most friars who, coming themselves from the ranks of the urban population, were able to supply the needs of young societies in a process of transition. Moreover, entrusted with the apostolic blessing, they provided the frustrated and the disoriented with a prodigious social task, which turned the Franciscans into the medieval ancestors of modern social workers and rewarded them with a suitable place within the safe wings of Catholic orthodoxy. They represented, in brief, the vitality of the monastic movement, which, whether through the Military Orders in the Levant or through the Franciscans, adapted its schemes to the changing needs of sociopolitical life.

The crucial socioreligious role played by the friars dictated the supporting policy of the papacy, as the Franciscans appeared as the most suitable answer that the orthodox camp could offer in the face of heresy. The papal support was also intended to stress the gap between the Franciscans and other groups, which, by overstressing generally accepted principles, were eventually declared by the Church as heretical. The Franciscans' commitment to evangelical poverty and to mysticism, indeed, made the close papal contact with the Order quite imperative, as both principles had been the touchstone that brought sectarians out of the ranks of orthodoxy. After some hesitation, in 1209 Pope Innocent III gave Francis and his eleven companions a verbal approval of their Rule, a first stage followed by the formal sanction of the Franciscan Order and its Rule by Pope Honorius III in 1223. The close cooperation between the papacy and the friars was further strengthened by the appointment of a cardinal-protector who represented the Franciscans in the curia, thus implementing their immediate subordination to the papacy.[6]

Thirteenth-century popes, however, were forced to invest much energy in endless efforts to restore peace among the restless brothers, namely, between those who contended most strictly for Francis's teachings, later known as *Spirituals*, and those more willing to accept the institutionalization process, the *Conventuals*.[7] Pope Gregory IX, himself a former protector of the Order, made the first of a long list of attempts to reconcile the quarreling parties while exerting the full force of papal authority on them both. His bull, *Quo elongati* (September 28, 1230), restricted Francis's command to live *sine proprio*, without goods of their own, to formal ownership, which nevertheless enabled the brothers to enjoy the use of elementary goods, such as furniture or clothing.[8] Pope Nicholas III went a stage further and defined in detail the concept of use in order to prevent any contravention of the Franciscan primary exigency of absolute poverty. The decretal *Exiit qui seminat* (August 14, 1279) drew a distinction between *usus juris*, use by right of ownership, and *usus facti*, the actual use of goods. The pope formally declared that the friars had no rights to own the goods they used, but only "the actual use" of food, clothes, books, and all material things they needed for their living and for the pursuit of wisdom. This *usus facti* was restrained to the most elementary things and would not justify any infringement of the Rule. The pope formally took upon himself the ownership of all Franciscan property, whose distribution and control he delegated to the ministers and the custodians of the Order. They were still encouraged to take into consideration the particular needs of individual friars, as long as this would not infringe the basic obligation of all

Franciscans to absolute poverty. Besides, the pope forbade, under threat of excommunication, the modification or cancellation of any clause established in the decretal *Exiit*, thus establishing, in fact, the immutability of the Rule forever.[9]

The endless conflicts between opposing frictions thus called for a continuous papal intervention; this, nevertheless, failed to restore peace within the Franciscan ranks. The pontificate of Pope John XXII could be considered, in this regard, as another, more radical attempt on behalf of the papacy to put an end to the internal conflicts and to enforce ecclesiastical discipline on the Franciscan Order as a whole. Being originally from southern France, the pope was well acquainted with the spread of heresy and was more aware than his predecessors of the ill-defined line between the meritorious Franciscan ideals and the commitment of heretics to the same ideas, which had brought about their expulsion from orthodoxy.[10] On the other hand, Pope John was less familiar than his predecessors with the Franciscan creed and, consequently, was more critical of their fanatical adherence to absolute poverty. His energetic handling of the Franciscan crisis between 1316 and 1318 indicates the major importance he ascribed to the reform of the Franciscan Order as a part of his efforts to bring about the implementation of the papal monarchy in the ecclesiastical order as a whole.[11] Yet the question still remains, to what degree John's approach was concomitant with the Franciscan spirit, or whether the radical tenets of the Spirituals had indirectly made the whole Order suspected of heresy. The gradual escalation of the papal policy seems to hint at the second alternative, while reverting his pontificate into a turning point in the relations between the heir of St. Peter and the heirs of St. Francis.

The appeals of both Spirituals and Conventuals for immediate papal intervention at the beginning of his pontificate provided John with the chance of reorganizing the Franciscan Order in a manner that would restore peace and discipline among the restless friars. The papal policy at this stage suggests that the main concern of John XXII did not focus on the Franciscans' divergent interpretations of their commitment to absolute poverty but, rather, on the readiness of all friars, Spirituals and Conventuals alike, to submit to papal authority. The pope expressed this creed quite plainly in his appeal to the Provençal Spirituals when he claimed that "poverty is important, but unity is more, and most is obedience, particularly if observed meticulously." He accordingly ordered the Spirituals to submit to their masters and to acknowledge their regulations in respect of food and clothing.[12] John XXII's attitude clearly indicates the papacy's scale of priorities at the time. Without dealing with the value of evangelical poverty and the implementation of the Rule, the crystallization of the papal monarchy encouraged the popes to focus attention on the readiness of ecclesiastical sectors to submit to the papal directives or, in John's words, their degree of discipline. One can argue that the pope's approach was motivated by the challenge presented by the diffusion of heresy on the one hand and the endless conflicts in the Franciscan Order on the other. Yet John's policy actually reversed the Franciscan freedom and voluntary spirit and channeled them into the more rigid patterns of ecclesiastical discipline.

The papal policy with regard to the Provençal Spirituals was well-received outside the Franciscan Order, as it relieved the tension that had developed in wide ecclesiastical sectors against the Franciscans in general and the Spirituals in particular. Some sources pointed out the hypocrisy inherent in the fictitious distinction between ownership and use, and "many came to the conclusion that the Franciscan way of life was not one of repentance, nor was it a holy way of life, but a pretentious one, devoid of reason."[13] John of Paris severely criticized the presumption of the Spirituals of Narbonne, who looked upon the ecclesiastical establishment as the Synagogue of Satan.[14] English chroniclers criticized them more severely still, referring to the Spirituals as "those that go the way of the Nicolaitans, whom God hates." They poured scorn on the Franciscans' preference of St. Francis to Christ Himself as a model of perfection.[15] Reactions from Germany and Italy were similar and, denying all values of the Franciscan ideals, pointed out the hypocrisy and pretentiousness behind them. Guido Cavalcanti declared in one of his *Canzone*: "Poverty! What a cloak you are for anger, jealousy, and every kind of dissension!"[16] A pamphlet issued about this time referred to the Spirituals as "not so much superstitious as pernicious, pestiferous, apostate and heretical, the breeders of new heresies and imitators of old ones." According to the anonymous author, the Spirituals refused to obey the pope whenever he interpreted the Rule contrary to their opinion, they wore a different kind of habit and they submitted themselves to all kinds of unnecessary austerities. And, worse still, they continued to regard the "heretical" Olivi as their leader.[17]

These reactions hint at the prevailing attitudes among the clergy toward radical groups, which, in one way or another, had become a threat to their own status and income in the traditional framework. The basic rivalry between the Franciscans and the secular clergy—as they both appealed to the same audiences, and the success of the friars resulted in many cases in the failure of prelates,—encouraged the readiness of the clergy to sustain the papal coercive policy against the Order, which provided a channel for old feuds. Besides, the tendency to identify the Spirituals with the Nicolaitans (Rev. 2:14), a Gnostic sect that had come to an end by about AD 200, suggests anew the inability of medieval commentators to face a new reality and their tendency to regress to the patterns of the past. Furthermore, the papal criticism of the Spirituals' fanatical adherence to absolute poverty created a defense mechanism for prelates and monks alike, as they both had to deal with the paradox inherent in the gap between the evangelical dictates and the Church's practices. The support of the papal policy as voiced by contemporary chroniclers, however, does not reveal the climate of opinion in differing social strata, nor could it be regarded as an expression of new attitudes of the laity toward the friars, whose services they still enjoyed. It reflects the traditional approach of privileged classes who did not actively participate in the process of change and, consequently, opposed also the entente between the Franciscans and the urban population, whose needs the friars were more capable to fulfill in daily practice.[18] This conflict of opinion left its mark on the ideological sphere as well, while the Franciscans provided the bourgeoisie with the ideological justification they were looking for, by replacing the ideal image of St. Louis by St. Francis, the son of a rich merchant of Assisi. This state of affairs suggests the threat inherent in the Franciscan creed from the perspective of the papacy, whose ideological foundations were also based on the same traditional patterns of the anointed monarchy and the Crusade, which had brought about the beatification of Louis IX. By elevating the image of St. Francis as the most perfect implementation of the evangelical way of life, the Franciscans also elevated the potentiality inherent in urban societies looking for socioreligious legitimization. Although the basic tension between the Franciscan creed and the institutionalized papal monarchy was already latent at the very beginnings of the Order, it became more acute at the start of the fourteenth century in light of the socioeconomic process of change, which was undermining the very foundations of the traditional society. In this regard, the drastic measures adopted by John XXII seem justified to some degree by the greater danger represented by the Franciscan voluntary spirit at a time when the traditional society was collapsing.

The prevailing support of ecclesiastical sectors on the papal coercive policy encouraged John XXII to deal even more severely with the Spirituals' challenge. On April 22, 1317, three cardinals wrote, on the pope's behalf, to the Spirituals of Narbonne and Béziers, demanding their immediate submission to the minister general. Only three days later, municipality officials issued sixty-five Spirituals with a papal summons to come to Avignon for further interrogation. Angelo da Clareno, possibly an eyewitness, reported in his *History of the Seven Tribulations of the Order of St. Francis* the grim, clipped dialogue between the pope and the Spirituals, when John's sardonic tongue actually shifted to intimidation.[19] The Spirituals, however, declared anew their complete allegiance to the Rule, which they identified with the Gospels and denied the pope's right to intervene in the affairs of the Order. Under inquisitorial pressure, forty Spirituals submitted to the pope while agreeing to obey the minister general and to accept his tenets with regard to food and clothing. Those who remained refractory were handed over to the Inquisitors, an act that led to twenty additional retractions. One of the five Spirituals who stood by their basic tenets was condemned to life imprisonment, and the others were burned at the stake in Marseille (May 7, 1318).[20]

Although most friars were not prepared to praise the papal radical measures, they still refrained at this stage from opposing John's policy against the Spirituals. The General Chapter held at Marseille in 1319, even condemned the writings of Olivi, a step aiming to come to terms with John XXII.[21] But reactions outside the Franciscan Order were quite different. According to the testimony of Bernard Gui, the Beguins of Provence recorded the Spirituals' names as those of martyrs, and kept the anniversary of their burning as a feast of the Church.[22] If this report is indeed accurate, it demonstrates anew the ill-defined line between the Franciscans as a whole and the heretical movements at the time, while the Spirituals condemned to death enlarged the ranks of those regarded by sectarians as martyrs of the true faith. The coercive policy of the papacy thus achieved the very opposite of its declared goals, as it closed the ranks among all those persecuted by the establishment and encouraged their identification as the chosen people of God. On the other hand, the absence of any criticism of the events at Marseille from the ranks of orthodoxy seems to point to a "conspiracy of silence" surrounding John's anti-Spiritual policy, at least among contemporary chroniclers. A number of factors helped to make the pope's policy quite acceptable at this stage: there was no political body at the time that could profit from the Spiritual struggle, nor was any factor within the ecclesiastical order willing to support them. Moreover, the endless tension between the Franciscans and other Orders prevented wide monastic circles from supporting the Spirituals, despite the fact that the basic obligation to evangelical poverty was at the very heart of all Monastic Orders, especially of those sustained by the Reformer papacy. Besides, the papal policy against the Spirituals followed the accepted patterns of apostolic prerogatives, the pope being entrusted with the surveillance of orthodoxy throughout Christendom, especially within the Church. The events at Marseille, however, brought about only momentary relief. The controversy over evangelical poverty was renewed in 1321, when it was no longer limited to the fringe groups of the Order, but included its leaders as well; it thus involved John XXII in an ideological conflict in which there was no room for compromise.

Following the profession of faith of a Beguine of Provence that Christ and the Apostles owned no property either individually or collectively, Berenger of Talon, a Franciscan lecturer at the University of Carcassonne, was asked for his opinion by the Dominican Inquisitor, John de Belna. To everyone's surprise, Berenger supported the Beguine, basing himself on the decretal *Exiit* of Nicholas III, which had conferred apostolic legitimacy on the principle of total poverty. Despite inquisitorial pressure, he refused to retract and, instead, appealed to the papal curia in Avignon (1321).[23] His appeal provided Pope John XXII with another symptom of the problematic nature of the Franciscan allegiance to poverty and its further implications in the field of heresy. The pope, however, refrained from immediate action but opted for a full investigation of the Franciscan creed. The bull *Quia nonnumquam* (March 26, 1322) raised the question of the Franciscan concept of poverty, despite the prohibition of the decretal *Exiit* to reopen the issue. Aware of the unprecedented nature of his move, John reasserted his right to alter the edicts of his predecessors in his opening remarks: "Because sometimes, what conjecture believed would be of profit, subsequent experience has shown to be harmful; it ought not to be thought reprehensible, if the legislator takes steps to revoke canons issued by himself or his predecessors, if he sees them to be harmful rather than profitable."[24] From John's perspective, the endless crisis in the Franciscan Order justified his intervention, despite the prohibition laid down by Nicholas III in the decretal *Exiit*. Without questioning the Franciscan creed, John's attention focused, first and foremost, on his own rights to be freed from papal precedents, against which he raised the claim of the common good. The complete freedom of action the pope claimed for himself represents one of the most advanced stages toward papal absolutism in the Middle

Ages, a state of affairs against which William Ockham reacted vehemently. From the perspective of most friars, however, the implications of the pope's declarations focused on his attempts to demolish with one stroke the shelter behind which they had been protected for generations: the decretal *Exiit*. Furthermore, John XXII's disregard of Nicholas III's edict was regarded as a contravention of the Rule and, as such, one that could undermine the unconditional adherence of all friars to absolute poverty.

The dangers implied in the papal policy encouraged both Spirituals and Conventuals to close their ranks to counteract the designs of their former ally, the pope. Although they differed over the ways to implement the ideal of poverty, all friars believed in the complete identification of their poverty with that of Christ and the Apostles, as recorded in the Gospels. The papal bull relegated old feuds between Spirituals and Conventuals to a secondary role, as it appeared as a threat to the very foundations of the Franciscan creed. The progressive erosion of the original Franciscan faith became the main topic on the agenda of the Chapter held in Perugia in May, 1322. The Perugia Chapter did not conceal its absolute belief in total poverty, arguing that in pointing the way to perfection, Christ and the Apostles owned no goods, either individually or collectively. The ideal of poverty, they stressed, was not peculiar to the Franciscan creed but, rather, was essential to the Church's teachings. This had been faithfully explained in the decretal *Exiit*, which had been confirmed by Clement V and even by John XXII himself in his bull, *Quorundam exigit* of October, 1317, a mere five years earlier. In other words, against the pope's assertions of his own plenitude of power, the Franciscans reiterated the immutability of the Franciscan doctrine of the poverty of Christ on which they were not ready to compromise.[25] An encyclical in this spirit was addressed to all the faithful in two versions: the shorter, expected to reach wide sectors of the clergy, and the more extensive, intended for the intellectual elite. More than 100 copies of each version were distributed through Christendom before the end of the year. This publicity hints at the Franciscans' search of favorable public opinion within the ecclesiastical order, which would strengthen their position in counteracting the papal offensive. The appeal to public opinion corroborates the gap between the Franciscans and the papal perspectives: against the traditional position of the pope, supported by the concept of the papal plenitude of power that could not be questioned, the friars posed the opinion of wide sectors in the Church which, they hoped, would force the pope to change his policy.

The Franciscans' refusal to submit to John XXII or to allow his reconsideration of their adherence to poverty actually transferred the conflict between the friars and the pope to the precarious path separating orthodoxy and heresy. The decision of the Perugia Chapter that deprived John XXII of the right to reconsider the decretal *Exiit* repealed the freedom of action of each individual pope, thus considerably restraining the papal prerogatives. On the other hand, it implied also the Franciscan recognition of *papal infallibility*, as former papal decisions could not be changed by anyone or for any reason whatsoever. The concept of papal infallibility conferred on the papal monarchy the prerogative Christ had promised to the universal Church (Luke 22:33). However, the support of papal infallibility was not in line with the papal monarchy but, rather, served as a means to give perpetual force to the decretal *Exiit*, while ultimately achieving infallibility for the Franciscan ideals. The ideological gains of papal infallibility in the field of doctrine thus appeared to bar the further development of the papal monarchy since each pope was compelled to submit to the precedents laid down by his predecessors. The later attempts to brand John XXII with the stigma of heresy illustrate the problematic nature of the infallibility principle. Though it conferred the virtue of eternal justice on the decisions of the papacy, the Franciscans still reserved to themselves the right to determine the orthodoxy of each individual pope. This eventually justified Ockham's later claim to bring about the immediate deposition of John XXII since the pope was suspected of heresy.

The Franciscan support of papal infallibility thus resulted from the belief that the Rule was a completion of the Gospels and, as such, not subject to question, as well as from the ambivalent relations between the Franciscan Order and the papacy.[26] The attitude of the Perugia Chapter further indicates the challenge that faced Pope John XXII: the Franciscans' claims on the immutability of the Rule and their attempt to manipulate public opinion among the clergy against the papal policy actually undermined both the apostolic authority in matters of dogma and the pope's prerogatives within the Church. The dangers implied in the concept of papal infallibility were clear enough to John XXII so as to lead him to renounce this doubtful prerogative.[27] By renouncing infallibility, the pope gained complete legal sovereignty and, as claimed by contemporary Canonists, his decisions won "the force of law."[28] Throughout his pontificate, John implemented these concepts in actual practice while he applied full authority to bring about the immediate suppression of any act of insubordination, even from the side of those who could be regarded as potential allies in his struggle against the Franciscans.[29]

John XXII reacted to the Perugia declarations with a series of bulls that delivered a frontal attack on the Franciscan "presumptuousness" and stated anew the papal plenitude of power. The bull, *Ad conditorem canonum*, which was affixed to the doors of the Avignon Cathedral on December 8, 1322, opened with a dogmatic declaration that "To the legislator there is no doubt that it pertains to provide new laws when he sees that those of himself or his predecessors are harmful rather than helpful." The pope further stressed that Christian perfection did not focus on the arbitrary rejection of all ownership but, rather, depended upon Christian charity. He severely criticized the fictitious distinction between use and ownership that had enabled the Franciscans to maintain their allegiance to poverty on purely technical grounds. Consequently, he revoked the former agreements by which his predecessors had assumed the ownership of the Franciscan property. These "stratagems" had imposed an excessively heavy burden on the papal curia, which derived no benefit from such a legalistic arrangement. Accordingly, John XXII renounced "such a fruitless and hazardous ownership" and discharged the legates who had managed the Franciscan property on behalf of the papal curia for generations.[30] The bull *Cum inter nonnullos* (November, 1323) was the pope's reply to the pronouncement of the Perugia Chapter regarding evangelical absolute poverty, whose very accuracy the pope now questioned. John XXII argued that such pronouncements actually contradicted the spirit of Holy Scriptures, and even suspected of heresy the friars' insistence on the total poverty of Christ and the Apostles.[31] This was the first time in the whole controversy that the pope had made a fully dogmatic assertion and, as such, it appeared as a grievous blow to the Franciscan creed. Within one single year, by apostolic sanction, the Franciscans thus became owners of vast property, and, in addition, a charge of heresy was leveled against them, as the pope questioned the very orthodoxy of their mystic marriage with "Lady Poverty."

The extremism inherent in the apostolic policy brought about the first rifts in the large support that the pope had enjoyed hitherto. John XXII's disregard for accepted beliefs and his indirect rejection of the ideal of poverty were regarded as too revolutionary, even if they had come from the acknowledged representative of God on earth. Some chroniclers expressed a fear that the pope's decision, if put into practice, would jeopardize the salvation of souls.[32] Dominican sources, however, took a different view and exploited the timing to expose the hypocrisy of their hated rivals: a contemporary Dominican etching depicted Christ with a moneybag, counting dinars on the cross, thus making a mockery of the Franciscan teachings.[33]

The papal legislation faced all friars, Spirituals and Conventuals alike, with the difficult alternatives of whether to submit to the papal dictates while abandoning their own creed, or to defend their principles and way of life against the official representative of Catholic orthodoxy. This course of events was to test the balance of power between the papal monarchy and the Franciscan Order, which, by the force of circumstances, was gradually suspected of heresy. The letter of William Ockham to the Franciscan Chapter of Assisi (1334) reflected the view of the minority, ready to fight to the end in the defense of the Franciscan creed. The English scholar minced no words in his condemnation of John XXII's pronouncements on evangelical poverty, which he described as "heretical, erroneous, ignorant, ridiculous, absurd, insane and, (as such), defaming the orthodox faith, the good traditions, and natural reason."[34] The Holy Scriptures, Canon Law, and ecclesiastical traditions supplied him with a solid dogmatic basis to refute the pope's views and prove the heresy of his premises. According to Ockham, Christ's call "If thou wilt be perfect, go and sell that thou hast, and give to the poor" (Matt. 19:21) pointed the way to perfection,[35] which led to the complete renunciation of all property, both individually and collectively. The transfer of all goods to the poor confirmed the rightness of Christ's total renunciation on property.[36] By using only essential goods and contributing all the property He was given to the poor, Christ gave His personal example and taught His Disciples the commitment to absolute poverty.[37] The Apostles also followed Christ's example while relinquishing all property and preaching the renunciation of unnecessary pomp, contenting themselves with the most essential things for their subsistence.[38] Their adherence to absolute poverty allowed the return to the pure way of life that anteceded sin, when the forefathers of humanity lived in a state of innocence. Ownership thus resulted from sin and followed the expulsion from Eden. It contradicts both Natural Law and the divine will, as expressed in Genesis, and could only jeopardize the salvation of souls.[39]

Beyond his general defense of the Franciscan approach to absolute poverty, Ockham systematically exposed every fault and flaw in the pope's stance. He asserted anew the Franciscan claim of having received apostolic sanction to their renunciation of all property while they limited themselves to the minimal use of goods necessary for their subsistence.[40] John's claims on evangelical poverty were in direct opposition to those of Nicholas III, Clement V, and even his own, as set forth in the bull *Quorundam exigit*, in which he had praised the decisions of his predecessors.[41] Still, Ockham refrained from a maximalist concept of poverty and did not conceal his reluctance to accept the Spirituals' tenets. Although the pursuit of perfection had dictated the Franciscan adherence to absolute poverty, this in itself could not guarantee the achievement of perfection. The Old Testament depicted the existence of kings who, though enjoying wordly goods, had still committed themselves to God's service and achieved perfection.[42] If used for charity and other meritorious acts, property could foster the way to salvation, though in actual practice it often harmed such a chance.[43] By ascribing to the friars the identification of poverty with perfection, Pope John had merely given further proof of his complete ignorance of the Franciscan creed.[44] Neither poverty nor riches could guarantee the attaining of perfection but only the very intention that stood behind them.[45]

Even if one takes into consideration Ockham's relatively moderate stance—his unwillingness to equate poverty with perfection, and his restriction of the duty of poverty to monks alone[46]—there was still a wide gap between his point of view and that of the institutionalized papal monarchy. Ockham's attitudes to evangelical poverty, moreover, reflect the questions and soul-searching, which spread beyond the Franciscan Order and were also common to some prelates. Gilles of Rome, archbishop of Bourges and very close to the papal curia, though recognizing the pope's rights to ownership over all property, saw fit to suggest that the pope should voluntarily renounce the use of this property in order to devote himself to his spiritual tasks.[47] Reasonings of this kind indicate the tension inherent in the ideal of poverty. The possession of property by the Church was a fact and had played a significant role in the consolidation of the papal monarchy, but it had not yet been legitimized in the framework of Catholic dogma, nor was it justified by large sectors of the clergy itself. The conflict between John XXII and the Franciscans, however, did not make things easier, but rather sharpened the problematic aspects inherent in the ideal of absolute poverty and its sequels in the field of dogma. In those areas where the Spiritual creed was deeply rooted, especially in France and Italy, the prevailing unrest in the Franciscan Order found an outlet in mysticism. In her confession before the Carcassonne Inquisition in 1326, Na Prous Boneta declared that John XXII was actually a reincarnation of Caiphas, Simon Magus, and Herod who, on account of their numerous moral shortcomings, were dismissed from their holy offices. The Holy Spirit had revealed to her God's intention to replace John by a Franciscan pope, in order to revert the actual deplorable situation.[48] Declarations of this kind underlined the mystical trend behind the Franciscan ideals and emphasized anew the tenuous dividing line between the friars' orthodoxy and the heresy ascribed to sectarian movements at the time. To some degree, they could also justify the escalation in the papal policy against the Franciscan leaders, who, at this stage, showed no significant inclination to cooperate further with John XXII.

In 1327, the minister general Michael of Cesena was summoned to Avignon in an attempt to weaken his leadership and prevent any cooperation with elements hostile to the papal curia, particularly, the claimant to the Empire, Lewis the Bavarian.[49] Michael of Cesena came to Avignon, where he met Bonagratia of Bergamo and William Ockham, who had also been summoned to the papal curia in order to defend some thesis he had delivered at Oxford.[50] Their common resolution to escape from the papal curia in May, 1328, eventually expelled all three from the safe wings of orthodoxy, as they openly questioned the authority of its uncontested leader, Pope John XXII.[51] The bull *Quia vir reprobus* (November 16, 1329) asserted at length the papal refutations of the Michaelists' tenets. At this stage, the confrontation between the Franciscan dissenters and Pope John XXII no longer centered upon different concepts of poverty, but upon the more fundamental issues of the papal dogma, which involved a reexamination of the papacy and the relations between the pope and his predecessors on the Holy See. The Spirituals' challenge was obliterated since no one within Cesena's group sympathized with their views nor was suspected of doing so. Michael of Cesena himself had assisted Pope John in the suppression of the Spirituals in Provence between 1317 and 1318 and had forced them to recant. Bonagratia of Bergamo had vehemently expressed the Conventuals' position in the Council of Vienne, while Ockham came from England, a Franciscan province not touched by Spiritual influence, nor could this be detected in his many writings on the issue of poverty. The Michaelists, however, saw themselves as the remnant of the true Franciscan Order facing an heretical pope who harmed not only the Rule, but the very foundations of Christian orthodoxy as well.

The escape from Avignon ultimately led the dissenter Franciscans to ally themselves with Lewis the Bavarian. It is said that at their first meeting, Ockham declared: "O Emperor, protect me with your sword and I will protect you with my pen!" The further alliance between Lewis and the Michaelists provided the dissenters with the political backing they were seeking, while the contendent for the Empire benefitted from a firm ideological platform. Their rebellion against the pope, however, caused Cesena and his companions to be relegated from the Order, as most Franciscans opted for cooperation with the pope in order to ensure their survival within the ranks of orthodoxy. At the Perpignan Chapter of 1331, Gerald Odonis was elected minister general in place of Cesena, who, in the meantime, had been excommunicated. In retrospect, the dissenting Franciscans seem a voice crying in the wilderness, the voice of a group of extremists excommunicated by the pope and expelled from their own Order, fighting desperately to survive. Ultimately they failed because they did not succeed in gaining wide support for their cause within the ecclesiastical order and among wide social strata. According to papal testimony, however, there were still in Italy, especially in the March of Ancona, members of the nobility including the bishop of Camerino, who followed the Michaelists' tenets for long periods and maintained that John XXII and his successors were actually heretics.[52] But sporadic support could not change the balance of power nor bring about the legitimization of the Michaelists' tenets in Christendom. On the contrary, early in 1348, Ockham, the last surviving member among the dissenters, handed over the Franciscan seal to Pope Clement VI and asked for apostolic pardon and readmission to the Order. Although the pope was ready to give Ockham absolution under certain circumstances, we lack satisfactory data as to whether this reconciliation ever took place or whether Ockham died earlier, during the Black Death (1349). His death or submission actually brought an end to the most formidable rebellion against the papacy from within the Franciscan Order, one which lasted for twenty-one consecutive years. The failure of the Franciscan leaders to achieve favorable public opinion for their views assumes problematic perspectives, bearing in mind both the Franciscans' communication experience and, in parallel, the success of heretical movements, less skilled, in spreading similar ideas among wide sectors of medieval society. Further insight into the Michaelists' tenets helps to understand this propaganda failure or, conversely, the success of the pope in advancing his policy without encountering any substantial opposition.

The imperative to restrain the pope's arbitrariness brought the Franciscan dissenters to reconsider the papal plenitude of power while paving the way for an original approach whose implications were developed in full by William Ockham. He extended the papal plenitude of power to include the sum of authority accorded to St. Peter and his heirs in leading Christendom in the true faith and the common good, but he made it subservient to the Divine and the Natural Law.[53] The restrictions of the Divine and the Natural Law, however, seemed to him too abstract to be used as a means of counterbalancing the heresy he ascribed to John XXII or for combating the papal coercive policy.[54] This brought Ockham to lay down a further restriction: the pope had no right to harm any institutions outside the Church, some of which had existed before the papacy, and were protected by the *Ius gentium*, the customary law that resulted from the usages and practices of a relevant group of people, whose will and consent to adhere to a particular practice had turned it into a binding rule.[55] The tendency to restrict the pope's freedom of action paved the way for Ockham's creed that *the Christian religion was one of perfect freedom*. Restraining the pope by means of Divine and Natural Law alone would furnish the pope with unlimited freedom of action, thus undermining the principle of Christian freedom. All Christians, their leaders first and foremost, would be turned into the pope's liegemen forever, as the pope would impose upon them the laws of servitude. Ockham also quoted the

Aristotelian argument of the common good. The papal plenitude of power elevated the pope's good before the common good, turning the pope into a mercenary whose only concern was his own welfare. This contradicted the will of God Who had elevated the pope to serve the common good and not his own. It also contradicted the essence of Christian freedom, making a minimalistic conception of the Petrine heritage unavoidable.[56] Hence, the pope had to be denied absolute authority in the spiritual sphere as well, and for two major reasons: the implications of the spiritual on the temporal, and its further priority over the material.[57] Moreover, though recognizing the subordination of the clergy to the pope, Ockham regarded the Church as the congregation of all the faithful which, as such, embraced the laity as well.[58]

The main contribution of Ockham's philosophy, moreover, lay not in the antithesis he posed to the norms propagated by the papal curia, but rather in his developing a new approach to the papal institution. In his exegesis of "Feed my lambs" (John 21:15), Ockham showed the functions included in Christ's mandate to St. Peter: to inculcate a doctrine leading to salvation, to exemplify a holy life, and to redeem the sinners. These duties were common to all the Apostles and did not elevate St. Peter above the others.[59] It follows that Christ did not raise Peter above the other Apostles, but they had chosen him to be their leader, and Peter's dignity of prince of the Apostles suggests only his superiority in charity.[60] The denial of the divine source of the papacy led to the acknowledgment of Emperor Constantine as the architect of papal primacy, thus further undermining the implications of the Petrine heritage.[61] These reasonings encouraged Ockham in his attempts to find a suitable substitute for papal authority: the Holy Scriptures, logic, and the decrees of the universal Church all served as a fitting curb on papal arbitrariness. A combination of the three should guarantee the fulfillment of Christ's promise directed to the universal Church through St. Peter: "But I have prayed for thee, that thy faith fail not" (Luke 22:32).[62] These declarations were much influenced by the conflict with Pope John XXII and Ockham's belief in the pope's heresy, which brought about his claim that any Christian could dismiss a heretical pope, but in such a contingency this task would be allocated to the emperor, thus turning the secular ruler into the keeper of Catholic orthodoxy.[63]

These reasonings hint at the critical implications of the Franciscan crisis in the plan of Catholic dogma. The endless conflicts within the Order and the complete opposition of the Spirituals to reach a way of compromise with their own leaders had justified John XXII's intervention in its first stages. As pointed out by Lambert, "John offered realism against the Spirituals' opposition to change."[64] Still, the prevailing view of modern researchers remains critical of the pope. Brampton argues that "with regard to religion, John XXII had a faith so simple that sometimes he failed to understand the difficulties of other people."[65] Referring to the papal bull *Quorumdam exigit*, Manselli points out that "it was a decision that reflected his characteristic psychology as a jurist together with the most complete insensitivity to the religious values he was touching."[66] Yet despite the fact that the papal policy did not actually bring about a change in the Franciscan approach to absolute poverty, it still led to the implementation of papal authority, thus heralding the final crystallization of the papal monarchy in the fourteenth century. The success of the papal monarchy in overcoming the Franciscan challenge acquires important dimensions from the perspective of communication, as the Franciscans were the most successful communicators in medieval society who nevertheless failed to safeguard their own creed against the papal offensive. Ockham's ultimate readiness to submit to papal authority reflects the weaknesses inherent in his individualistic approach, which failed to achieve wide support, nor did it change the prevailing normative attitude toward the Church's institutions. The question thus still remains: why did the Franciscans fail when less skilled communicators such as the sectarians succeeded? A tentative answer to this question could be given on two different, but complementary, levels, which deal with the essence of the Michaelists' message and the channels they used to spread it. The use of classical concepts such as the "common good" and "complete freedom," so dear to the modern reader, were rather strange to the ordinary clergyman, who was not yet ready to change his customary ways of thought for more abstract concepts quite alien to a traditional society. On the other hand, the intellectual elite, more familiar with such concepts, opted for a way of compromise with John XXII instead of beginning a hopeless struggle against the acknowledged leader of Catholic orthodoxy, the pope. This state of affairs reflects the gap between the Franciscan leaders and wide sectors of fourteenth-century prelates and monks who remained faithful to ancestral norms. This gap, however, could have been overcome if the Franciscans had looked for the support of wide social sectors *outside* the Ecclesiastical Order, especially of the townsmen who were their natural allies. Such an appeal could free the Franciscans from their isolation and bring them the large audiences to which the heretical movements appealed, and ultimately gained, for their cause. Yet with the exception of the Perugian encyclical, the Franciscans refrained from a propaganda campaign outside the ecclesiastical order, thus leaving their struggle in the traditional framework that gave the pope all the advantages. The appeal to Lewis the Bavarian, too, brought them back to the old patterns of the Investiture Contest, hardly relevant to the actual state of affairs. By turning the ecclesiastical order into the main target of their message, though allowing the Franciscans to remain faithful to the Church, it nevertheless prevented them from exploiting their propaganda skills to the fullest. It did not advance their cause, but that of Pope John XXII, the acknowledged leader of Christendom and the representative of Catholic orthodoxy. One can conclude, quite paradoxically, that the Franciscan failure was due both to the friars' conservative policy and to their too-novel message—conservative in that they followed the traditional patterns of discussion within the ecclesiastical order, which a priori gave the pope a place of preference. Moreover, the novelty inherent in the ideas of absolute freedom, common good, and *Ius gentium* acted against them, as fourteenth-century prelates and monks were not yet ready to accept such a message. This paradox reflects the many obstacles facing a minority group when attempting to survive within the framework of a traditional society in which the scope of propaganda was often restrained by coercion.

13

Political Propaganda and the Heresy of the Knights Templar

The refusal to tolerate the very existence of heresy and the resulting determination to annihilate its harmful consequences dictated the close cooperation between the leading powers of Christian society, namely, the pope and the secular monarchs. They both committed their swords to the uncompromising struggle against any hint of heresy and its sympathizers. In most cases, the Church's definition of a doctrine or a sect as heresy was followed by royal penalties, which ranged from corporal punishment to confiscation of all goods and the death sentence by burning of the so-called heretics. The lasting entente between priest and king over the suppression of heresy left little room for public opinion as this usually followed the path of events and in most cases lacked real influence on the policy of the establishment. Such a state of affairs, however, could radically change when the basic agreement between priest and king was broken; this was the case when Christian princes hindered the Church's antiheresy policy and lent their support to groups which had been formally declared as heretics or, conversely, when kings attempted to take the initiative in matters of heresy and persecuted groups that enjoyed ecclesiastical immunity. The support of the dissenter Franciscans by Lewis the Bavarian from 1327 onward faced the candidate to the empire against the papal policy, since Pope John had excommunicated them and condemned their ideology. The arrest of the Knights Templar by Philip the Fair in 1307 exemplified the other alternative, when the "Most Christian King" arrested all knights in France and confiscated their possessions. In both cases, contemporary reactions acquired maximal importance in light of kings and popes seeking maximal support, each hoping to fill the vacuum left by the rupture with former partners. Research into the reactions to the Templars' arrest and the abolition of the Order further reveals the scope of royal propaganda in fourteenth-century society and its contribution to the process of state-building in the kingdom of France.

The emergence of the Templar Order in 1119 combined the two dearest ideals of Christian Europe: knighthood and monasticism. Born amidst the struggle for survival of the young Crusader kingdom, Hugues de Payns, Godefroi de St. Omer, and a few knights vowed before the Patriarch of Jerusalem to follow a quasi-monastic life of poverty, chastity, and obedience, and to devote themselves to the defense of Christian pilgrims on their way from Acre to the various centers of pilgrimage across the Holy Land. The first stages of the new Order were quite humble: the nine original knights lived from alms and were forced to share their few horses. King Baldwin II of Jerusalem allowed them to use a wing of his former palace in the Dome of the Rock (mosque al-Aqsa), known to the Crusaders as the Temple of Solomon.[1] This, eventually, gave its name to the Order which came to be known as *Pauperes commilitiones Christi templique Salomonici*, the Poor Knights of Christ and the Temple of Solomon. It appealed to all those who "scorn to follow their own will and who desired to fight in high and true purity of soul for the king, in that they might choose to assume and fulfill the noble armour of obedience, fulfilling and preserving (it) with the most eager care." From this came into being the elaboration of a peculiar monastic code, which, though rejecting "immoderate abstinence," called for moral strength, chastity, and integrity.[2] In the eyes of Bernard de Clairvaux, their virtuous way of life turned the Templars into the most reliable knighthood of Christ, real candidates for martyrdom, as they were always ready to offer their lives in the holiest of wars: the Crusade. This resulted in the gap between the Templars, presented as the *militia Dei*, the knighthood of God, and the secular knights, who represented all the badness (*malicia*) of earthly life.[3]

Being the most original outcome of the Crusader Kingdom of Jerusalem, the knights played a crucial role in the defense of fortresses, small strongholds, towers, and turrets along the main roads. Together with the Hospitallers, they became the only regular army in the Holy Land and its most experienced military force. In the Battle of Hattin (1187), both Orders recruited some 600 knights, who represented about half of all the Christian forces in the battlefield.[4] The Templars' commitment to the defense of pilgrims and Christian citadels in the Holy Land made them beneficiaries of papal generosity, while the Order, its members, and property were placed under apostolic tutelage. They were accorded all property rights over any booty they won from the Moslems and were exempted from paying tithes on all territories conquered from the Infidel; this clause was gradually expanded with regard to all lands, pasture and agricultural, which came into the hands of the knights in whatever way. The Templars were allowed to build their own chapels and cemeteries and to enjoy the services of their own clerics who submitted to the authority of the Order's masters alone. Successive popes built the foundations of the Templars' power on an international scale.[5] As pointed out by Riley-Smith, the large papal privileges accorded to the Templars, as to the other Military Orders, resulted from the tendency of the papacy to turn them into instruments of its will.[6] Submitting to the Holy See and independent of any outside interference, secular or ecclesiastical, the Military Orders could advance the interests of the papal monarchy throughout Christendom, while becoming an important tool in the hands of the Reformer papacy.

Entrusted with ecclesiastical legitimacy, the Templar Order provided the most suitable framework for pious laymen seeking an outlet for their religious impulses. They employed their bellicosity against the Infidel, while channeling the spiritual effervescence of the times into most convenient targets. Their distinctive symbol gave visual confirmation of the virtuosity of their mission: professed knights were granted the white habit, on which a large red cross was sewn. The same cross appeared also on the Templar banner, known as *Baucent* (black-white), because of the black and white lines that appeared in the background. Nobles and knights who were not ready to relinquish all material goods joined the Templars sometimes for a specific period, or with a particular status that did not enforce on them all monastic duties. The increasing number of knights improved the economic situation of the Order, due to the considerable sums of money and lands they contributed at their official reception. Christian princes also became vigorous supporters of the Order and, beside generous political and economic grants, they fostered the establishment of Templar houses all over Europe. In the 1120s, Henry I encouraged their establishment in Normandy, by 1129 a Templar fortress was built in Castile as well as in La Rochelle (1131), Languedoc (1136), Rome (1138), and later in England and Germany.

In addition to their military tasks, a significant number of knights became involved in financial transactions. Originally, the Order had lent money to pilgrims who were unable to afford the heavy expenses of the long journey to Jerusalem. These loans were usually backed by mortgages on lands and involved interest, openly contravening the Christian prohibition on usury. In addition, the establishment of Templar houses on both sides of the Mediterranean turned the Order into a most convenient intermediary in financial transactions between Europe and the Levant.[7] The status of *asylreicht* accorded to all houses and their building as fortresses, encouraged people to entrust their goods to the knights, while strengthening their power of financial maneuvering.[8] The Templars thus became a leading force in international trade and finance, one that could also offer the best services to popes and princes. Philip August entrusted the keeping of the royal treasure into the Templar house of Paris when he departed for the Third Crusade.[9] The centralizing policy of western monarchs and their growing needs for money strengthened their cooperation with the Templars, since the knights were able to fulfill most difficult administrative functions and to lend large sums of money.[10] The spread of the Templars throughout Europe and the involvement of many knights in financial transactions did not undermine their basic obligations to the Holy Land, nor did they bring about significant changes in the military organization of the Order. Yet they brought about a change of emphasis, while much of the Templars' power was translated westward: against the 500 knights reported in Acre during the final Moslem offensive, there were some 15,000 all over Europe. Nevertheless, if the Military Orders served mainly as a means to advance papal interests, as maintained by Riley-Smith, after 1291 the Order was further able to serve the papal policy, especially the current plans for a reneval of the Crusade and the immediate recovery of the Holy

Land. Although the fall of the Crusader Kingdom encouraged the reform of the Military Orders, it could hardly be regarded as a catalyst for the dissolution of the Templar Order some twenty years later. The sudden arrest of all knights in France on charges of heresy further appears in stark contrast against the traditional papal support of all Military Orders, and as a bad omen with regard to the cooperation between priest and king in matters of heresy.

Modern scholars who have written about the dissolution of the Order of the Knights Templar have integrated it into the process of state-building, which encouraged the gradual suppression of feudal barriers while paving the way for the centralizing policy of Western kings. The suppression of the Order was also evaluated as another symptom of the power struggle between priesthood and kingship, which by the later Middle Ages turned the balance of power in favor of the kings.[11] It has been also argued that the loss of the Crusader kingdom had considerably harmed the further existence of a universal Order whose *raison d'être* focused on the fight against the Infidel. In contrast with other Military Orders, which had already channeled their military efforts into new areas, such as the Hospitallers in Rhodes or the Teutonic Knights in Prussia and Poland, the Templars had not yet created such a substitute for the kingdom of Jerusalem; they became, therefore, more vulnerable to criticism and royal pressure. The opposition of the last master, James of Molay, to the reform of the Military Orders had further hindered the adaptation of the Templars in the new political reality while emphasizing the anachronistic nature of the Order. On the other hand, the financial activities of the Templars could no longer be justified on the grounds of their assistance to the Crusader kingdom and, in fact, became an additional factor of resentment against the Order. All these argumentations could explain the dissolution of the Templar Order from a historical perspective, but they did not give a satisfactory answer as to the charges of heresy leveled against the knights and the unprecedented propaganda campaign promoted in this regard by the Capetian court. From the perpective of medieval communication, the affair of the Templars reflects two additional facets, namely, the advanced communication stage of fourteenth-century society, and the awareness of royal communicators regarding the importance of public opinion in the process of state-building.

> In the early hours of the morning, on Friday, 13 October, a strange event occurred, the likes of which have never been heard of before. The Grand Master of the Order was imprisoned in the Temple house of Paris and, on the same day, all the Templars in France were incarcerated in various prisons. It seemed to everyone that this action could have been taken only by order of the Roman Curia in consultation with the king, and carried out by the knights William of Nogaret and Raynaud of Roy. . . . On the following day a public meeting was held in the king's gardens, at which the reasons for the action were explained to the people and the matter was expounded, first by the Dominicans and then by the king's officers, so that the people might not be too violently shocked by this sudden action, since the Templars were all men of great wealth and standing.[12]

The testimony of John of Paris reflects the propagandistic challenge that faced the Capetian court following the arrest of the Templars, both because there was no precedent for such an action and because the knights were so wealthy and highly respected. The royal court sensed the prevailing climate of opinion and, from the very beginning, aimed at refuting any seeds of criticism. Already, at the order of arrest, Philip the Fair suggested previous consultations "with our most holy father in Christ, the pope,"[13] thus vindicating himself from any charges of having violated the ecclesiastical immunity of the Order.

Papal correspondence, however, does not confirm the royal version, but suggests, rather, the pope's embarrassment arising from the affair:

> But you, our beloved son, setting aside all rules, while we are far from you, laid hands upon the souls and possessions of the Templars; you placed them under arrest, and what grieves us further, you added much suffering to their pain. . . . You took away the bodies and possessions of people directly dependent on the Roman Church. . . . Your impulsive behaviour is seen as an insult to us and to the Roman Church.[14]

Clement V's reluctance to deal with the charges of heresy attributed to the Templars and, instead, his focusing on the prevailing climate of opinion, hint at the waves of criticism aroused by the royal move and its main implications from the papal perspective. Charges of greed and avarice had previously been leveled against the Templars, who had often disregarded the prerogatives of other ecclesiastical sectors for their own profit. Besides, the secrecy that surrounded their proceedings fostered rumors about magical practices, sorcery, and sodomy. It seems quite probable that these charges had been previously discussed between Clement V and Philip the Fair and encouraged the papal tendency to reform all Military Orders. Nevertheless, such charges, whose reliability had never been satisfactorily proved, had not yet brought about any papal radical measures against the Order, nor could they justify the royal initiative. By arresting the Templars on charges of heresy without asking first for the papal sanction, Philip the Fair made a mockery of the traditional alliance between priest and king and, actually, faced the pope with a fait accompli.

Royal propaganda, nonetheless, significantly succeeded in the initial stages in spreading the version of the knights' heresy. The charges against the Templars ran to over 127 articles, which can be summarized as follows: every member received by the Order was requested to deny Christ and, sometimes, also the Holy Virgin and the saints, and to deny the holiness of the cross three times, while spitting upon it or upon a crucifix or the image of Christ. New members were also required to kiss their masters on all parts of their body and, though committed to observe celibacy, were encouraged to fulfill each other's physical needs, thus fostering homosexual practices. Not all Templars believed in the sacraments, and their priests omitted the words of consecration during Mass. They believed that the master and other leaders of the Order could hear confessions and absolve them from sin, despite the fact that many of them were laymen who had not received holy orders. The charges of sacrilege were further elaborated on the grounds of their alleged worship of an idol, Baphomet, whom the knights supposedly believed to be God, though only a few later confessed to have actually seen it. They sought gain for the Order by whatever means came to hand, whether lawful or not. Chapter meetings and receptions were held at night, in secret, under heavy guard, and only professed knights were allowed to attend. Any knight who dared to reveal to an outsider anything of the proceedings was punished by imprisonment and, eventually, death.[15] Royal propagandists developed these basic religious charges in the political language of the times, while giving further legitimacy to the royal policy. According to Pierre Dubois, the Templars' crimes of apostasy and homicide justified the king's intervention against them, because their sin was much more serious than that of ordinary heresy.[16] William of Plaisans linked the Christian defeat at Acre with the Templars, and accused the knights of being the main reason for the loss of the Holy Land, as they had carried out secret conversations with Saladin and other Moslem leaders.[17]

The charges leveled against the Templars were readily accepted in the kingdom of France. The secrecy of their meetings had fed the popular imagination and directed the many fears of contemporaries against the Order, which, only one year after the expulsion of all Jews, provided a most convenient target to channel the prevailing unrest. Moreover, the many frustrations resulting from the loss of the Holy Land could now be conveniently directed to the path that distinguished between orthodoxy and heresy, the heresy of the Templars providing the perfect answer for the troublesome questions arising from the final victory of the Infidel. Royal propagandists developed these premises to justify the arrests and further claimed that if speedy action against the knights were not taken, the wrath of God would also fall on the kingdom of France for sheltering the perpetrators of such unnatural acts. The Church and the king had to eradicate this danger most expeditiously. The royal policy was depicted as the most virtuous deed a Christian king could perform in his capacity as *advocatus ecclesiae*, defender of the Church, thus blurring the king's disregard of the Order's ecclesiastical immunity. The prevailing readiness to accept royal propaganda and, consequently, the widespread belief in the Templars' heresy, encouraged in France the further recognition of the king's initiative *de iure* or *de facto*.[18] After the initial shock,[19] most French chroniclers regarded the arrest of the Templars as an exercise of the king's initiative and a display of his power, thus overshadowing the Church's magistracy in matters of heresy.[20] They dutifully echoed the royal version concerning the Templars' treachery, and linked their heresy with the defeat of the Holy Land and the Crusades.[21] Yet being acquainted with the royal independent policy regarding the Church, most of the French chroniclers did not accept the royal version, which implied the papal initiative, nor did they find the papal sanction as a formal requisite to justify the royal policy. On the other side of the Channel, however, the perspective of most chroniclers was quite different. Although the heresy of the Templars was widely accepted in

England as well, following the version spread by the Capetian court, the arrest of the knights was considered the result of the pope's and not the king's initiative as, in England, heresy was still considered as coming under the sole jurisdiction of the pope.[22]

Criticism to the Templars' arrest grew with the distance from the Ile de France. Opposition to the policy of Philip the Fair was particularly strong among German and Italian chroniclers, who emphasized the royal disregard of ecclesiastical prerogatives without making any attempt to defend the Order or its members.[23] Envy and greed were believed to be the main factors behind the policy of Philip the Fair,[24] who intended one scion of the Capetians to be crowned as the future king of Jerusalem.[25] European rulers also reacted with restraint. Albert of Hapsburg expressed regret at the Templars' behavior as, "although a crime of such evil infamy ought to be reprehensible and damnable in all persons, it still became more reprehensible among the religious, who ought by the splendor of their life to be mirror for others and an example." Yet he expressed no readiness to follow the example of the king of France at once, but made the arrest of the knights in the empire conditional upon the explicit request of the pope.[26] Edward II was more incensed against what seemed him to be another "French plot." In his letters to various Christian princes, the king of England minced no words in condemning the reprehensible motives, which, in his view, had led Philip the Fair to arrest the knights.[27]

The different reactions to the arrest of the Templars reflect the conflict of opinion at the time. The charges of heresy leveled against the knights were widely accepted in the sphere of influence of Philip the Fair and, when discussed, they did not lead to noticeable attempts to defend the Order but, rather, to the criticism of the motives behind the royal move. This state of affairs hints at the varying acceptability of heresy charges in spite of the fact that they still appeared as a formidable propaganda weapon to be used against potential adversaries. Although considered "a work of genius,"[28] Capetian propaganda was not therefore received equally by all social strata, and the sociopolitical elite remained quite reluctant to support the policy of the king of France. The charges of heresy leveled by French propaganda were not universally regarded as satisfactory to revoke the ecclesiastical immunity of the Order, in complete disregard of the papal prerogatives. This position was clearly sustained by the lecturers of theology at the University of Paris. Without explicitly criticizing the royal policy, the university members denied the king's right to act independently in matters of heresy or to make use of the Templars' property without receiving the papal sanction.[29]

Despite his original hesitations, only two months after the arrest, Pope Clement V came to the help of the Capetian king. The bull *Pastoralis praeminentiae* (November 22, 1307), advanced the dissolution of the Order by one further step and bestowed on Philip the Fair the apostolic sanction he needed. After quoting the charges of heresy developed by royal propaganda, supposedly corroborated by the confessions made by the Templars under arrest, Clement required all Christian princes to arrest the Templars and to confiscate their property until he would pronounce judgement on them in the forthcoming Council of Vienne. The inquisition of the knights was entrusted to local bishops, while the investigation of the Order and its leaders was commissioned to legates especially appointed by the pope to this task.[30] The papal order brought about the arrest of the Knights Templar all over Europe at the beginning of 1308, but it hardly brought all conjectures to an end; nor did it bridge the differences of opinion concerning the division of labor between priest and king on matters of heresy. The interrogation of the Templar leaders by the papal legates revealed many weaknesses in the "Holy War" promoted by the king of France, and thus further underlined the pope's difficult position. Many knights defended the innocence and integrity of their Order so fearlessly that the doubts concerning the legitimacy of their arrest actually increased. Furthermore, the reluctance of Capetian officers to follow the papal edict and transfer the Templars' property to the curia damaged even further the critical relations between Philip the Fair and Clement V. The papal order to stop all procedures against the Templars in France (March, 1308)[31] thus appears as a belated attempt to change the course of events and restore to the Holy See the initiative in matters of heresy while improving the apostolic prestige. Clement's change faced the Capetian court with the disappointing alternatives of either following the papal order while abandoning its achievements, or finding some other partner, one more ready to support the royal commitment to the persecution of heresy. The Assembly of the three estates appeared as the most suitable partner the king could find, the unquestionable loyalty of its members bestowing on the royal policy the juridical basis of consent.

The Assembly of Tours (April-May, 1308) reflects the attempts to turn the representatives of the three estates into active participants in the uncompromising royal struggle against heresy. The summons to the Assembly appealed to the religious feelings of the nobility, the clergy, and the bourgeoisie to which the king added a national character. Philip stressed the ancestral Capetian devotion to the true faith, which had led him to submit all his forces "for the glory of God, for the salvation of our mother the Holy Church, and for the defence of the rights and liberties of the Church of France."[32] The absolute devotion to the Church and the Catholic faith was also the leading motif in the speeches of the Capetian lawyers headed by Dubois, Nogaret, and Plaisans.[33] Their vigorous defense of the royal policy against the Templars gained the support of the representatives of the bourgeoisie, 700 in all, who fell an easy prey to royal indoctrination.[34] The conspicuously large number of absentees from among the prelates and nobles, however, indicates their reservations concerning the dissolution of the Order. Besides the personal involvement of many nobles with the knights, the dissolution of the Order appeared to them as a further concentration of power in the king's hands, to the detriment of their own privileged status. The pressure exerted by the royal court, however, was more effective than class considerations and the large majority of the Assembly members (four-fifths) supported the resumption of negotiations with the pope along the lines already adopted in the Templars affair.

From then on, Philip could rely on the consent of the realm, a state of affairs that was well known to his subjects. Evaluating the aims pursued at Tours, one chronicler asserted that "the king has acted wisely, for he sought the support of all his people, and of all the estates, so that none could criticize his policy in the future."[35]

The Assembly of Tours thus created a new balance between Philip the Fair and Clement V, as the king succeeded in presenting his policy as the embodiment of the religious zeal of all his subjects, one which placed the Vicar of God in a rather embarrassing situation. This was the backdrop for the meeting with the pope at Poitiers (May, 1308), where Philip attended with a "great multitude" of supporters.[36] The Poitiers meeting indicates the success of royal communicators in creating a new conjucture, as the issue focused no longer on the Templars' heresy and the papal tendency to protect the Order but, rather, the guilt of the pope, condemned by the representatives of the French people. The speech of William of Plaisans faithfully reflects, right from its opening remarks, the aggressive stance of royal propaganda, as William appealed to the customary motto of *Christus vincit, Christus regnat, Christus imperat!* Plaisans referred to Christ's victories in the past and compared them to the similar victory achieved in his own times by the king of France against the Templars. He particularly stressed the religious zeal of the king, the nobles, the clergy, and all the French people, and praised their devotion to the Church and to the faith. Plaisans also cleared the king of any suspicion of coveting the property of the Templars since it would be held by the Church and dedicated to the glory of God and the propagation of the Holy Faith. This praise of Philip eventually placed him above the pope, and Plaisans added insult to injury by reminding Clement of the duties of his office. As for the Templars, Plaisans not only corroborated the charge of heresy, but added a further charge, namely, their responsibility for the loss of the Holy Land. In the name of the king and the people of France, Plaisans called upon the pope to acknowledge without delay the guilt of the Templars, since, otherwise, the French people, "the most zealous keepers of the Catholic Faith," would execute God's judgment with their own hands.[37] Royal pressure led the pope to agree to resume the inquisition of the Templars in France, while the matter of their property was left to the Council of Vienne.[38] The pope's sanction actually gave the royal officers a free hand, which was exploited to harm the knights' capacity to defend themselves while fostering the gradual suppression of the Order. In the following months, those Templars who confessed to heresy were immediately set free, while those who protested their innocence were sentenced to life imprisonment. Fifty-four Templars who retracted their former confessions of heresy were burned at the stake in Sens. But the ruthless attitude of the kings's officers against the Templars had an adverse effect on public opinion, engendering seeds of criticism against the royal policy. The fearless persistence with which many knights declared their innocence aroused wonder and admiration. Even those chroniclers who had formerly accepted the charges of heresy found it difficult to reconcile the diabolical image spread by royal propaganda with the heroic martyrdom of many Templars at the stake.[39]

The course of events in France reveals the power of royal propaganda and its impact in contemporary society. In its reluctance to support the royal policy, the nobility and the University of Paris represented the point of view of a minority whose opinions were not ultimately shared by wide social strata. The bourgeoisie, together with the low clergy and the monks who voiced their opinion in the chronicles, willingly or otherwise, supported Philip's attack on the Order and his disregard of ecclesiastical privileges. These propagandistic achievements cannot be explained by means of a coercive policy alone; it seems, rather, that the use of concepts deeply rooted in traditional patterns helped to make the royal propaganda more acceptable to many as the Templars themselves had aroused the contempt of the laity and the clergy alike. In open contrast to the virtuous ideals pursued by the first knights, the expansion of the Order through the twelfth and thirteenth centuries had, in the view of many, brought the Templars closer to the interests and vicious of the secular knights, thus exposing them to criticism. Even the empathy between the Templars and the lay aristocracy had its dangers, for when it began to believe that the Templars fell short of their original ideals, its reaction became hostile as its initial reception had been enthusiastic. In France, the Capetian policy against the Templars thus revived old feuds and ensured the wide reception of the charges of heresy.

In other areas of Christendom, however, the prevailing attitudes were less aggressive and reflect diverging reactions to both the charges of heresy and the moral validity of the papal policy. At the beginning of the fourteenth century, there were between 144 to 230 Templars in England, Scotland, and Ireland who, at their arrest, owned the relatively small sum of 36 pounds, 12 shillings, and 6 pence in cash. In addition, they had an annual land income of about 4,800 pounds, which equaled that of the Hospitallers in the area.[40] These figures could hardly encourage much hostility against the Order or motivate royal greed. Nevertheless, the papal order brought about the arrest of all Templars in England, Scotland, and Ireland between January and February, 1308. Following this second wave of arrests, all English chroniclers still repeated in full the French version of the Templars' heresy and justified the papal decree that required the cooperation of the secular arm, thus formally legitimizing the arrest of the knights in England as well.[41] On the other hand, King Edward II and wide sectors of the ecclesiastical elite of England seem to have been more sceptical regarding the charges of heresy as well as the motives behind the papal policy. The king's critical attitude is reflected in the very favorable treatment accorded to the knights after their arrest, while the royal treasury granted them a daily allowance to solve their immediate needs. Moreover, despite papal pressure to conduct the inquisition along the same lines developed in France, the prelates of York, for example, refrained from using torture and expressed considerable reservation concerning the charges of heresy leveled against the Order.[42] Only after lengthy deliberations did they force the Templars to swear an oath of self-purification, which further reflects the prevailing lack of faith in the charges of heresy spread by Capetian propaganda and echoed in the papal bulls. Instead of forcing the knights to acknowledge their guilt, thus giving legitimacy to the papal policy, the Templars of the province of York were actually forced to declare their innocence as follows: "I, brother R., acknowledging myself defamed by the articles sustained in the papal bull from which I cannot purify myself, I therefore submit myself to the grace of God and the Council's decision."[43] The version of the Templars' declaration provided by the chronicler Walter of Hemingburgh also leaves no room for doubt concerning the critical attitude of the northern prelates: "I reject and condemn, before these Holy Four Gospels, all the heresies and particularly those contained in the papal bull by which *I was defamed* (my emphasis). And I promise to observe in the future the Catholic and Orthodox faith, as it is held, taught and preached by the Holy Roman Church."[44] The twenty-four knights who complied were granted indulgence and sent to various monasteries to complete the penitential process.[45]

The reluctance of the York prelates to follow the papal directives seems somewhat surprising in view of the endless rivalry between the secular clergy and the knights. The English prelates could have gained some immediate benefit from the dissolution of the Order, which, throughout its existence of over 200 years, had shown very little inclination to cooperation. The dissolution of the Templar Order would have meant the removal of a hated rival that had consistently undermined both the authority and the income of prelates. However, this squaring of old accounts did not determine the attitude of the York prelates but rather their fears that the dissolution of the Order would strengthen the papal monarchy, a possibility further complicated by the close entente between Clement V and Philip the Fair. Such an attitude was clearly indicated by Langland who claimed that:

> Bothe riche and religious. that Rode thei honoure,
> Hat in grotes is ygraue. and in golde nobles,
> For coneityse of that crosse. men of holykirke
> shul tourne as templeres did. the tyme approcheth faste.[46]

In the province of Canterbury, however, the prelates adopted a different approach and, after subjecting all knights to the full force of inquisition through torture, the Council of London (1311) compelled them to admit the charges of heresy.[47] This was one of the last sequences in the dissolution of the Order, which took place within the framework of the Council of Vienne.

On May 16, 1311, the Council of Vienne opened its sessions, which focused on the fate of the Templar Order and its property. Fewer than 150 prelates were present at this stage, a relatively small number in comparison to the attendance in the previous century. Allocating the work to committees reduced to three the total number of general sessions, the last taking place on May 4, 1312. The committee, which was appointed to decide the fate of the

Templars, consisted of fifty members from England, Spain, Germany, and France, this international composition restraining the actual influence of the king of France on the proceedings. One month after the opening of the council, the Aragon representative wrote to King Jayme II:

> Concerning the Templars' Order, the results of the investigations that have been carried out so far are being studied. On the basis of what we have heard from cardinals and other prelates, it is not possible to condemn the Order as a whole, since there is no evidence of guilt on the part of the Order. . . . However, since most of the knights have been found guilty of serious transgressions, the pope will issue a decree for the dissolution of the Order. . . . and the setting up of a new Order to be based overseas.[48]

Other reports strengthened the opposition of many prelates in following the dictates of the French-papal entente while bringing about the immediate dissolution of the Order: "Most of the prelates stood by the Templars, except for the French who, it would seem, did not dare to act otherwise for fear of the king, the source of all this scandal."[49] This view was supported by other English as well as Aragonian and Italian chroniclers who testify to the prevailing readiness to give the Templars the chance of voicing their defense.[50] According to papal correspondence, some 900 knights arrived in Vienne in order to defend themselves and their Order from all charges and to give the prelates a more accurate picture, freed from the pressure and torture on which the inquisitorial reports had been based.[51]

The alarming possibility of further delays or, even worse, of a new trial that might eventually prove the innocence of most Templars, led to the sudden arrival at Vienne of Philip the Fair with a large committee. This ultimately put an end to all speculations and paved the way for the final dissolution of the Order. The bull *Vox in excelso* (April, 1312) gave apostolic sanction to the suppression of the Templar Order :

> It would seem wiser and more expedient for the glory of God and the Christian Faith, and for the welfare of the Holy Land. . . . to comply with the apostolic decree and exercise apostolic caution, which we trust will best prevent scandal and danger. . . . therefore, without a definitive sentence but by means of a provision or apostolic order, we hereby dissolve, for evermore, the Order of the Templars, its regulations, its dress, and its name. We forbid all men to enter this Order, to wear its habit or to claim to belong to the Order. And if anyone should dare to disobey this decree he shall be excommunicated by the Church.[52]

The bull *Ad providam* (May, 1312) transferred the Templars' property "to those who are ever placing their lives in jeopardy for the defence of the faith beyond the seas." Thus, the Hospitallers inherited "all (goods) that were in the possession of the Templars from the month of October 1307 when they were arrested in France."[53]

The dissolution of the Templar Order, however, did not lead to universal support for Clement V, neither was it connected with the charges of heresy that had been so widely accepted a mere five years before. A sense of consternation is evident even in the French sources on account of the way in which this prestigious Order had been abolished.[54] The *Chronology of the French Kings*, however, tried to dispel the doubts concerning the Templars' guilt by appealing to the confessions forced from the knights before the University of Paris and the Synods of Reims and Sens. It also abolished the influence of torture in extracting the confessions on the grounds that "the Templars are brave men who could not be easily intimidated."[55] Yet only a few English sources remained faithful to the original charges leveled against the knights and connected the abolition of the Order with their heresy.[56] Most English authors were well aware of the drift of opinion in the council, and openly expressed their disapproval of Philip's motives and the weight of his influence on the pope and the prelates.[57] Reports from other countries described the dissolution of the Order as an act of injustice calculated to serve the unbridled ambition of Philip the Fair.[58] The treatise of Bernard de Clairvaux in support of the Templars inspired a German author to confront Clement V with Abraham's challenging question: "Wilt thou also destroy the righteous with the wicked?" (Gen. 18:23).[59] According to Dante Alighieri, the suppression of the Templar Order was yet another act of the "New Pilate" (Philip the Fair) to undermine further the foundations of Christian society.[60] The chronicle of Pistoia went even further and considered the dissolution of the Order as one of the three main causes that later brought about the Black Death.[61] In the course of the years, resentment against the Templars abated and their memory was sanctified by an aura of martyrdom. Their leaders' brave death at the stake (1313–1314) bestowed a mythical halo over the memory of all Templars, and their martyred bodies were considered as holy relics.[62]

Reactions of this kind indicate that the doubts regarding the reliability of the charges leveled against the Templars not only persisted but actually increased after the pope sanctioned the final suppression of the Order. Although in the early stages of the trial Capetian propaganda fostered the belief in the Templars' heresy, between 1308 and 1312 the opposition to the royal policy strengthened and, consequently, the number of those who still supported the dissolution of the Order significantly diminished. The Capetian propaganda had not succeeded in convincing wide social circles outside France of the genuineness of the French king's commitment to maintain the purity of the faith. The growing opposition to the suppression of the Order, moreover, was not a result of a belief in the Templars' integrity but, rather, of the reprehensible motives behind the policy of both Philip the Fair and Clement V. The entente between priest and king seemed a bad omen to all those who cherished ecclesiastical freedom, particularly the ecclesiastical elite, that mostly disapproved of the pope's shameful submission to the Capetian interests. The failure of Philip's propaganda outside France, however, could not neutralize its impact in the French kingdom, but rather reflects the new scale of priorities of the political communication system at the time. Against the former focus of royal diplomacy in foreign countries, by the fourteenth century, the emergence of a centralizing monarchy turned the royal subjects into the main target of the king's message and the very basis of his increasing power. The incorporation of townsmen into the Assembly from 1302 onward faithfully reflects the tendency of royal communicators to turn the *tiers état* into the main audience of the royal message, as this represented the most communication-oriented sector in contemporary society. To the traditional perspectives of both the clergy and the nobility, the townsmen provided royal propaganda with a receptive public, which proved again and again its commitment to the king. The maximal scope of royal propaganda during the Templar affair was therefore due both to the emergence of the centralizing monarchy and, in parallel, to the decline of the traditional society. The townsmen played the leading role in this process, further emphasized by the appearance of different social classes, each having its own particular interests. Yet in the lack of legitimization for *Realpolitik* or *raison d'état*, Philip the Fair and his communicators were forced to use traditional norms in order to make their novel policy more acceptable to their contemporaries. The charges of heresy leveled against the Templars appeared as lip service to ancestral beliefs whose effectivity was still conditioned a priori by the existence of an embryonic national awareness. The reluctance of Italian, German, and Aragonian chroniclers to accept the Capetian version of the Templars' heresy further corroborates the weight of the national awareness, which actually conditioned the very success of royal propaganda. Furthermore, even in the kingdom of France, the response to royal propaganda was significantly influenced by class considerations; this forced royal communicators to direct their message to more specific social strata. In this regard, the arrest of the Templars and, ultimately, the abolition of the Order fostered one of the first propaganda campaigns in the modern sense, because of the well-defined nature of both the audience and the royal message that focused on one single issue: the guilt of the Templars. The royal policy during the Templar affair appeared, therefore, as another stage in the process of state-building, in which the emergence of a centralizing monarchy was established on the basis of favorable public opinion. In a society not acquainted with nuances but which faced reality in well-defined poles of good and bad, orthodoxy and heresy, Philip the Fair played the same rules; he tried and to some degree succeeded in harming the right of the Templars to survive in the national context. The intensive use of propaganda and compulsion, dogma and torture, *Weltanschauung* and manipulation ultimately paved the way for the centralizing monarch who, rather parodoxically, could further legitimize his policy through the general consent of the community of the realm.

EPILOGUE

The basic premise of this study was that communication is a product of history no less than of technology. Though influenced by the technological evolution, the relationship between the media and the message has been dictated by the changing needs of a given society in space and time. The emergence of communication systems in the Central Middle Ages thus appears as one crucial stage in the development of medieval society. This raises another premise: that communication systems are not an innovation of modern society but, rather, the outcome of the modernization process of Western society from the eleventh century onward. The consolidation of the Church's power, the emergence of centralizing monarchies, the gradual evolution toward a monetary economy, and the prosperity of commerce and towns all served as catalysts for the development of a more communication-oriented society. The process of integration of Western society in the political and socioeconomic levels, in turn, brought about the disintegration of the traditional society and the feudal system, which had structured medieval society for generations. It also encouraged the sophistication of the means of communication while facilitating the emergence of communication systems. The appearance of media in the modern sense appears, therefore, as one facet in the decline of the traditional society, notably, of the corporate structure inherent in the feudal system. Once medieval people had divorced themselves from the familiar confines of the corporate framework, they became more exposed to propaganda and manipulation. The way for the *Leviathan* was thus paved.

The crucial role of political communication in the Central Middle Ages dictated the three sections of this study. At the institutional level, differentiation has been made between three different systems, that of the Church, the monarchy, and the so-called heretical movements, the development of which followed the different stages in the disintegration process of the traditional society. In the eleventh century, the Church undoubtedly appears as the leading factor in the use of propaganda to consolidate its power. It thus introduced a new dynamics in European politics and provided a precedent for the development of political communication that has lasted up to this day. The consolidation of the Church's power through an integrated communication system provided an example for the state to follow in challenging the monopoly while appropriating the means. The heretical movements also used alternative means to challenge not directly, as the state did, but by undermining the Church's monopoly indirectly, through opposing the message. Once the emerging state and the heretical movements began to challenge the Church's power, they adapted the myths, symbols, and ideas of the Church to their own needs.

The interaction between the Church, the monarchy, and the heretical movements was indicated by their indiscriminate use of the image of the "chosen people," while each institution ascribed to itself the monopoly of the *Vox Dei*. God thus remained the only source of legitimization in medieval communication. The common reliance on the Holy Scriptures and the wealth of symbols attached to them, both by the Church and by those who challenged its monopoly, facilitated the reception of new messages as their basic concepts had been well-known for generations. On the other hand, it asked for a higher degree of sophistication, which, in turn, complicated the reception of new messages among traditional audiences. The propaganda failure of the dissident Franciscans during the pontificate of John XXII suggests but limited room for the maneuvering of propaganda campaigns when they remained within the traditional frameworks of the Franciscan Order or the ecclesiastical establishment as a whole. Although the Franciscan message relied on old concepts, once developed by Ockham it became too sophisticated for large sectors of the clergy. These limitations eventually became more pronounced when larger masses of the public were concerned. The affair of the Templars exemplifies, in this regard, one of the most spectacular propaganda campaigns in the Middle Ages. Philip the Fair aimed then to legitimate his unprecedented move against a Military Order, which enjoyed ecclesiastical immunity by fostering favorable public opinion to his policy. Yet research into the different reactions it aroused reflects the heterogeneous nature of medieval public opinion and the necessity to extend the focus of research beyond the cultural and political elites of the times. This necessity became further justified because of the appeal to the masses inherent in medieval propaganda campaigns during the Crusades or the Hundred Years War, the success of which was conditioned by the massive identification with the goals they represented.

Despite the normative essence of medieval society, however, old symbols were continuously adapted to fulfill new needs. The evolution of the papal Crusade policy, as well as the approach to death and the world-to-come, reflects the high awareness of the Church to the changing climate of opinion and, consequently, the capability of the clergy to adapt its message according to the circumstances. The monarchy's use of the Crusade theme, as well as of national stereotypes, corroborates the dynamics inherent in medieval political communication and the continuous process of change it involved. Themselves an outcome of change, the heretical movements appear to be the most capable of exploiting the new historical context by turning merchants and craftsmen into their main communicators while focusing their propaganda among the urban populations. As the socioeconomic process of integration reached more advanced stages and, in parallel, the communication channels became more sophisticated, heresy reached its zenith in the Early Modern period with the Reformation. The vernacularization of learning, and the competition for public opinion it promoted, then became the basis of humanism.

As a result of the disintegration of the traditional society, the emergence of communication systems and the more intensive use of manipulation and propaganda heightened the alienation between the individual and society. On the other hand, the emergence of communication systems was closely related to the growing gap between this state of alienation and the still-existing social needs of medieval man. The historical approach to communication thus reflects further two divergent roles it accomplished according to the interests of the institution that promoted its use. As with all human functions, the question still stands as to the goals served by the communication systems in the Central Middle Ages. The divergent alternatives were clearly reflected in the opposing communication policies of the emerging state and of the heretical movements. Medieval kings hastened the disintegration of the traditional society as an important stage in the emergence of a political solidarity countrywide which in time became the basis for a national consciousness. Only after medieval man had divorced himself from the corporate framework was he able to embrace the idea of a national power, a historical and cultural background and a territory common to all its inhabitants. The heretical movements, on the other hand, provided new social frameworks for the urban populations, which had already dissociated themselves from the traditional society. The strong solidarity between the members of the heretical group thus replaced the former corporation, while giving its members the sense of security and the legitimization of their existence they had been looking for—thence the contribution of the historical approach to a better understanding of communication, and the sociopolitical conditions from which the communication systems emerged, and which they reciprocally influenced. The medieval heritage provides, therefore, not only a precedent for modern communication but also new insight into the conditions that favored or opposed the emergence of communication systems.

The wide use of stereotypes and symbols taken from the fauna and the flora by the emerging state is another important facet of the medieval heritage and its lasting influence. The atomization of modern society and the dissociation of modern man from the natural world did not weaken the impact of symbols taken from the flora and the fauna; on the contrary, they became an integral part of propaganda campaigns as well as of political manipulation. The Nazis' depiction of rats to symbolize the Jews provides one of the most shocking examples of antisemitic propaganda in the twentieth century (as in the film *The Eternal Jew*), while the success it achieved in manipulating old fears and new hates indicates both the universal scope and the efficiency of medieval concepts. This state of affairs brings us to the conclusion that both in the Central Middle Ages and in our own days, the success of communicators depends on their manipulation of fears, stereotypes, hates, and prejudices prevailing in the social group whose approval they are seeking. A *Newsweek* Poll (April 3, 1989, p. 41), in which 750 American adults were interviewed by the Gallup Organization (December, 1988), unambiguously corroborates the longevity of some medieval concepts: over three-quarters of those surveyed believe that there is a heaven and that they have a good to excellent chance of getting there. Once they get here, they believe heaven will be peaceful (ninety-one percent), will have humor (seventy-four percent), and will be populated by people they know (seventy-seven percent) along with God (eighty-three percent). Fifty-eight percent also believe there is a hell.

Medieval man, whose knowledge of science and technology was virtually nil, also feared God, the Last Judgment, and Hell. Though to a great extent these fears are shared by modern individuals, they do not result from the illiteracy prevailing in the Middle Ages, but from the state of alienation in which our contemporaries find themselves. The role of medieval preachers and those of today epitomize the similarities and diversities between communications then and now. The messages of both the medieval preachers and their modern heirs were legitimized by the *Vox Dei* and their content is often quite similar. Yet beyond these similarities of content, medieval preachers and those of today are actually diametrically opposite. The irreconcilability between them results not only from the different channels of communication at their disposal, but, first and foremost, from the different social environment to which they appeal. The preacher of the Middle Ages emerged from within the corporate framework of the medieval village whose members knew each other well, while the Church provided them with the basic cells of their social life. On the other hand, the need for the modern preacher whose sermons are broadcast nationwide on television, results from the state of alienation of his public among whom communication delusions actually replace real social links. The field of action of this modern preacher is no longer the well-known medieval corporation but the *global village* of Marshall McLuhan. The very fact that today's preachers use the same ideas elaborated by their medieval ancestors more than 600 years ago surely indicates the relationship between the message and the media then and now, and the need to elucidate the origins, mechanism, and nature of the modern one. One can further emphasize the adaptation of the *Vox Dei* by the nation-state as well as by the Church of today. The invocation of God, from the controversies about school prayer and the Pledge of Allegiance ("one nation, under God") to the coins, documents, and speeches of the realm ("God bless") suggest that the *Vox Dei* is still a legitimating force in (at least American) politics and, no doubt elsewhere as well. As Robert Bellah's notion of a "civil religion" suggests, the *Vox Dei* still resonates in the civil society as well as the religious groups of today.

ABBREVIATIONS

A.H.R.	American Historical Review.
B.E.C.	Bibliothèque de l'école des Chartes.
C.I.C.	Corpus Iuris Canonici, ed. Aemilius Friedberg, 2 vols., (1879, reprint Graz: Akademische Druck-U. Verlagsanstalt, 1959).
E.H.R.	English Historical Review.
Foedera	Foedera, Conventiones, Literae. . . . inter Reges Angliae ab ineunte saeculo duodecimo ad nostra usque tempora, ed. Thomas Rymer, 3rd ed., 40 vols. in 10 (1739–1745, reprint Hants: Gregg Press, 1967).
F.R.G.	Fontes Rerum Germanicarum, ed. Johann F. Böhmer, 4 vols. (1843–1868, reprint Aalen: Scientia Verlag, 1969).
H.F.	Recueil des historiens des Gaules et de la France, ed. M. Léopold Delisle et al., 25 vols. (1869–1904, reprint Farnborough, Hants: Gregg, 1967–1968).
Mansi	Sacrorum Conciliorum Nova et Amplissima Collectio, ed. Giovanni Mansi, 59 vols. (1759–1798, reprint Graz: Akademische Druck-u.Verlagsantalt, 1960–1961).
M.G.H.	Monumenta Germaniae Historica.
P.G.	Patrologia Graeca, ed. J. P. Migne, 162 vols. (Paris: Turnhout, 1857–1868).
P.L.	Patrologia Series Latina, ed. J. Migne, 17 vols. (Paris, 1844–1901).
R.E.J.	Revue des études Juives.
R.H.	Revue historique.
RHC Occ.	Recueil des historiens des Croisades—Historiens Occidentaux, éd. Académie des inscriptions et belles lettres, 5 vols. (Paris: Imprimerie royale, 1845–1906).
R.H.E.F.	Revue d'histoire de l'église de France.
R.I.S.	Rerum Italicarum Scriptores, ed. L. A. Muratori, new ed. Giosue Carducci et al., 114 vols. (Citta di Castello: S. Lapi, 1900–1975).
R.Q.H.	Revue des questions historiques.

R.S. Rolls Series, Rerum Britannicarum Medii Aevi Scriptores, 251 vols. (1858–, Kraus reprint, 1965–1968).

Raynaldus Annales Ecclesiastici auctore Caesare Baronio, ed. Odorico Rinaldi, 24 vols. (Luca 1738–1750).

Vitae Vitae Paparum Avenionensium, ed. Etienne Baluze, 2 vols. (Paris: F. Muguet, 1693).

NOTES

INTRODUCTION

1. *Epistola 132 seu capitulare admonitionis ad eundem Carolum,* Alcuini *Epistolae Karolini aevi*, ed. Ernst Dümmler et al., 5 vols. (*M.G.H.*: Berlin, 1895), 4:198–99. Walter Map, a scholar and courtier of Henry II, shared this opinion while expressing his reluctance to involve the mob in the diffusion of the *Vox Dei*: "Shall then the pearl be cast before swine, the word be given to the ignorant whom we know to be unfit to receive it, let alone to pass on what they have received? Away with such a thought; root it out! From the head let ointment go down to the beard and thence to the clothing; from the spring let water be led, not from puddles in the streets." Humbert de Romans, general master of the Dominican Order, claimed, in a similar vein, that the laity should endorse the faith without inquiring into its secrets. See Walter Map, *De Nugis Curialum*, I, 31, ed. and trans. Montague R. James (London: Honourable Society of Cymmrodion, 1923), pp. 65–66. Humbert de Romans, "*Ad omnes laicos,*" *De modo prompte audendi sermones*, ed. M. de La Bigne (Lyon: Maxima Bibliotheca Veterum Patrum, vol. 25, 1677), p. 491.

2. Colin Morris, *Medieval Media: Mass Communication in the Making of Europe* (Southampton: University of Southampton, 1972), p. 14.

3. Charles H. Haskins, "The Spread of Ideas in the Middle Ages," *Speculum* 1 (1926): 19.

4. Joseph R. Strayer, "The Historian's Concept of Public Opinion," in *Common Frontiers of the Social Sciences*, ed. Mina Komarovsky (Glencoe, Ill.: Free Press, 1957), pp. 264–67.

5. Robert A. Kann, "Public Opinion Research: A Contribution to Historical Method," in *Quantitative History*, ed. D. Karl Rowney and James Q. Graham, Jr. (Homewood, Ill.: The Dorsey Press, 1969), pp. 65–73; see also, William O. Aydelotte, "Quantification in History," *Ibid.*, p.4; Lee Benson, "An Approach to the Scientific Study of Past Public Opinion," *Ibid.*, p. 25; Henry David, "Opinion Research in the Service of the Historian," *Common Frontiers*, pp. 277–8.

6. Lucian Pye, "Communication and Political Development," in *Public Opinion Propaganda and Politics*, ed. N. Kies (Jerusalem: Hebrew University, 1971), pp. 368–70.

7. Michel Zink, "Détachement du monde et soumission au monde dans la prédication et la littérature édifiante en français du XIIe et du XIIIe siècle," in *Idéologie et propagande en France*, ed. Myriam Yardeni (Paris: Picard, 1987), p. 43.

8. Haskins, "Spread of Ideas," p. 21.

9. Morris, *Medieval Media*, pp. 12–14.

10. On the concept and its evolution, see, Roger Chartier, s.v. "Populaire," *La*

nouvelle histoire, ed. Jacques le Goff et al. (Paris: Retz, 1978), pp. 458–60; Jean C. Schmitt, "Religion populaire et culture folklorique," *Annales E.S.C.* (1976): 941–53.

11. Georges Duby, "The Diffusion of Cultural Patterns in Feudal Society," *Past and Present* 39 (1968): 3–4.

12. Wilber Schramm, "Two Concepts of Mass Communication," in *Public Opinion*, p. 262.

13. Alfons Huber and Alphons Dopsch, *Österreichische Reichsgeschichte* (Vienna: Neudruck der Ausgabe, 1901), pp. 32–61.

14. Nancy Harper, *Human Communication Theory: The History of a Paradigm* (Rochelle Park: Hayden Book Co., 1979), p. 262.

15. C. Cherry, *On Human Communication: A Review, a Survey, and a Criticism* (Cambridge, Mass.: M.I.T. Press, 1978), p. 9.

16. Carol Harding, "Acting with Intention: A Framework for Examining the Development of the Intention to Communicate," in *The Origins and Growth of Communication*, ed. Lynne Feagans et al. (New Jersey: Ablex Publishing Corporation, 1984), p.124.

17. Hanno Hardt, "Communication and History: The Dimensions of Man's Reality," in *Approaches to Human Communication*, ed. Richard W. Budd and Brent D. Ruben (Rochelle Park: Hayden Book Company, 1972), p. 148.

18. David Windlesham, *Communication and Political Power* (London: J. Cape, 1966), pp. 17–30.

19. James N. Rosenau, "The Opinion Policy Relationship," in *Public Opinion*, pp. 196–200.

Chapter 1

1. George G. Coulton, *Medieval Village, Manor and Monastery* (New York: Harper & Brothers, Torchbooks, 1960), p. 15.

2. Johan Huizinga, *The Waning of the Middle Ages*, trans. F. Hopman (1924, reprint ed. Middlesex: Penguin Books, 1982), p. 9.

3. Joseph Benzinger, "Zum Wesen und zu den Formen von Kommunikationen und Publizistik im Mittelalter," *Publizistik* 15 (1970): 301.

4. John of Salisbury, *Policraticus*, VI, 24, ed. Clemens I. Webb, 2 vols. (1909, reprint Frankfurt: Minerva G.M.B.H., 1965), 2:67.

5. See, for instance, the reports sent by Venetian merchants in Egypt about the political situation in the area, R. Morozzo della Rocca, ed., *Lettere di mercanti a Pignol Zucchello (1336–1350)* (Venice: Comitato per la pubblicazione delle fonti relative alla storia di Venezia, 1957).

6. Wilhelm Scherer, *Geschichte der deutschen Litteratur*, 3 vols. (Berlin: Weidmann, 1905), 1: 59.

7. Thomas Stapleton, "A Brief Summary of the Wardrobe Accounts of 10, 11, 14 Edward II," *Archaelogia* 26 (1836): 321, 326.

8. C. A. J. Armstrong, "Some Examples of the Distribution and Spreed of News in England at the Time of the Wars of the Roses," in *Studies in Medieval History Presented to F. M. Powicke*, ed. R. W. Hunt, et al. (Oxford: Clarendon Press, 1948), pp. 444–45.

9. Tacitus, *Annales*, 4, ed. Henry Furneaux (Oxford: Clarendon Press, 1956), p. 23.

10. Gordon W. Allport and Leo J. Postman, "The Basic Psychology of Rumor," in *Public Opinion*, pp. 170–75.

11. Silvia Schein, "Gesta Dei per Mongolos 1300: The Genesis of a Non Event," *E.H.R.* 94 (1979): 805–19.

12. Giovanni Bocaccio, *The Decameron*, trans. Richard Aldington (New York: Garden City, 1930), p. 222, 363; Geoffrey Chaucer, *The Canterbury Tales*, "The Reve," "The Clerk of Oxenforde," ed. Paul G. Ruggiers (Oklahoma: Oklahoma University Press, 1979), p. 202, 713.

13. Henry S. Lucas, "The Machinery of Diplomatic Intercourse," in *The English Government at Work, 1327–1336*, ed. James F. Willard and William A. Morris, 3 vols. (Cambridge, Mass.: The Medieval Academy of America, 1940), 1: 303.

14. *Acta Aragonensia*, ed. Heinrich Finke, 3 vols. (Berlin: W. Rothschild, 1908–1922), 3: 185.

15. On the school of the Annales and its contribution to modern historiography, see T. Stoianovich, *French Historical Method: The Annales Paradigm* (Ithaca-London: Cornell University Press, 1976); *La nouvelle histoire*, s.v. "Annales," pp. 26–32.

16. Brian Stock, *The Implications of Literacy: Written Language and Models of Interpretation in the Eleventh and Twelfth Centuries* (Princeton: Princeton University Press, 1983), p. 85.

17. Etienne Delaruelle, "La culture religieuse des laïcs en France aux XIe et XIIe siècles," in idem. *I Laici nella Societas Christiana dei secoli XI e XII*, Miscellanea del Centro di Studi Medievali, vol. 5 (Milano, 1968), pp. 568–69.

18. See Bernard Guenée's instructing research on the number of manuscripts and their spread in *Histoire et culture historique dans l'Occident médiéval* (Paris: Aubier Montaigne, 1980), pp. 250 f.

19. Nancy Harper, *Human Communication*, p. 74.

20. Jacques Toussaert, *Le sentiment religieux, la vie et la pratique religieuse des laïcs en Flandre maritime et au "West Hoeck" de langue Flamande aux XIVe, XVe et début du XVIe siècle* (Paris: Librairie Plon, 1963), pp. 60–66, 85–87.

21. Emmanuel Le Roy Ladurie, *Montaillou, village occitan de 1294 à 1324* (Paris: Gallimard, 1975), pp. 357–65.

22. Much research has been done on medieval universities. For the purposes of this study, the following will be mentioned: Alan B. Cobban, *The Medieval Universities, their Development and Organization* (London: Methuen, 1975); Jacques Verger, *Les universités au Moyen Age* (Paris: Presses universitaires de France, 1973); P. Glorieux, "L'enseignement au moyen âge: Techniques et méthodes en usage à la faculté de théologie de Paris au XIIIe siècle," *Archives d'histoire doctrinale et littéraire du moyen âge* 43 (1968): 90–95.

23. Giovanni Villani, *Cronica*, 4. 11. 94, ed. Franco G. Dragomanni, 4 vols. (Firenze: S. Coen, 1844–1845), pp. 324–25.

24. Accordingly, Giles Constable saw in the eleventh and twelfth centuries the "Golden Age" of medieval epistolography; see Giles Constable, *Letters and Letter—Collections*, in *Typologie des sources du moyen âge occidental*, ed. L. Genicot, vol.17 (Turnhout: Brepols, 1976), p. 31.

25. Ambrosius, *Epistola* 66 (*P.L.* 16.1225).

26. *Foedera*, II–2, p. 985.

27. F. Guterbock, *Die Gelnhauser Urkunde* (Berlin, 1920), pp. 23 f.

28. Eugène Déprez, *Les préliminaires de la guerre de cent ans* (Paris: A. Fontemoing, 1902), p. 407.

29. Walter of Hemingburgh, *Chronicon*, ed. H. Hamilton, English Historical Society (London, 1848–1849), p. 373.

30. Boccaccio, *Decameron*, p. 210; Chaucer, *Canterbury Tales*, "The Frankeleyn," p. 614, v. 837–40.

31. Philippe de Novare, *Les quatre âges de l'homme*, ed. M. de Fréville (Paris: F. Didot, 1888) §24–25; see also A. A. Heutsch, *La littérature didactique au moyen âge* (Halle, 1903), pp. 53–54, 101, 151. On women's literacy in the Middle Ages, see Shulamit Shahar, *The Fourth Estate: A History of Women in the Middle Ages*, trans. Chaya Galai (London: Methuen, 1983), pp. 214–19.

32. Giraldus Cambrensis, *Speculum duorum*, ed. Yves Lefèvre and R. B. Huygen (Cardiff: University of Wales Press, 1974), pp. 249, 251.

33. Jean Leclerq,"Deux témoins de la vie des cloîtres au moyen âge," *Studi Medievali* 12 (1971): 989–92.

34. Jean Leclerq, "Modern Psychology and the Interpretation of Medieval Texts," *Speculum* 48 (1973): 476–90.

35. Jean Leclerq, "Lettres de S. Bernard: Histoire ou littérature?," *Studi Medievali* 12 (1971): 1–74.

36. Quoted by F. Guterbock, *Die Gelnhauser Urkunde*, p. 23.

37. Hugh de St. Victor, *De institutione novitiorum* (*P.L.* 176. 925–52). On the essay and its author, see Jean Claude Schmitt, "La geste, la cathédrale et le roi," *L'arc* (numéro consacré à G. Duby, 1978): 9–12.

38. Jacques Le Goff and Jean Claude Schmitt, "Au XIIIe siècle: Une parole nouvelle," in *Histoire vécue du peuple chrétien*, ed. Jean Delumeau, 2 vols. (Paris: Privat, 1979), 1: 263.

39. See for instance, Charles Swan and W. Hooper, ed. and trans., *Gesta Romanorum* (New York: Dover Publications, 1959), p. 11, 126. Chaucer, *Canterbury Tales*, "The Millere," p. 173, v. 3375; Boccaccio, *Decameron*, p. 231, 302.

40. Ruth Crosby, "Oral Delivery in the Middle Ages," *Speculum* 11 (1936): 88.

41. Michael Curschmann, "Oral Poetry in Medieval English, French, and German Literature: Some Notes on Recent Research," *Speculum* 42 (1967): 36–52.

42. See, for instance, Bede, *Historia ecclesiastica*, ed. Robert Hussey (Oxford: Oxford University Press, 1846), pp. 1–2.

43. Colin Morris, *Medieval Media*, pp. 7–10.

44. Augustine, "The Confessions," trans. Edward B. Pusey, Thomas Symons, ed., *Medieval Classics* (London-New York: Nelson, 1953), pp. 46–47, 51–52.

45. Petri Cantoris, *Verbum abbreviatum*, c. I (*P.L.* 205. 25); Thomas Aquinas, *Breve principium in Sacram Scripturam Opusculum*, ed. Pierre Mandonnet (Paris: Desclèe, de Brouwer et Cie., 1938), 4: 494.

46. Lynn Thorndike, "Public Readings of New Works in Medieval Universities," *Speculum* 1 (1926): 101–3.

47. Giraldus Cambrensis, *De rebus a se gestis*, in *Opera*, ed. J. S. Breuer, 8 vols. (London: *R.S.*, 1861), 1: 72–73.

48. Froissart, *Chroniques*, ed. Léon and Albert Mirot, 17 vols. (Paris: C. Klincksieck, 1952–), 11: 85.

49. Robert K. Root, "Publication Before Printing," *Publications of the Modern Language Association* 28 (1913): 421–23.

50. Clair C. Olson, "The Minstrels at the Court of Edward III," *Publications of the Modern Language Association* 56 (1941): 601–11.

51. See, for instance, the interesting collection of Barbara Leonie Picard, *French Legends, Tales and Fairy Stories* (London: Oxford University Press, 1960).

52. *Mansi*, 19. 544.

53. *Mansi*, 23. 830, 837.

54. Gerald R. Owst, *Preaching in Medieval England: An Introduction to Sermon Manuscript (1350–1450)* (Cambridge: Cambridge University Press, 1926), pp. 145–46.

55. G. Meersseman, O.P., "La prédication dominicaine dans les congrégations mariales en Italie au XIIIe siècle," *Archivum Fratrum Praedicatorum* 18 (1948): 149–51.

56. Alani de Insulis, *Summa de arte praedicatoria* (*P.L.* 210. 111).

57. Augustine, *On Christian Doctrine*, IV, 4, ed. and trans. Durant W. Robertson (New York: Liberal Art Press, 1958).

58. Gerald R. Owst, *Literature and Pulpit in Medieval England* (Oxford: Basil Blackwell, 1966), p. 526.

59. Michael Richter, "Kommunikationsprobleme im lateinischen Mittelalter," *Historische Zeitschrift* 222 (1976): 43 f.

60. Possidonius, *Vita Augustini*, 5, ed. Mary M. Muller and Roy J. Deferrari, The Fathers of the Church, vol. 15 (New York, 1953).

61. Quoted by Michel Zink, *La prédication en langue romane avant 1300* (Paris: Honoré Champion, 1976), p. 142.

62. Evelyn Underhill, *The Life of the Spirit and the Life of To-day* (New York: E. P. Dutton, 1922), pp. 141 f.

63. Eunice Cooper and Marie Jahoda, "The Evasion of Propaganda: How Prejudiced People Respond to Anti-Prejudice Propaganda," in *Public Opinion*, pp. 176–79.

64. Alani de Insulis, *Summa de arte praedicatoria*, cols. 184–98.

65. Humbert de Romans, *De eruditione Religiosorum Praedicatorum libri duo*, 1. 4. 17, (Rome: A. de Rubeis, 1739), p. 35.

66. *Caesarii Heisterbacensis monachi ordinis Cisterciensis Dialogus Miraculorum*, 4. 36, ed. Josephus Strange, 3 vols. (Cologne: J. M. Heberle, 1851–1857).

67. Johan Huizinga, *The Waning*, p. 173.

68. *Summa Praedicantium*, quoted by Owst, *Preaching*, p. 183.

69. *Mansi*, 15. 72.

70. *The Exempla or Illustratives Stories from the Sermones Vulgares of Jacques de Vitry*, ed. Thomas F. Crane (1890, reprint New York: B. Franklin, 1971), pp. 7–8, 27.

71. Harry Caplan, "The Four Senses of Scriptural Interpretation and the Mediaeval Theory of Preaching," *Speculum* 4 (1929): 283.

72. Frederick H. Dudden, *Gregory the Great: His Place in History and Thought* (London-New York: Longmans, Green & Co., 1905), p. 193.

73. A. Lecoy de la Marche, ed., *Anecdotes historiques . . . d' Etienne de Bourbon* (Paris: Librairie Renouard, 1877), pp. 4–5.

74. See the interesting collection edited by Thomas Wright, *Political Songs of England from the Reign of John to that of Edward II,* Camden Society, vol. 6 (London, 1839).

75. Zehava Jacoby, "Le chapiteau allégorique et le sermon: deux voies parallèles dans le processus créateur de l'imagerie romane," *Atti del Convegno Romanico mediopadano-romanico europeo* (Parma, 1982): 382–90.

76. For the basic concepts of medieval art, see the classic study of Emile Mâle, *The Gothic Image: Religious Art in France of the Thirteenth Century*, trans. Dora Nussey (1913, New York: Harper & Row, 1958).

77. Georges Duby, *Le temps des cathédrales: L'art et la société 980–1420* (Paris: Gallimard, 1976), pp. 18–19.

78. *Sancti Gregorii Magni Registrum Epistolarum*, letter III. 19, ed. D. Norberg,

Corpus Christianorum Series Latina, 140 A. (Turnhout, 1982), p.165. See also Daniel Arasse, "Entre dévotion et culture: Fonctions de l'image religieuse au XVe siècle," *Faire Croire: Modalités de la réception et de la diffusion des messages religieux du XIIe au XVe siècles* (Rome: Ecole Française de Rome, 1981), pp. 131–46.

79. Nurith Kenaan, "The Sculptural Lintels of the Crusader Church of the Holy Sepulchre in Jerusalem," in *Jerusalem in the Middle Ages,* ed. Benjamin Z. Kedar (Hebrew) (Jerusalem: Yad Ben Zvi, 1979), pp. 316–26.

80. Emile Mâle, *L'art religieux de la fin du moyen âge en France* (Paris: A. Colin, 1908), pp. 3–4.

81. William C. Jordan, "The Lamb Triumphant and the Municipal Seals of Western Languedoc in the Early Thirteenth Century," *Revue Belge de Numismatique* 123 (1977): 213–19.

82. G. T. Salusbury-Jones, *Street Life in Medieval England* (Sussex: Harvester Press, 1975), pp. 167–78.

83. Jacques Le Goff, "Au moyen âge: Temps de l'église et temps du marchand," *Annales E.S.C.* 15–3 (1960): 426.

84. The synodal statutes of Cambrai in 1277 et 1301 already mentioned the dominical procession.

85. Jean Delumeau, *La peur en Occident (XIVe—XVIIIe siécles)* (Paris: Fayard, 1978), p. 124.

86. Boccaccio, *Decameron*, pp. 1–4.

87. *Les Continuateurs de Guillaume de Nangis (H.F.* 20. 614); see also, *Chronique rimée attribuée à Geffroi de Paris (H.F.* 22. 160–61, v. 7359–80).

88. *Annales Paulini*, ed. W. Stubbs, *Chronicles of the Reigns of Edward I & II*, 2 vols. (London: *R.S.*, 1882), 1: 278.

89. On the rich literature in the subject one should mention, Ronald C. Finucane, *Miracles and Pilgrims: Popular Beliefs in Medieval England* (London: J. M. Dent & Sons, 1977); John H. Bernard, ed., *The Pilgrimage of S. Silvia of Aquitania to the Holy Land* (New York: AMS Press, 1971); Sidney H. Heath, *Pilgrim Life in the Middle Ages* (New York: Kennikat Press, 1971); Jonathan Sumption, *Pilgrimage: An Image of Medieval Religion* (London: Faber and Faber, 1975).

90. *Sancti Eusebii Hieronymi Epistulae*, Epistula 58 ad Paulinum, ed. Isidorus Hilberg, 3 vols., *Corpus Scriptorum Ecclesiasticorum Latinorum*, vols. 54–56 (Vienna: 1910), 54:533. See also Aryeh Grabois, "Medieval Pilgrims, the Holy Land and Its Image in European Civilization," in *Pillars of Smoke and Fire*, ed. Moshé Sharon (Johannesburg: Southern Books Pub., 1987): 65–79.

91. J. J. Jusserand, *English Wayfaring Life in the Middle Ages*, trans. Lucy Toulmin Smith (1889, London: Ernest Benn, 1950), p. 234.

92. The Church's reluctance to allow the clergy's uncontrolled leave to pelegrinage was faithfully expressed by the decree of Pope Alexander II, see *Decretum, C.I.C.*, 16. 1. 11; see also Giles Constable, "Monachisme et pélerinage au moyen âge," *R.H.* 258 (1977): 3–27.

93. *Statuts of Lincoln Cathedral*, ed. Henry Bradshaw and C. Wordsworth (Cambridge: Cambridge University Press, 1892), pp. 56–57.

94. Elisa M. Ferreira-Priegue, "Circulación y red viaria en la Galicia medieval," in *Les communications dans la péninsule Ibérique au moyen âge*, Actes du colloque tenu à Pau les 28–29 Mars 1980 sous la direction de P. Tucoo-Chala (Paris: Centre national de la recherche scientifique, 1981), pp. 65–67; Antonio Ubieto Arteta, "Los caminos que unían a Aragón con Francia durante la Edad Media," *Ibid.*, pp. 21–27.

95. *Miracula S. Thomae auctore Willelmo Cantuariensi*, in *Materials for the History of Thomas Becket*, ed. James Craigie Robertson, 7 vols. (London: *R.S.*, 1875), 1: 320.

96. Jean Gobi, *Scala Coeli*, ed. Marie Anne Mercoyrol (Ph. D. diss., Ecole des Hautes Etudes 1984–1985), *"confessio,"* nos. 247, 249.

97. See, for instance, C. Desimoni, "Actes passés à Famagouste de 1299 à 1301 par devant le notaire Lamberto di Sambuceto," *Revue de l'Orient Latin* 1 (1893): 333–34; Georg I. Bratianu, ed., *Actes des notaires génois de Pera et de Caffa de la fin du treizième siècle (1281–1290)* (Bucarest: Cultura nationala, 1927), pp. 276–77.

98. Charles E. Dufourcq, "Les communications entre les royaumes chrétiens ibériques et les pays de l'occident musulman dans les derniers siècles du moyen âge," *Les communications*, pp. 29, 43.

99. *Lettres de Jacques de Vitry*, ed. R. Huygens (Leyden: E. J. Brill, 1960), pp. 80–81.

100. Sire Jean de Joinville, *Histoire de St. Louis*, ed. J. Natalis de Wailly (Paris: Renouard, 1872), pp. 347–48.

101. "Placides et Timeo ou Livre des Secrets aus Philosophes," ed. Charles Langlois, *La vie en France au moyen âge*, 4 vols. (Paris: Hachette, 1926), 3: 318.

102. *Ibid.*, p. 127.

103. Bernard Bischoff, "The Study of Foreign Languages in the Middle Ages," *Speculum* 36 (1961): 210.

104. Hugo of Trimberg, "Der Renner," v. 3633–35, ed. G. Ehrismann, *Bibliothek des Literarischen Vereins*, vol. 247 (Tubingen, 1908), pp. 149 f.

105. Giraldus Cambrensis, "*De principis institutione*," ed. George F. Warner, *Opera*, 8 vols. (London: *R.S.*, 1861–1891), 8: 7, 42.

106. Arno Borst, *Der Turmbau von Babel: Geschichte der Meinungen über Ursprung und Viefalt der Sprachen und Völker* (Stuttgart: A.Hiersemann, 1957), II–2, p. 690.

107. *Ibid.*, p. 756.

108. Froissart, *Chroniques*, ed. H. Kervyn de Lettenhove, 25 vols (Brussels, 1867–1877), 2: 419.

109. *Clementinae*, 5. 1. 1. 2, *C.I.C.*, cols. 1179–80.

Chapter 2

1. Richard W. Southern, *Western Society and the Church in the Middle Ages* (1970, Middlesex: Penguin Books, 1972), pp. 15–16. For the historical background of this chapter, see Walter Ullman, *A Short History of the Papacy in the Middle Ages*. 2nd ed. (London: Methuen & Co., 1974).

2. Georges Duby, *Fondements d'un nouvel humanisme, 1280–1440* (Genève: Skira, 1966), p. 89.

3. The priority ascribed to the ecclesiastical order in the Reform schemes of Gregory VII was clearly reflected in his *Dictates of the Pope* (1075); see *Das Register Gregors VII*, II, 55a., ed. E. Caspar, 1929, (reprint Berlin: *M.G.H.*, 1967), *Epistolae Selectae*, 1: 202–8.

4. Dom Jean Leclercq, *L'idée de la royauté du Christ au moyen âge* (Paris: Cerf, 1959), pp.36–37; see also, Etienne Delaruelle, "La pietá populare nel secolo XI," *idem.*, *La piété populaire au moyen âge* (Turin: Bottega d'Erasmo, 1975), p. 18.

5. Much research had been done on the struggle between Boniface VIII and Philip the Fair. For the purposes of this study, see Georges Digard, *Philippe le Bel et le*

Saint Siége de 1285 à 1304 (Paris: Librairie du Recueil Sirey, 1936); Jean Rivière, *Le problème de l'église et de l'état au temps de Philippe le Bel* (Paris-Louvain: E. Champion, 1926), pp. 79 f.; Sophia Menache, "Un peuple qui a sa demeure à part: Boniface VIII et le sentiment national français," *Francia* 12 (1985): 193–208; P. Mury, "La bulle Unam sanctam," *R.Q.H.* 26 (1879): 41–130, 46 (1889): 253–57; Maurice Powicke, *The Christian Life in the Middle Ages* (Oxford: Clarendon Press, 1935), pp. 48–73.

6. *Extravag. Commun.*, 1. 8. 1., *C.I.C.*, 2: 1245–1246.

7. *Decretal. Gregor.*, 9. 1. 1. 1–2, *C.I.C.*, 2: 5–7.

8. On Manichaeism, a Persian gnostic sect uncompromisingly dualistic and its spread on medieval Europe, see, René Nelli, *La philosophie du catharisme: Le dualisme radical au XIIIe siècle* (Paris: Payot, 1978), pp.7–27.

9. Jacques de Viterbe, *De Regimine Christiano*, ed. H. X. Arquillière, *Le plus ancien traité de l'église* (Paris, 1926), pp. 118, 209–12.

10. P. Lapparent, *L'oeuvre politique de François de Meyronnes* (Paris, 1940–1942), pp. 60–61.

11. Johannes Quidort von Paris, *De Regia Potestate et Papali*, c. 3, ed. Fritz Bleienstein (Stuttgart: E. Klett, 1969), p. 81.

12. Aegidius Romanus, *De Ecclesiastica Potestate*, ed. Richard Scholz (Stuttgart: Druck der Union deutsche Verlagsgesellschaft, 1902), pp. 51, 77, 153.

13. Petrus Lombardi, *Liber de ecclesiasticis sacramentis*, 4. 1 (*P.L.* 192. 1089–91).

14. Donald M. Baillie, *The Theology of the Sacraments and Other Papers* (New York: Scribner, 1957), pp. 37–124.

15. *Mansi*, 24. 71.

16. Huguccio, *Summa*, ad D. 21, c. 2, quoted by J. Watt, "The Theory of Papal Monarchy in the XIIIth Century: The Contribution of the Canonists," *Traditio* 20 (1964): 257.

17. Jacques de Viterbe, pp. 204 f.; *Aegidius Romanus*, p. 109.

18. Raymundus Dreiling, *Der Konzeptualismus in der universalienlehre des franziskanerzbischofs Petrus Aureoli* (Munster: W. Aschendorff, 1913), p. 44.

19. *Decreti*, prima pars, 20. 1–3, *C.I.C.*, 1: 65–66.

20. Bryan Tierney, *Origins of Papal Infallibility (1150–1350)* (Leiden: E. J. Brill, 1972), pp. 15–50.

21. *Aegidius Romanus*, p. 194.

22. *Ibid.*, p. 152.

23. See the justification adduced by Pope Innocent III in 1202, "*Per Venerabilem,*" *Decretal. Gregor.* 9. 4. 17. 13, *C.I.C.*, 2: 716.

24. *Aegidius Romanus*, pp. 22–26, 28–30.

25. H. Leclercq, s.v. *"Ancilla Dei," Dictionnaire d'archéologie chrétienne et de liturgie*, 1–1: 1973–93; *idem.*, s.v. *"Servus servorum Dei," Ibid.*, 15–1: 1360–63.

26. "*Famuli vestrae pietatis*," Ep. 12 to Anastasius, *Decretum Gratiani*, 10. 96, *C.I.C.*, 1: 340.

27. Ep. n. 13, *Epistolae Henrici IV*, ed. Carl Erdmann, Deutches Mittelalter, Kritische Studien Texte, vol. 1, (Weimar: *M.G.H.*, 1950), p. 19.

28. Bernard de Clairvaux, *De Consideratione*, 4. 3. 7 (*P.L.* 182. 776).

29. Hugh de St. Victor, *De Sacramentis Christianae Fidei*, 2. 2. 4 (*P.L.* 176. 417).

30. John of Salisbury, *Policraticus*, 4. 3, 1: 515.

31. Alanus Anglicus, gloss to c. 7, "si duobus," quoted by T. Eschmann O.P., "St. Thomas on the Two Powers," *Medieval Studies* 20 (1958): 186.

32. *Acta Imperii inedita saeculi XIII et XIV*, ed. Eduard Winkelmann (Innsbruck: Wagner'sche Universitats-Buchhandlung, 1885), 2: 698.

33. S. Bonaventura, *De perfectione evangelica*, 4. 3 (7), in *S. Bonaventurae Opera Omnia*, 11 vols. (Quaracchi: Collegii S. Bonaventurae, 1891), 5: 190.

34. Roger Bacon, *Opus Majus*, ed. John H. Bridges, (1897, London: Williams & Norgate, 1914), 2: 227–28.

35. Arnold of Villanova, "De tempore adventus Christi," ed. C. Mirbt and K. Aland, *Quellen zur Geschichte des Papsttums und des romischen Katholizismus* (Tubingen: Siebeck, 1967), p. 460. *Jacques de Viterbe, 2. 7, Ibid.*, pp. 232–33.

36. Augustinus Triumphus, *Summa de potestate papae*, *Quellen*, p. 462. These ideas are based on *The Celestial Hierarchy* of Dionysius (c. AD 500), a mystical theologian who argued that the relations between God and the world are established by a hierarchically graded series of beings, the nine choirs of angels.

37. *Aegidius Romanus*, 2. 4, pp. 49–52, 1. 7, p. 26.

38. St. Augustine, *De civitate Dei contra paganos*, 4. 4, in *Oeuvres de St. Augustin*, ed. Gustave Bardy, 37 vols. (Paris: Desclée de Brouwer, 1959–), 33: 540; see also Walter Ullmann, *Medieval Papalism: The Political Theories of the Medieval Canonists* (London: Methuen, 1949), pp. 160–61.

39. *Aegidius Romanus*, p. 15, 54, 59.

40. *"Novit Ille," Decretal. Gregor.* 9, 2. 1, *de iudiciis*, c. 13, *C.I.C.*, 2: 243.

41. *Codex*, 2, 12, 10, William W. Buckland, ed., *Textbook of Roman Law from Augustus to Justinian* (Cambridge: Cambridge University Press, 1921), p. 710.

42. *"Plenitudo potestatis est in aliquo agente, quando illud agens potest sine causa secunda, quicquid potest cum causa secunda," Aegidius Romanus*, p. 190.

43. *"Credimus quod papa, qui est Christi vicarius, potestatem habet non tantum super Christianos, sed etiam super omnes potestatem," Clementinarum*, 5. 2, *C.I.C.*, 2: 1180–81.

44. *Dictatus papae, Das Register Gregors VII*, II, 55a, ed. E. Caspar, 1929 (reprint Berlin, 1967), *M.G.H., Epistolae Selectae*, 1: 202–8.

45. From the rich bibliography on the subject, see W. McCready, "Papalists and Antipapalists: Aspects of the Church/State Controversy in the Late Middle Ages," *Viator* 6 (1975): 241–73. S. Hendrix, "In Quest of the *vera ecclesia*: The Crises of Late Medieval Ecclesiology," *Viator* 7 (1976): 347–78; Sophia Menache, *The Status of the Papacy and the Image of the Popes at the Beginning of the Avignonese Period (1335–1334)*, (Ph.D. diss., Hebrew University, 1980), pp. 8–18.

Chapter 3

1. Guilhelmus Durandus, "Speculum juris," rubric *de legato*, in V.E. Hrabar, *De Legatis et Legationibus Tractatus Varii* (Dorpat: Typographeo Mattieseniano, 1905), p. 31.

2. Christopher Cheney, "Cardinal John of Ferentino, Papal Legate in England in 1206," *E.H.R.* 76 (1961): 654–60; see also Gabriel Le Bras, *Institutions ecclésiastiques de la Chrétienté médiévale* in *Histoire de l'église*, ed. Augustin Fliche and V. Martin, vol. 12 (Tournai 1959–1964), 2: 556–57.

3. *Councils and Synods with Other Documents Relating to the English Church*, eds. F. Maurice Powicke and Christopher Cheney, 2 vols. (Oxford: Clarendon Press, 1964), 2: 1297, 1353–56; see also, *Mansi*, 25. 539.

4. *Flores historiarum*, ed. H. Luard, 3 vols. (London: *R.S.*, 1890), 3: 136, 154; see also, Matthaeus Parisiensis, *Chronica Majora*, ed. H. Luard, 7 vols. (London: *R.S.*,

1872–1883), ad a. 1213, p. 571; ad a. 1225, p. 98; ad a. 1237, p. 403, 412; ad a. 1239, p. 567; ad a. 1245, p. 441.

5. Walter of Hemingbourgh, p. 371; Adam Murimuth, *Continuatio Chronicarum*, ed. E. Thompson (London: *R.S.*, 1889), p. 16.

6. *Marcha di Marco Battagli da Rimini*, ed. A. Massera, 1913 *(R.I.S.* 16–3. 17).

7. *Chronicon Estense*, ed. Giulio Bentoni and Emilio Vicini, s.d. (*R.I.S.* 15–3. 72).

8. *Ibid.*, pp. 88–89; *Lettres communes de Jean XXII (1316–1334)*, ed. Guillaume Mollat, 16 vols. (Paris: Bibliothèque des écoles françaises d'Athènes et de Rome, 1921–1947), nos. 12270, 14197; Peter Partner, *The Lands of St. Peter* (Berkeley: University of California Press, 1972), p. 308.

9. *Lettres communes*, no. 41361; Petri Cantinelli, *Chronicon*, ed. F. Torraca, 1902 (*R.I.S.* 28–2. 97).

10. *Cronaca Senese di autore anonimo*, ed. A. Lisini and F. Iacometti, in *Cronache Senesi* (*R.I.S.* 15–6. 133).

11. *Matthew Paris*, ad a. 1216, p. 645; ad a. 1255, p. 514; see also W. N. Bryant, "Matthew Paris, Chronicler of St. Albans," *History Today* 19 (1969): 772–82.

12. John of Trokelowe, *Annales*, ed. H. Riley, *Chronica Monasterii S. Albani*, 12 vols. (London: *R.S.*, 1866), 3: 78; Walsingham, *Historia Anglicana*, ed. H. Riley, 2 vols., *Chronica Monasterii S. Albani*, 1: 133–34; *Councils and Synods*, 2: 1377.

13. At this time Edward II was involved in a conflict with both the Earl of Lancaster and the Scots, see, M. H. Keen, *England in the Later Middle Ages* (London: Meuthen, 1973), pp. 61 f. On the goals of the papal policy, see, *Lettres communes*, nos. 5148, 5150–61.

14. *Flores historiarum*, p. 179.

15. *Adam Murimuth*, pp. 27–28.

16. *Registrum Palatinum Dunelmense: The Register of Richard de Kellawe, Bishop of Durham (1311–1316)*, ed. T. Hardy, 4 vols. (London: *R.S.*, 1873–1878), 4: 394–95; *Eulogium historiarum sive temporis . . . a monacho quodam Malmesburiensi exeratum*, ed. F. Haydon (London: *R.S.*, 1858–1863), p. 232.

17. *Lettres communes*, nos. 25139–40, 27390; *Mansi*, 25. 131, 225, 511, 770, 777.

18. Charles J. Héfèle and H. Leclercq, *Histoire des conciles*, 16 vols. in 8 (Paris: Letouzey et Ane, 1913), 5–2: 1087. Also reports of the Council of Vienne stressed the participation of prelates from Spain, Germany, Dacia, England, Scotland and Ireland, *Ibid.*, 6–2: 645.

19. *Ibid.*, 5–2: 1092, 1317.

20. See, for instance, the reports to Jayme II, king of Aragon, written by his representatives at the Council of Vienne, *Papsttum und Untergang des Templerordens*, ed. Heinrich Finke, 2 vols. (Munster: Aschendorffsche Buchhandlung, 1907), 2: 245, 251, 258, 287, 299.

21. See the faithful record sent to the German prelates by an anonymous author who participated in the Fourth Council of Lateran, A. García y García and S. Kuttner, "A New Eyewitness Account of the Fourth Lateran Council," *Traditio* 20 (1964): 115–78.

22. *Mansi*, 22. 991–92.

23. Christopher R. Cheney, *English Synodalia of the Thirteenth Century* (London: Oxford University Press, 1940), p. 33.

24. On the peace movement, see, Loren C. Mackinney, "The People and Public Opinion in the Eleventh Century Peace Movement," *Speculum* 5 (1930): 181–206. On its contribution to the consolidation of the French monarchy, see Aryeh Grabois, "De

la trêve de Dieu à la paix du roi: Etude sur les transformations du mouvement de la paix au XIIe siècle," in *Mélanges offerts à René Crozet*, ed. Pierre Gallais and Yves-Jean Rion, 2 vols. (Poitiers: Société d'Etudes Médiévales, 1966), 1: 585–96.

25. *Mansi*, 1–4. 19, 23–27, 29–34.

26. *Ex delatione Corporis S. Juliani (H.F.*, 10. 361).

27. *C.I.C.*, Decretum, Dist. 80. 1 (*urbes*); Dist. 99. 1 (*provintiae*), 13. 1. 1 (*ecclesias*).

28. *C.I.C.*, Decretum, 10. 1. 9–12.

29. *Matthew Paris*, 4: 497.

30. *Annales de Dunstaplia*, in *Annales Monastici*, 3: 147.

31. *Registrum John Peckham*, ed. C. T. Martin, 3 vols. (London: *R.S.*, 1882), 1: 363–64.

32. Georges Huard, "Considérations sur l'histoire de la paroisse rurale des origines à la fin du moyen âge," *R.H.E.F.* 24 (1938): 5.

33. *C.I.C.*, Decretum, 12. 2. 28; 7. 2. 27.

34. Jean Gaudemet, "La paroisse au moyen âge: état des questions," *R.H.E.F.* 59 (1973): 13. See also Bernard Guillemain, "Chiffres et statistiques pour l'histoire ecclésiastique du moyen âge," *Le moyen âge* 8 (1953): 341–65.

35. John Pontissara, *Registrum*, ed. C. Deedes, Surrey Record Society vols. I and VI (London, Canterbury & York Society, 1915–1924), p. 209.

36. *Mansi*, 22. 1007–10.

37. Jean Gobi, *Scala Coeli, confessio*, nos. 247, 248, 249, etc.

38. Jeannine Horowitz, *La peur en milieu ecclésiastique: XIIIe—première moitié du XIVe siècle* (Ph. D. diss., Ecole des Hautes Etudes, 1985), pp. 189–90.

39. *Registrum Ralph Baldock, G. Segrave, R. Newport et S. Gravesend,* ed. R. C. Fowler (Canterbury and York Society, 1911), pp. 73, 145–46.

40. L. Gougaud, "La danse dans les églises," *Revue d'histoire de l'église* 15 (1914): 5–22.

41. *Le Livre de Guillaume le Maire*, ed. Célestin Port (Paris 1874), p. 293; Guillaume Durand le Jeune, *Tractatus de modo generalis Concilii celebrandi* (Paris: F. Clousier, 1671), p. 340.

42. William Rishanger, *Chronica et Annales Regnantibus Henrico Tertio et Edwardo Primo*, ed. H. T. Riley (London: *R.S.*, 1865), pp. 74–75.

43. *Mansi*, 23. 784, 1061; 25. 757.

44. Caesarius von Heisterbach, *Die Fragmente des Libri VIII Miraculorum* (Rome: Römische Quartalschrift für Christliche Altertumskunde und für Kirchengeschichte, 1901), 8: 1–17.

45. H. Thurston, "The Bells of the Mass," *The Month* 123 (1914): 389–401; see also Gerald R. Owst, "The People's Sunday Amusements in the Preaching of Medieval England," *Holborn Review* 68 (1926): 32–45.

46. Paul Adam, *La vie paroissiale en France au XIVe siècle* (Paris: Sirey, 1964), pp. 87–91.

47. Grosseteste, *Epistolae*, ed. H. R. Luard (London: *R.S.*, 1861), pp. 155–56.

48. David Wilkins, *Concilia Magnae Britanniae et Hiberniae*, 4 vols. (London: R. Gosling, 1737), 1: 573.

49. Jean Claude Schmitt, "Du bon usage du 'Credo'," in *Faire croire*, pp. 354–55.

50. W. A. Pantin, *The English Church in the Fourteenth Century* (1955, reprint ed., Notre Dame-Indiana: Notre-Dame University Press, 1962), pp. 195–201.

51. André Vauchez, "Présentation," in *Faire croire*, pp. 10–13.

52. Gabriel Le Bras, "Les confréries chrétiennes: problèmes et propositions," in

idem., Etudes de sociologie religieuse, 2 vols. (Paris: Presses Universitaires de France, 1956), 2: 432–33.

53. G. Meersseman, O.P., "Etudes sur les anciennes confréries Dominicaines," *Archivum Fratrum Praedicatorum* 20 (1950): 10; 21 (1951): 57; 22 (1952): 15–16.

54. F. Jegou, "Confrérie de Saint Nicolas de Guérande, statuts de 1350," *Revue de Bretagne et de Vendée* 6–2 (1874): 8.

55. G. Meersseman,"Confréries dominicaines," 20 (1950): 74.

56. Etienne Delaruelle, "La piété populaire en Ombrie au siècle des communes," *idem., La piété*, pp. 45 f.

57. G. Meersseman, "Confréries dominicaines," 22 (1952): 105.

58. Elizabeth Siberry, "Missionaries and Crusaders, 1095–1274: Opponents or Allies?," *Studies in Church History* 20 (1983): 103–4.

59. Marie D. Chenu, *Introduction à l'étude de Saint Thomas d'Aquin* (Montréal: Institut d'études médiévales, 1950), Introduction; see also Robert I. Burns, "Christian Islamic Confrontation in the West: The XIIIth Century Dream of Conversion," *A.H.R.* 76 (1971): 1386–1434.

60. Robert Chazan, "The Barcelona 'Disputation' of 1263: Christian Missionizing and Jewish Response," *Speculum* 52 (1977): 830; *Id.,* "From Friar Paul to Friar Raymond: The Development of Innovative Missionizing Argumentation," *Harvard Theological Review* 76:3 (1983): 290.

61. *Bullarium Franciscanum Romanorum Pontificum*, ed. Joannes H. Sbaralea, 5 vols. (Rome: Typ. Sacrae congregationis de propaganda fidei, 1759–1921), 1: 376.

62. Zvi Baras, "Jewish-Christian Disputes and Conversions in Jerusalem" [Hebrew], *Jerusalem in the Middle Ages*, pp. 27–38.

63. The most popular Christian works were the *Pugio Dei adversus mauros et iudaeos* of Raymond Marti and the monumental *Summa contra gentiles* of Thomas Aquinas; see B. Blumenkranz, "Les auteurs chrétiens latins du moyen âge sur les Juifs et le Judaïsme," *Revue des études Juives* 109 (1948–1949): 3–67; 111 (1951–1952): 5–61; 113 (1954): 5–36; 114 (1955): 37–90; 117 (1958): 5–58; E.E. Urbach, "Etudes sur la littérature polémique au moyen âge," *Revue des études Juives*, 100 (1935): 49–77; Robert Chazan, "A Jewish Plaint to Saint Louis," *Hebrew Union College Annual* 45 (1974): 287–305.

64. Isidore Loeb, "La controverse de 1240 sur le Talmud," *Revue des études Juives* 1 (1880): 247.

65. Yitzhak F. Baer, "Criticism of the Disputations of Rabbi Yehiel of Paris and Nahmanides" [Hebrew], in *idem., Studies in the History of the Jewish People*, 2 vols. (1931, reprint Jerusalem: The Historical Society of Israel, 1985), 2: 128–34.

Chapter 4

1. Joannis Chrysostomi, *Commentarius in Sanctum Matthaeum Evangelistam*, Homilia 31–32 (*P.G.* 57. 374–76).

2. Hyeronimus, *Epistula 77 ad Oceanum, Epistulae* 55:48.

3. *Ex actibus pontificum Cenomannensium (H.F.* 10. 385).

4. Dujka Smoje, "La mort et l'au-delá dans la musique médiévale," in *Le sentiment de la mort au moyen âge*, ed. Claude Sutto, Etudes présentées au cinquiéme colloque de l'Institut d'études médiévales de l'Université de Montréal (Montréal: L'Aurore, 1979), pp. 255–56.

5. *Inferno*, c. III, l. 9, Dante Alighieri, *La Divina Commedia*, ed. L. Magugliani, 3 vols. (Milan: Biblioteca universale Rizzoli, 1949), 1: 24.

6. Arieh Sachs, "Religious Despair in Mediaeval Literature and Art," *Mediaeval Studies* 26 (1964): 256.

7. Philippe Aries, *Essais sur l'histoire de la mort en Occident du moyen âge à nos jours* (Paris: Seuil, 1975), p. 41.

8. Gustavo Vinay, *Il dolore e la morte nella spiritualitá dei secoli XII e XIII* (Todi, 1967), pp. 13–14.

9. Julien of Vézelay, *Sermons*, ed. Damien Vorreux, Collections sources chrètiennes, vol. 13 (Paris: Cerf, 1972), pp. 450–55.

10. The Dance of Death was a literary or artistic representation of a procession or dance, in which both the living and the dead take part. The dead may be portrayed by a number of figures or by a single individual personifying death. The living members are arranged in some kind of order of precedence, such as pope, cardinal, archbishop or emperor, king, duke, etc; see James M. Clark, *The Dance of Death in the Middle Ages and the Renaissance* (Glasgow: Jackson, Son & Company, 1950), p. 1, 111. This motif is faithfully presented in the closing scene of Bergman's movie, *The Seventh Seal*.

11. Michel Hebert, "La mort: impact réel et choc psychologique," in *Le sentiment de la mort*, pp. 21–23.

12. Philippe Aries, *L'homme devant la mort* (Paris: Seuil, 1977), pp. 13–96.

13. Thomas Aquinas, *Summa Theologica*, suppl., q. 75, "Of the Resurrection," art. I and II, ed. Fathers of the Dominican Province, 3 vols. (New York 1920 reprint: Benziger Brothers Inc., 1948), 3: 2874–75; Sophia Menache, "La naissance d'une nouvelle source d'autorité: L'université de Paris," *R.H.* 268–2 (1982): 320–27; Claude Carozzi and Huguette Traviani-Carozzi, *La fin des temps: Terreurs et prophéties au moyen âge* (Paris: Stock, 1982), pp. 37–45, 133–37.

14. St. Benedict, the well-known father of Western monasticism, had regarded the cult of fear just as a preliminary step to the very goal of Christian life, the *amor Dei*, love of God; see *Benedicti regula*, ed. R. Hanslik (Vienne: Corpus Scriptorum Latinorum editum consilio et impensis Academicae Scientiarum Austricae 75, 1960), cf. 7. On the cult of fear in the Central Middle Ages, see Jean Delumeau, *Le péché et la peur: La culpabilité en Occident* (Paris: Fayard, 1983), pp. 363–69; Jeannine Horowitz, *La peur*, pp. 180–82.

15. Arnold de Liége, *Alphabetum Narracionum*, ed. Colette Ribaucourt (Ph. D. diss.: Ecole des Hautes Etudes, 1985), cf. *abbas*.

16. *Liber Exemplorum ad usum praedicantium saeculo XIII compositus,* ed. Andrew G. Little (Aberdeen: British Society of Franciscan Studies, 1918), pp. 116, 63–64.

17. *Ibid.*, pp. 115, 41, 8.

18. Bernard de Clairvaux, for instance, mocked the care of the dead body, which it called *cadaver*; see *Epistola* 460 (*P.L.* 182. 655).

19. J. C. Payen, "Le *Dies irae* dans la prédication de la mort et des fins dernières au moyen âge," *Romania* 86 (1965): 66–73.

20. Hans Lietzmann, *The Era of the Church Fathers*, trans. Bertram Lee Wolf (New York: Meridian Books, 1950), pp. 124 f.

21. Gerald R. Owst, *Literature*, p. 528.

22. *Alphabetum narracionum*, "Timor;" Jacques de Voragine, *Legenda aurea*, trans. J. B. M. Roze (Paris: Garnier, 1967), p. 230.

23. See also, Guy H. Allard, "Dante et la mort," in *Le sentiment*, pp. 211–27.

24. Jean-Claude Schmitt, "Le suicide au moyen âge," *Annales E.S.C.* 31 (1976): 4–18.

25. "Suffrages profit not the damned, nor does the Church intend to pray for them," Thomas Aquinas, *Summa*, suppl., qu. 71, "Of the Suffrages for the Dead," art. 5, pp. 2847–49.

26. Jacques Le Goff, *The Birth of Purgatory*, trans. Arthur Goldhammer (1981, Chicago : Chicago University Press, 1984), p. 285.

27. S. Petri Damiani, *epistola 20 ad B. Monachum (P.L.* 144. 403).

28. Jean Gobi, *Scala coeli*, s.v. "Cruce signatis," no.409.

29. Jacques Chiffoleau, "Sur l'usage obsessionnel de la messe pour les morts à la fin du moyen âge," in *Faire croire*, pp. 235–42. On the importance ascribed to Mass and donations to the Church, see Bernard Saint-Pierre, "Mourir au XVe siècle: le testament de Jeanne d'Entrecasteaux," in *Le sentiment*, pp. 79–96.

30. *Councils and Synods*, 2: 895–96.

31. Louis Bourgain, *La chaire française au XIIe siècle* (1879, reprint Genève: Slatkine, 1973) p. 61.

32. Michel Zink, *La prédication*, p. 269.

33. *Ibid.*, p. 460.

34. S. Ivonis Carnotensis episcopi, *Epistola 1 ad Humbaldo Antissiodorensis episcopo, (P.L.*, 162. 169).

35. Huizinga, *The Waning*, pp. 134–45; Philippe Aries, "Huizinga et les thèmes macabres," in *Bijdragen en mededelingen Betreffende de geschiendenis der Nederlanden*, Colloque Huizinga (Gravenhage, 1973) 88(2): 246–57.

36. Jacques Chailley, *L'école musicale de Saint-Martial de Limoges jusqu'á la fin du XIe siècle* (Paris : Les livres essentiels, 1960), pp. 153–57.

37. See, for instance, the declarations of D. Peregrinus (April 2, 1212) and D. Gil Malrique (October 28, 1238) at their reception to the Order of Calatrava, *Bullarium Ordinis Militiae de Calatrava*, ed. D. Ignatius J. de Ortega et Cotes et al. (1747–1761, reprint Barcelona: El Albir, 1981), pp. 451–52.

38. Jonathan Riley-Smith, *The Knights of St. John in Jersualem and Cyprus, c. 1050–1310* (London: Macmillan, 1967), pp. 236 f.

39. *Etienne de Bourbon*, ex. 405, p. 355.

40. Thomas Cobham, *Summa Confessorum*, 4. 1. 7., ed. F. Bloomfield (Louvain: Analecta medievalia Namurcensia, 1968), p. 261.

41. *Manuel des péchés: Etude de littérature religieuse anglo-normande (XIIIe siècle)*, ed. Emile J. Arnould (Paris 1940), v. 2949–90.

42. *Etienne de Bourbon, exemplum* 56, p. 62.

43. Jeannine Horowitz, *La peur*, pp. 291–92.

44. *Etienne de Bourbon, exemplum* 32, pp. 41–42.

Chapter 5

1. Stanley J. Kahrl, "Introduction," in *The Holy War*, ed. Thomas P. Murphy (Ohio: Ohio State University Press, 1976), p. 3.

2. Carl Erdmann, *The Origins of the Idea of Crusade*, trans. Marshall W. Baldwin and Walter Goffart (1935, Princeton: Princeton University Press, 1977), pp. 332–68; see also, John Gilchrist, "The Erdmann Thesis and the Canon Law, 1083–1141," in *Crusade and Settlement*, ed. P.W. Edbury (Cardiff: University College Cardiff Press, 1985), pp. 37–45.

3. H. E. J. Cowdrey, "Pope Gregory VII's 'Crusading' Plans of 1074," in *Outremer: Studies in the History of the Crusading Kingdom of Jerusalem Presented to Joshua Prawer*, ed. Benjamin Z. Kedar et al. (Jerusalem: Yad Ben Zvi, 1982), pp. 27–40.

4. Palmer A. Throop, *Criticism of the Crusade: A Study of Public Opinion and Crusade Propaganda* (1940, Philadelphia: Porcupine Press, 1975), p. 4.

5. H. E. Cowdrey, "Pope Urban II's Preaching of the First Crusade," *History* 5 (1970): 178–86.

6. E. O. Blake, "The Formation of the Crusade Idea," *Journal of Ecclesiastical History* 21–1 (1970): 18.

7. Dana Carleton Munro, "The Speech of Pope Urban II at Clermont, 1095," *A.H.R.* 11 (1906): 231–42.

8. Fulcher of Chartres, *Historia Iherosolymitana: Gesta Francorum Iherusalem peregrinantium* (*RHC Occ.* 3. 323–24).

9. Robert of Reims, *Historia Iherosolimitana* (*RHC Occ.* 3. 728).

10. *Mansi*, 20. 816. This is the version reported by Bishop Lambert of Arras. The canons of Clermont have not survived in any official transcript, but are extant only through private collections and in personal notes made by participants at the council.

11. Fulcher of Chartres, p. 324.

12. Guibert of Nogent, *Historia quae dicitur Gesta Dei per Francos* (*RHC Occ.* 4. 138, 140).

13. Robert of Reims, p. 730.

14. Guibert of Nogent, p. 138.

15. *Die Kreuzzugsbriefe aus den Jahren 1088–1100*, ed. H. Hagenmeyer, (Innsbruck: Wargner'sche Universitats-Buchhandlung, 1901), pp. 136–37.

16. Robert of Reims, p. 727.

17. Fulcher of Chartres, p. 324.

18. Louise and Jonathan Riley-Smith, *The Crusades: Idea and Reality 1095–1274* (London: Edward Arnold, 1981), pp. 39–40.

19. Hagenmeyer, *Die Kreuzzugsbriefe*, p. 138.

20. The leaders of the First Crusade addressed Urban II as one "who by your sermons made us all leave our lands and whatever was in them," Robert of Reims, pp. 728, 730.

21. Baldric of Bourgueil, *Historia Jerosolimitana* (*RHC Occ.* 4. 16).

22. The synod held in Rouen in February, 1096, for instance, did not even mention the Crusade, although many other resolutions at Clermont were indeed cited; see *Mansi*, 20. 921–26.

23. Norman Housley, *The Italian Crusades: The Papal Angevin Alliance and the Crusades Against Christian Powers, 1254–1343* (Oxford: Clarendon Press, 1982), p. 113.

24. Jean Richard, "La papauté et la direction de la première croisade," *Journal des savants* (1960): 53, 58.

25. A. Lecoy de la Marche, "La prédication de la croisade au treizième siècle," *R.Q.H.* 42 (1890): 25.

26. Raimundi de Aguilers, *Historia francorum qui ceperunt Iherusalem* (*RHC Occ.* 3. 295–302).

27. *De Expugnatione Lyxbonensi*, ed. and trans. Charles Wendell David (1936, New York: Octagon Books, 1976), pp. 146–58; see also Yael Katzir, "The Vicissitudes of the True Cross of the Crusaders," in *The Crusaders in their Kingdom: 1099–1291*, ed. Benjamin Z. Kedar (Jerusalem: Yad Ben Zvi, 1987), pp. 243–53.

28. A. Lecoy de la Marche, "Prédication," pp. 15–16.

29. Colin Morris, "Propaganda for War: The Dissemination of the Crusading Ideal in the Twelfth Century," *Studies in Church History* 20 (1983): 86.

30. A. Lecoy de la Marche, "Prédication," p. 20.

31. Joshua Prawer, *A History of the Latin Kingdom of Jerusalem*, 2 vols. (Jerusalem: Mosad Bialik, 1963) [Hebrew], 1:251–82. A different view of the Second Crusade, as a pilgrimage, in Aryeh Grabois, "The Crusade of King Louis VII: A Reconsideration," in *Crusade and Settlement*, pp.94–104.

32. Contemporary stories tell that in his second meeting with King Conrad at Speyer, Bernard reproached him with the words of Isaiah: "What could have been done more to my vineyard that I have not done in it?" (Is. 5:4). The king knelt in all humility on his knees and proclamed at once his readiness to render to Jesus Christ the service he owed Him.

33. Giles Constable, "The Second Crusade as seen by Contemporaries," *Traditio* 9 (1953): 244–45.

34. Alexandri III *Epistola 1505 ad archiepiscopis, episcopis, abbatibus (P.L.* 200. 1297).

35. E. Caspar, "Die Kreuzzugsbullen Eugens III," *Neues Archiv. d. Gesch. f. altere deut. Geschichtskunde* 45 (1924): 285–86; see also the testimony of Pope Alexander III in 1265, *Epistola 360 ad principes, comites, barones et universos Dei fideles (P.L.* 200. 383–86).

36. C. J. Tyerman, "Philip VI of France and the Recovery of the Holy Land," *E.H.R.* 100 (1985): 40.

37. Jonathan Riley-Smith, *The First Crusade and the Idea of Crusading* (London: Athlone Press, 1986), pp. 63–79.

38. Robert of Reims, p. 730.

39. Guibert of Nogent, pp. 124, 140; see also Paul Alphandery and Alphonse Dupront, *La Chrétienté et l'idée de croisade*, 2 vols. (Paris: Albin Michel, 1954–1959), 1: 60–70.

40. Steven Runciman, *The First Crusade* (Cambridge: Cambridge University Press, 1980), pp. 59–64.

41. The exclusion of King Philip I, however, was related to his former excommunication for adultery, which at the time, was being reconsidered by the pope.

42. Robert of Reims, p. 731; see also E. O. Blake and C. Morris, "A Hermit goes to War: Peter and the Origins of the First Crusade," *Studies in Church History* 22 (1985): 79–108.

43. G. Miccoli, "La crociata del fanciulli del 1212," *Studi medievali* 2 (1961): 407–43; Malcolm Barber, "The Crusade of the Shepherds in 1251," *Proceedings of the 10th Annual Meeting of the Western Society for French History* (Lawrence, 1984), pp. 1–23.

44. C. F. Beckingham, *Between Islam and Christendom: Travellers, Facts and Legends in the Middle Ages and the Renaissance* (London: Variorum Reprints, 1987), pp.77f.

45. Baldric of Bourgueil, p. 16.

46. *Annales Herbipolenses*, ed. Georgius H. Pertz (*M.G.H., Scriptores*, 16. 3).

47. Benjamin Z. Kedar, "The Passenger List of a Crusader Ship, 1250: Towards the History of the Popular Element in the Seventh Crusade," *Studi Medievali* 13 (1972): 267–79.

48. *Ekkehardi Chronicon Universale*, ed. D.G. Waitz and P. Kilon (*M.G.H., Scriptores*, 6. 213).

49. *Chronica Sigeberti Gemblacensis*, ed. D. L. C. Bethmann (*Ibid.* 6. 367).

50. Jonathan Riley-Smith, *What were the Crusades?* (London: MacMillan Press, 1977), p. 15, see also Maureen Purcell, "Changing Views of Crusade in the Thirteenth Century," *The Journal of Religious History* 7 (1973): 8.

51. Riley-Smith, *What Were the Crusades?*, pp. 57–58.

52. Norman Cohn, *The Pursuit of the Millennium*, 2nd ed. (New York: Harper Torchbooks, 1961), p. 44.

53. Baldric of Bourgueil, pp. 16–17; Guibert of Nogent, p. 142.

54. Cornelius Tacitus, *De Germania*, ch. 14, ed. Henry Furneaux (Oxford: Clarendon Press, 1894), pp. 64–65.

55. Sophia Menache, "Un ideal caballerezco en el medioevo tardío: Don Alonso de Aragón," *Revista de la Universidad de Alicante* 6(1987): 9–29.

56. Henry Chadwick, *The Early Church* (London: Penguin Books, 1967), pp. 152–73.

57. Gregorius Turonensis *Historiae Francorum*, ed. Henri Omont (Paris: Berthaud, 1905), pp. 58–60.

58. Leo IV, *Epistola* 28 (*M.G.H., Epistolae*, 5. 601). John VIII, *Epistola* 150 (*M.G.H., Epistolae*, 7. 126 f).

59. According to St. Augustine, any war became a "Just War" if it was declared by a recognized authority, if there was some logical or ethical justification, if there was no other alternative to solve the problems that had brought about the war, and if it was carried out by accepted weapons; see Fernando Ortega, "La paz y la guerra en el pensamiento augustiniano," *Revista española de derecho canónico* 20 (1965): 5–35.

60. H. E. J. Cowdrey, "The Genesis of the Crusades: The Spring of Western Ideas of Holy War," in *The Holy War*, p. 16.

61. It has to be noted, however, that the warrior saints had achieved holiness not because of their military profession, but because of their pious behavior; see Carl Erdmann, *Idea of Crusade*, pp. 3–25.

62. *Das Register Gregors VII*, 2. 54, 1, pp. 199 f.; see also James A. Brundage, "Holy War and the Medieval Lawyers," in *The Holy War*, pp. 104–5.

63. The theological basis of indulgences, however, was consolidated later on, during the pontificate of Eugene III.

64. Jonathan Riley-Smith, *What Were the Crusades?*, p. 29.

65. M. Bennet, "The First Crusaders' Image of Muslims: The Influence of Vernacular Poetry," *Forum for Modern Language Studies* 22 (1986): 101–22.

66. Guibert of Nogent, pp. 138, 241.

67. Fulcher of Chartres, p. 328.

68. Baldric of Bourgueil, p. 14.

69. Robert of Reims, p. 763.

70. Raymond of Aguilers, p. 241.

71. J. H. and L. Hill, ed., *Historia de Hierosolymitano Itinere* (Paris: Paul Geuthner, 1977), p. 120.

72. Sylvia Schein, "The Crusades as a Messianic Movement," in *Messianism and Eschatology*, [Hebrew], ed. Zvi Baras (Jerusalem: Israel Historical Society, 1984), pp. 182–89.

73. Robert of Reims, pp. 739, 881–82.

74. *Gesta Francorum expugnantium Iherusalem* (*RHC Occ.* 3. 515).

75. *Ibid.*, p. 516; Robert of Reims, p. 723.

76. Bernard de Clairvaux, *De Laude Novae Militiae ad milites Templi* (*P.L.* 182. 924).

77. Orderic Vitalis, *Historia ecclesiastica*, ed. Marjorie Chibnall, 7 vols. (1969–1979), 5: 132.

78. Iohannes de Tulbia, *De Iohanne Rege Ierusalem*, in *Quinti Belli Sacri Scriptores Minores*, ed. R. Röhricht (Genève: Société de l'Orient latin, 1879), p. 122; *Liber duelii Christiani in obsedione Damiate, Ibid.*, p. 147.

79. Bernard de Clairvaux, *De Consideratione (P.L.* 182. 743–44).

80. Bernard McGinn, "Iter Sancti Sepulchri: The Piety of the First Crusaders," in *Essays in Medieval Civilization*, ed. R. E. Sullivan, et al., (Austin-London: University of Texas Press, 1978), p. 52.

81. Joshua Prawer, "Jerusalem in the Christian and Jewish Perspectives of the Early Middle Ages," *Settimane di Studi Sull' Alto Medio Evo* (Spoleto, 1980): 1–8.

82. "*Ordinacio de predicatione S. Crucis in Anglia*," ed. Röhricht, *Quinti belli*, p. 8.

83. R. Röhricht, "Die Pilgerfahrten nach dem Heiligen Lander vor den Kreuzzugen," *Historisches Taschenbuch*, fifth series, vol. 5 (Leipzig, 1875), p. 376.

84. Guibert de Nogent, p. 180, 211, 237; Fulcher of Chartres, p. 341; Robert of Reims, p. 727; see also Paul Alphandery, "Les citations bibliques chez les historiens de la première croisade," *Revue de l'histoire des religions* 99 (1929): 139–57; L. Siedl, "Images of the Crusades in Western Art: Models as Metaphors," in *The Meeting of Two Worlds* (Kalamazoo, 1986), pp. 377–92.

85. M. Markowski, "*Crucesignatus*: Its Origins and Early Usage," *Journal of Medieval History* 10 (1984): 157–65.

86. Jean Gobi, *Scala Coeli*, s.v. "*Cruce signatis*," nos. 401–4, 409.

87. Ernst H. Kantorowicz, " *Pro patria mori*," p. 481; Ivo of Chartres shared the same opinion; see *Decretum*, 10.87 (*P.L.* 161. 720).

88. "*Ahi, amors*," trans. Colin Morris, "Propaganda," p. 97.

89. A. Lecoy de la Marche, "La prédication," p. 7.

90. Georges Duby, *Cathédrales*, p. 70.

91. Etienne de Bourbon, "De diversis materiis praedicabilibus," 3. 7, in *Anecdotes historiques*, p. 172.

92. *Les registres d'Urbain IV*, ed. Joseph Guiraud, 5 vols. (Paris: Bibliothèque des écoles françaises d'Athènes et de Rome, 1901–1906), n. 471.

93. *Les registres d'Innocent IV*, ed. Elie Berger, 4 vols. (Paris: Bibliothèque des écoles françaises d'Athènes et de Rome, 1884), n. 2636.

94. There is a wide consensus between Innocent III, Urban IV, Gregory IX, Nicholas IV, and Boniface VIII, see, (*P.L.* 216. 433); O. *Raynaldus*, 22: 148–49, 23: 95–96; *Les registres de Boniface VIII*, ed. G. Digard et al., 4 vols. (Paris: Bibliothèque des écoles françaises d'Athènes et de Rome, 1884–1891), n. 868.

95. Innocent III, *Epistola 28 ad universis Christi fidelibus per Maguntinensem provinciam constitutis (P.L.* 216. 817).

96. Humbert de Romans, *Opus Tripartitum* 1. 16., ed. C. Brown, *Appendix ad fasciculum rerum expetendarum* (London 1690), 2: 196.

97. A. C. Krey, "Urban's Crusade: Success or Failure?," *A.H.R.* 53 (1948): 248–50.

98. Roger Bacon, *Opus majus*, pp. 120–22.

99. Humbert de Romans, *Opus tripartitum*, p. 195.

100. Hagenmeyer, *Kreuzzugsbriefe*, pp. 174–76.

101. *Mansi*, 22. 1062.

Chapter 6

1. Tacitus, *De Germania*, ch. XIV, trans. Alfred J. Church and William Jackson Brodribb (London, Macmillan, 1882), p. 712.

2. *Ibid.*, p. 716.

3. Jordanes, *Getica*, ed. Theodorus Mommsen, *Auctores antiquissimorum* (*M.G.H.*, Berlin: Weidmann, 1882, 5. 76).

4. "*Coronatio Philippi I seu ordo qualiter is in regem coronatus est*," (*H.F.* 11. 32–33).

5. E. Berger, "Les préparatifs d'une invasion anglaise et la descente de Henry III en Bretagne (1229–1230)," *B.E.C.* 54 (1893): 5–44.

6. Louis Halphen, "La place de la royauté dans le système féodal," *R.H.* 172 (1933): 256.

7. *Layettes du Trésor des chartes*, eds. Alexandre Teulet et al., 2 vols. (Paris: H. Plon, 1863–1909), 1: 559.

8. *Les Olim, ou registres des arrêts rendus par la cour du roi*, ed. Arthur Beugnot, 4 vols. (Paris: Collection de documents inédits sur l'histoire de France, 1839–1848), 2: 616.

9. *The Treatise on the Laws and Customs of the Realm of England Commonly Called Glanvill*, VII, 10, ed. G. D. Hall (London: Nelson, 1965), p. 84.

10. Hugh de Fleury, *Tractatus de regia potestate et sacerdotali dignitate*, ed. Ernest Sackur (*M.G.H., Libelli de Lite*, 2: 465–66), trans. Ewart Lewis, *Medieval Political Ideas*, 2 vols. (London: Routlege & Paul, 1954), 1: 166–68.

11. John L. McKenzie, *The Power and the Wisdom* (Milwaukee: Bruce Publishing Company, 1965), pp. 233–42.

12. Joannis Chrysostomi, *Commentarius in Epistolam ad Romanos (P.G.* 60. 615).

13. Augustine, *De Resurrectione mortuorum (P.L.* 39. 1600).

14. *De necessariis observantiis scaccarii dialogus*, ed. A. Hughes, C. G. Crump and Charles Johnson (London, 1902), p. 55.

15. Ranulphe d'Homblières, for instance, referred to God as the king of France, His bailiffs and provosts as priests, while demons were commissioned to execute His sentences in hell. The saints are members of His court, and St. Mary, as any queen, could always interfere and influence the king's sentence; see Nicole Beriou, "L'art de convaincre dans la prédication de Ranulphe d'Homblières, in *Faire croire*, p. 56. On the propaganda elements of the royal unction, see Aryeh Grabois, "La royauté sacrée au XIIe siècle: manifestation de propagande royale," in *Idéologie et Propagande en France*, pp. 35–41.

16. Barrows Dunham, *Heroes and Heretics: A Political History of Western Thought* (New York: Alfred A. Knopf, 1964), p. 7.

17. Ernst H. Kantorowicz, *Laudes Regiae: A Study in Liturgical Acclamations and Mediaeval Ruler Worship* (Berkeley: University of California Press, 1958), pp. 62–81.

18. Ernst H. Kantorowicz, *The King's Two Bodies: A Study in Mediaeval Political Theology* (Princeton: Princeton University Press, 1957), pp. 42–60.

19. Percy Ernst Schramm, *Herrschaftszeichen und Staatssymbolik, Schriften der M.G.H.* (Stuttgart, 1956), 1: 316 ff.

20. T. S. R. Boase, "Fontevrault and the Plantagenêts," *Journal of the British Archeological Association* 24 (1971): 1–10.

21. Elizabeth M. Hallam, "Royal Burial and the Cult of Kingship in France and England, 1060–1330," *Journal of Medieval History* 8 (1982): 359–80.

22. Gabrielle M. Spiegel, "The Cult of St. Denis and Capetian Kingship," *Journal of Medieval History* 1 (1975): 43–69.

23. Petri Blesensis, *Epistola ad clericos aulae regiae (P.L.* 207. 440).

24. Marc Bloch, *The Royal Touch*, trans. J. E. Anderson (1923, London: Routledge & K. Paul, 1973), pp. 3–12.

25. *Roberti regis diplomata* (*H.F.* 10. 612).

26. Marc Bloch, *The Royal Touch*, p. 1.

27. Norman Cohn, *Millenium*, pp. 81–97.

28. *"Reges a regendo et recte agendo,"* Isidori Hispalensis Episcopi *Etymologiarum,* I, XXIX–3, ed. Wallace M. Lindsay, 2 vols. (Oxford: Clarendon Press, 1911).

29. *Li proverbe au vilain,* ed. Charles V. Langlois, *La vie en France*, 2: 148, 209.

30. Augustine, *De civitate Dei contra paganos*, p. 140

31. Hugh of Fleury, above n. 10.

32. During the reign of Philip the Fair, the king's counselors, such as Enguerrand de Marigny, Pierre Flotte, Guillaume de Plaisans, and Guillaume de Nogaret provided the main target for the criticism of the royal policy; see Sophia Menache, "Philippe le Bel: Génèse d'une image," *Revue Belge de philologie et d'histoire* 72–4 (1984): 699.

33. *Dialogus de Scaccario*, 1: 59–60.

34. Manegold of Lautenbach, *Ad Gebehardum Liber*, ed. K. Francke (*M.G.H., Libelli de Lite*, 1. 365).

35. Iohannis Saresberiensis Episcopi Carnotensis *Policratici*, IV, c. 2, 1: 237f.

36. Kantorowicz, *The King's Two Bodies*, pp. 99–155; Gaines Post, "Law and Politics in the Middle Ages: The Medieval State as a Work of Art," in *Perspectives in Medieval History*, ed. Katherine Fischer Drew and Floyd Seyward Lear (Chicago: University of Chicago Press, 1963), p. 60.

37. Charles Wood, "The Mise of Amiens and Saint Louis' Theory of Kingship," *French Historical Studies* 6 (1970): 305–10.

38. Joinville, *The Life of St. Louis*, ch. CXLII, in *Chronicles of the Crusades*, ed. and trans. M. Shaw (Baltimore: Penguin Books, 1963), p. 342.

39. Numa Denis Fustel de Coulanges, *Saint Louis et le prestige de la royauté* (Colombes: Martin Morin, 1970), pp. 15–29.

40. Robert Folz, "La sainteté de Louis IX d'après les textes liturgiques de sa fête," *R.H.E.F.* 57 (1971): 31–40; William C. Jordan, *Louis IX and the Challenge of the Crusade* (Princeton: Princeton University Press, 1979), pp. 185–95.

41. Matthaei Parisiensis, *Chronica Majora*, 5: 466.

42. Salimbene de Adam, *Cronica*, ed. G. de Scalia (Bari: G. Laterza, 1966), pp. 429, 629.

43. *Chronicon Girardi ab Arvernia (H.F.* 21. 215).

44. *Majus Chronicon Lemovicense a Petro Coral et aliis conscriptum (H.F.* 21. 777).

45. J. Reviron, *Les idées politico-religieuses d'un évêque au IXe siècle et le "De Institutione Regia"* (Paris: J. Vrin, 1930), pp. 84–85, 140–41; see also Charlemagne's *Capitulare missorum generale (M.G.H., Legum*, 1. 91 f).

46. J. De Pange, *Le roi très chrétien* (Paris: Sireil, 1949), pp. 374–78.

47. *Continuatio chronici Odoranni monachi (H.F.*, 10. 165).

48. *Clausula de unctione Pippini*, ed. Arndt (*M.G.H., Scriptores Rerum Merovingiarum*, 1. 465–66).

49. Charles Wood, "Queens, Queans and Kingship: An Inquiry into Theories of Royal Legitimacy in Late Medieval England and France," in *Order and Innovation in*

the Middle Ages: Essays in Honor of Joseph Strayer, ed. William C. Jordan et al. (Princeton: Princeton University Press, 1976), pp. 396–400.

50. *English Constitutional Documents: 1307–1485*, ed. Eleanor Lodge and Gladys Thornton (New York: Octagon Books, 1972), pp. 11–12.

51. William Huse Dunham, Jr. and Charles Wood, "The Right to Rule in England: Depositions and the Kingdom's Authority, 1327–1485," *A.H.R.* 81 (1976): 738–60.

52. *Select Charters and Other Illustrations of English Constitutional History . . . to the Reign of Edward the First*, ed. William Stubbs, 9th ed. (1870, Oxford: Clarendon Press, 1929), p. 158.

53. *English Historical Documents: 1189–1327*, gen. ed., David C. Douglas, 11 vols., vol. 3, ed. Harry Rothwell (London: Eyre & Spottiswoode, 1975), p. 525; H. G. Richardson, "The Coronation Oath in Medieval England: The Evolution of the Office and the Oath," *Traditio* 16 (1960): 111–202.

54. See above note 10.

55. Walter Ullmann, "A Medieval Document on Papal Theories of Government," *E.H.R.* 61 (1946): 192–94.

56. Gaines Post, "Two Notes on Nationalism in the Middle Ages," *Traditio* 9 (1953): 289–91.

57. Ernst H. Kantorowicz, "*Pro patria mori* in Medieval Thought," *A.H.R.* 56 (1951): 488.

58. Gaines Post, "Two Notes," 281–89.

59. Johannes Quidort von Paris, *De regia potestate et papali*, pp. 111–13.

60. Sophia Menache, "Un peuple," 198–208.

61. Joseph Strayer, "France, the Holy Land, the Chosen People and the Most Christian King," in *Medieval Statecraft and the Perspectives of History* (Princeton: Princeton University Press, 1969), pp. 3–16.

Chapter 7

1. On the historical background of this chapter see, Charles Petit Dutaillis, *The Feudal Monarchy in France and England*, trans. E. D. Hunt (1936, London: Routledge & Kegan, 1966) and the rich bibliography it contains.

2. V. Krause, "Geschichte der Missi dominici," *Mitteilungen des Instituts für osterreichische Geschichtsforschung* 11 (1890): 193–214; Francois L. Ganshof, "Merowingisches Gesandtschaftswesen," in *Aus Geschichte und Landeskunde.* Foreschungen und Darstellungen. Franz Steinbach zum 65. Geburstag gewidmet von seinen Freunden und Schulern (Bonn, 1960), pp. 166–83.

3. Gustave Alef, "The Origin and Early Development of the Muscovite Postal Service," *Jahrbucher für Geschichte Osteuropas* 15 (1967): 1–15.

4. Frederick Furnivall, *King Edward II's Household and Wardrobe Ordinances, 1323* (London: Chaucer Society, 1876), p. 46.

5. Mary C. Hill, "King's Messengers and Administrative Developments in the Thirteenth and Fourteenth Centuries," *E.H.R.* 61 (1946): 315–28.

6. Georges d'Avenel, *L'évolution des moyens de transport* (Paris, 1919), p. 142.

7. *Regesta Imperii*, ed. Johann F. Böhmer, 11 vols. (Innsbruck: Wagner, 1893), 2–2: 7.

8. Jerónimo Zurita, *Anales de la Corona de Aragón,* 19. 44, ed. Angel Canellas Lopez, 8 vols. (Zaragoza: Fernando el Católico, 1977), 8: 179

9. *Foedera*, 1–2. 194.

10. *Foedera*, 1–1. 44.

11. K. Edwards, "The Political Importance of the English Bishops During the Reign of Edward II," *E.H.R.* 59 (1944): 311–47.

12. Among the many English bishops who actively participated in royal diplomatic missions in the early-fourteenth century, these could be mentioned: Anthony Beck (Durham, 1283–1311), Henry Burghersh (Lincoln, 1320–1340), Richard of Bury (Durham, 1333–1325), Thomas Cherleton (Hereford, 1327–1344), William Gainsborough (Worcester, 1302–1307), John Grandisson (Exeter, 1327–1369), Walter Langton (Coventry and Lichfield, 1296–1321), William Melton (York, 1316–1340), and Adam Orleton (Hereford, Worcester, Winchester, 1317–1345), see Sophia Menache, *Papacy at the Beginning of the Avignonese Period*, pp. 212–18.

13. Henry S. Lucas, "Diplomatic Relations Between England and Flanders, 1329–1336," *Speculum* 11 (1936): 59–87.

14. Henry S. Lucas, "The Machinery of Diplomatic Intercourse," pp. 303–4, 320–23.

15. *Foedera*, 1–3. 139–40.

16. Donald E. Queller, *The Office of Ambassador in the Middle Ages* (Princeton: Princeton University Press, 1967), p. 13.

17. *Codex*, 2, 12, 10, *Roman Law*, p. 710.

18. Helen Maud Cam, "Medieval Representation in Theory and Practice," *Speculum* 29 (1954): 347–55.

19. In a document of 1235, King Henry III declared that procuration was given "because many things can arise unexpectedly which need a rapid solution."

20. *Flores historiarum*, 2: 223.

21. Sophia Menache, "Isabelle of France, Queen of England: A Reconsideration," *Journal of Medieval History* 10 (1984): 110–12.

22. *Acta Imperii Selecta*, ed. Johann F. Böhmer (Innsbruck: Wagner, 1870), p. 422. On the historical background and the consequences of the *Romfahrt*, see W. Bowsky, *Henry VII in Italy: The Conflict of Empire and City-State* (Lincoln: University of Nebraska Press, 1960), pp. 25–42.

23. H. Offler, "Empire and Papacy: The Last Struggle," *Transactions of the Royal Historical Society*, 5 ser. 6 (1956): 21–47.

24. *Foedera*, 1–2. 29–30; Jocelyne G. Dickinson, "Blanks and Blank Charters in the Fourteenth and Fifteenth Century," *E.H.R.* 66 (1951): 375–87.

25. J. E. Neale, "The Diplomatic Envoy," *History* 13 (1928): 204.

26. François L. Ganshof, *Le Moyen Age, Histoire des relations internationales*, vol. 1 (Paris: Hachette, 1953), pp. 43–44, 278–79.

27. Philippe de Comynes, *Mèmoires*, III, ch. 8, ed. Joseph Calmette, 3 vols. (Paris: Les Classiques de l'histoire de France, 1924–1925), 1: 218–19.

28. Garret Mattingly, "The First Resident Embassies: Mediaeval Italian Origins of Modern Diplomacy," *Speculum* 12 (1937): 439.

29. Vivian H. Galbraith, *The Making of Domesday Book* (Oxford: Clarendon Press, 1961); Rex Welldon Finn, *The Domesday Inquest and the Making of Domesday Book* (London: Longmans, 1961).

30. According to Einhard, the Merovingian kings "used to ride in a cart, drawn by a yoke of oxen, driven, peasant-fashion, by a ploughman," see Einhard, *The Life of Charlemagne*, ch. 1, ed. and trans. Sidney Painter (Michigan: University of Michigan Press, 1960), p. 24.

31. Lady Stenton, "Communications," in *Medieval England*, ed. Austin Lane

Poole, 2 vols. (Oxford: Clarendon Press, 1958), 1: 200–201. On the itinerary of the Norman kings see the detailed research of W. Farrer, "An Outline Itinerary of Henry I," *E.H.R.* 24 (1919): 303–82, 505–79.

32. Frank Stenton, "The Road Systems of Medieval England," *Economic History Review* 7 (1936–1937): 1–21; James F. Willard, "Inland Transportation in England During the Fourteenth Century," *Speculum* 1 (1926): 361–74.

33. Grace Stretton, "The Travelling Household in the Middle Ages," *Journal of the British Archaeological Association*, n.s. 40 (1940): 75–103.

34. Peter of Blois, *Opera omnia*, ed. John A. Giles, 4 vols. (Oxford: Patres Ecclesiae Anglicanae, 1843–1848), 1: 195.

35. *Richeri Historiarum libri III*, ch. II, ed. Georgius Waitz (Hanovre: Hahnianus, 1839); The very fact that Pope John XXII repeated the same advice before Edward III in 1331 hints at the lasting gap between theory and practice; see Rainaldus, *Annales ecclesiastici*, ad a. 1331, p. 520.

36. William Stubbs, *Select Charters*, p. 370, 383.

37. *Recueil des actes de Philippe Auguste*, ed. Henri F. Delaborde et al., 2 vols. (Paris: Académie des Inscriptions et Belles Lettres, 1916), 1: 416–20.

38. Albert Beebe White, "Was There a 'Common Council' Before Parliament?," *A.H.R.* 25 (1919): 17.

39. Humbert de Romans, *De eruditione praedicatorum*, II, ii, c. 86, p. 559.

40. Roger of Howden, *Gesta Regis Henrici Secundi*, ed. William Stubbs, 1: 107, 139, 336.

41. Antonio Marongiu, *Medieval Parlements*, trans. S. J. Woolf (London: Eyre & Spottiswoode, 1968), pp. 45, 80–93.

42. Guillaume de Nangis, *Gestae Sanctae Memoriae Ludovici Regis Franciae (H.F.* 20. 428).

43. Thomas N. Bisson, "The General Assemblies of Philip the Fair, Their Character Reconsidered," *Studia Gratiana* 15 (1972): 537–64.

44. Ada E. Levett, "The Summons to a Great Council, 1213," *E.H.R.* 31 (1916): 85–90.

45. Benedict of Peterborough, *Gesta regis Henrici secundi*, ed. William Stubbs, 2 vols. (London: *R.S.*, 1867), 2: 178.

46. *English Historical Documents*, 3: 396. See the original document in W. Stubbs, *Select Charters*, pp.441–42.

47. J. Taylor, "The Manuscripts of the *Modus Tenendi Parliamentum," E.H.R.* 83 (1968): 673–88.

48. On the parallel function in France, see Felix Aubert, "Les huissiers du parlement de Paris, 1300–1420," *B.E.C.* 47 (1886): 370–93.

49. *English Historical Documents*, pp. 924–27.

50. A. F. Pollard, "History, English and Statistics," *History* 11 (1926–1927): 23.

51. J. S. Roskell, "The Problem of the Attendance of the Lords in Medieval Parliaments," *Bulletin of the Institute of Historical Research* 29 (1956): 155.

52. Thomas N. Bisson, "Consultative Functions in the King's Parliament (1250–1314)," *Speculum* 44 (1969): 353.

53. Vincent de Beauvais, *Speculum Quadruplex, sive Speculum Maius Naturale, Doctrinale, Morale, Historiale*, 4 vols. (reprint Graz 1964–1965), 3: 1287.

54. Elizabeth Brown, " 'Cessante Causa' and the Taxes of the Last Capetians: The Political Applications of a Philosophical Maxim," *Studia Gratiana* 15 (1972): 565–87.

55. Gaines Post, "A Roman-Canonical Maxim: "Quod omnes tangit" in Bracton and Early Parliaments," *Traditio* 4 (1946): 197–251; Antonio Marongiu, "Il principio

della democrazia e del consenso: "*quod omnes tangit ab omnibus approbari debet*" nel secolo XIV," *Studia Gratiana* 8 (1962): 555–75.

56. Hilary C. Jenkinson and Mabel H. Mills, "Rolls from a Sheriff's Office of the Fourteenth Century," *E.H.R.* 43 (1928): 21–32; Helen Maud Cam, "Shire Officials: Coroners, Constables and Bailiffs," in *The English Government at Work 1327–1336*, 3: 143–183.

57. *De necessariis observantiis scaccarii dialogus, qui vulgo dicitur Dialogus de Scaccario*, p. 84.

58. *English Historical Documents*, 3: 463.

59. H. Gravier, "Essai sur les prévôts royaux du XIe au XIVe siècle," *Nouvelle revue historique du droit français et étranger* 27 (1903): 648 f.

60. *Ordonnances des roys de France de la troisième race*, ed. E. Laurière, 21 vols. (1727–1734, reprint Farnborouth: Gregg Press, 1967–1968), 1: 358, 12: 449.

61. *Recueil des actes de Philippe Auguste*, loc. cit., (note 37).

62. Joseph Strayer, *The Reign of Philip the Fair* (Princeton: Princeton University Press, 1980), pp. 143, 100 f.

63. Sophia Menache, "Contemporary Attitudes Concerning the Templars' Affair: Propaganda Fiasco?," *Journal of Medieval History* 8 (1982): 137.

64. Matthew Paris, *Chronica Majora*, 4: 638–39.

65. Sire de Joinville, *Crusade of St. Louis*, pp. 322–23.

66. *Ibid.*, chap. 143; Lester K. Little, "Saint Louis Involvement with the Friars," *Church History* 33 (1964): 133–34.

67. Léopold Delisle, "Preuves de la Préface," (*H.F.* 24. 318).

68. Charles V. Langlois, "Doléances recueillies par les enquêteurs de Saint Louis et des derniers Capétiens directs," *R.H.* 100 (1909): 63–95.

69. John B. Henneman, "Enquêteurs-réformateurs and Fiscal Officers in Fourteenth Century France," *Traditio* 24 (1968): 309–49.

70. Alexander Wyse, "The enquêteurs of Louis IX," *Franciscan Studies* 25–1 (1944): 62.

Chapter 8

1. Paul Kecskemeti, "Propaganda," *Handbook of Communication*, ed. Ithiel de Sola Pool et al. (Chicago: Rand McNally 1973), pp. 845–48.

2. Jacques Ellul, *Propaganda: The Formation of Men's Attitudes*, trans. Konrad Kellen and Jean Learner (New York: A. A. Knopf, 1965), p. 4.

3. Harold D. Lasswell and Dorothy Blumenstock, *World Revolutionary Propaganda* (New York: A. A. Knopf, 1939), p. 5.

4. Ellul, *Propaganda*, p. 4.

5. Frederick, E. Lumley, *The Propaganda Menace* (New York-London: The Century Co., 1933), p. 44.

6. Uzi Elyada, *Manipulation et Théatralité: Le père Duchesne, 1788–1791* (Ph. D. Diss.: Ecole des hautes études, 1984–1985), pp. IX–XII, 348.

7. Garth S. Jowett, "Propaganda and Communication: The Re-emergence of a Research Tradition," *Journal of Communication* 37–1 (1987): 97–98.

8. Jurgen Habermas, *Communication and the Evolution of Society*, trans. Thomas McCarthy (Boston: Beacon Press, 1979), pp. 178–83.

9. On the historical background, see, Joseph Strayer, *Philip the Fair*, pp. 314 ff.; M. H. Keen, *England*, pp. 27 ff.

10. Fulcher of Chartres, p. 468.

11. Bernard Guenée, "Etat et nation en France au moyen âge," *R.H.* 237 (1967): 17–30.

12. The accounts of numerous cross-taking ceremonies and the significance of Crusade in contemporary legacies, corroborate indeed the undiminished popularity of the Crusade at the early-fourteenth century.

13. Simon Lloyd, "Political Crusades in England c. 1215–1217 and c. 1263–1265," in *Crusade and Settlement*, p. 113.

14. *Selected Letters of Pope Innocent III Concerning England (1198–1216)*, ed. Christopher R. Cheney and W. H. Semple (London: Th. Nelson, 1953), pp. 207–9.

15. *Ibid.*, p. 227.

16. Anthony Luttrell, "The Crusade in the Fourteenth Century," in *Europe in the Late Middle Ages*, ed. J. R. Hale et al. (London: Faber and Faber, 1965), p. 126.

17. The tithe was an income tax of ten percent or less, enforceable by law, which was of three sorts: praedial, of the fruits of the ground; personal, of the profits of labor; and mixed, arising partly from the ground and partly from the industry of men. Tithes fell especially upon agriculture, livestock, oil, wine, fish, game, mills, ovens, and on personal income. Under this last category, the tithe touched most human activities. Industry, commerce, and mines came under it; see Paul Viard, *Histoire de la dîme ecclésiastique principalement en France jusqu'au décret de Gratien* (Dijon: Jobard, 1909), pp. 10 f.

18. See the bulls of Honorius III, Gregory IX and John XXII in this regard, *Bullarium Ordinis Militiae de Calatrava*, pp. 55, 57, 73, 177–9; Peter Linehan, "The Church, the Economy and the *reconquista* in Early XIVth Century Castile," *Revista española de teología* 43 (1983): 275–303.

19. H. Deighton, "Clerical Taxation by Consent, 1279–1301," *E.H.R.* 68 (1953): 161–92.

20. Robert I. Burns, "A Medieval Income Tax: The Tithe in the Thirteenth Century Kingdom of Valencia," *Speculum* 41 (1966): 444–45.

21. William E. Lunt, *Financial Relations of the Papacy With England to 1327* (Cambridge, Mass.: Medieval Academy of America, 1939), pp. 307–19.

22. Charles Samaran and Guillaume Mollat, *La fiscalité pontificale en France au XIVe siècle* (1905, reprint Paris: E. de Boccard, 1968), pp. 12–16.

23. C. J. Tyerman, "*Sed nihil fecit*?: The Last Capetians and the Recovery of the Holy Land," in *War and Government in the Middle Ages: Essays in Honour of J. O. Prestwich*, ed. J. Gillingham and J. C. Holt (Woodbridge: Boydell Press, 1984), p. 172.

24. C. J. Tyerman, "Philip VI and the Recovery," p. 42.

25. *Clément VI: Lettres cioses, patentes et curiales*, ed. Eugène Deprez et al., Bibliothèque des écoles françaises d'Athènes et de Rome, vol. 3 (Paris, 1901), no. 914.

26. *Foedera* 1–4. 56.

27. *Ibid.*, p. 68.

28. *Regestum Clementis Papae V*, ed. A. Tosti et al., 8 vols. (Rome: Typographia Vaticana: 1885–1892), nos. 243, 244.

29. Already in the mid-twelfth century, Hugh of St. Victor claimed that when reason and necessity concurred, royal authority had the right to tax the clergy; see Gaines Post, "Law and Politics," p. 170.

30. Joseph Strayer, "France, the Holy Land," p. 308.

31. Elizabeth Brown, "Taxation and Morality in the XIIIth and XIVth Centuries: Conscience and Political Power and the Kings of France," *French Historical Studies* 8–1 (1973): 26–28; see also Joseph Strayer and C. H. Taylor, *Studies in Early French Taxation* (Cambridge, Mass.: Harvard University Press, 1939), pp. 151 f.

32. Ernst H. Kantorowicz, "*Pro patria mori*," pp. 478–79.

33. Renè Nelli, *Ecrivains anti-conformistes du moyen âge occitain* (Paris: Phebus, 1977), pp. 357–61; J. Robert Wright, *The Church and the English Crown, 1305–1334* (Toronto: Pontifical Institute of Mediaeval Studies, 1980), pp. 172–73; Louis Caillet, *La papautè d'Avignon et l'èglise de France* (Paris: Presses Universitaires de France, 1975), pp. 373f.

34. N. Housley, "Pope Clement V and the Crusades of 1309–1310," *Journal of Medieval History* 8 (1982): 41–42.

35. *Regestum Clementis papae V*, nos. 9941–62.

36. *Foedera* 2–2. 89.

37. *Lettres secrètes et curiales du pape Jean XXII relatives à la France*, ed. Auguste Coulon and S. Clémencet, Bibliothèque des écoles françaises d'Athènes et de Rome, 9 vols. in 4 (Paris: E. de Boccard, 1900–1960), nos. 74, 75, 1051; *Lettres communes*, no. 6808.

38. *The Declaration of Arbroath*, ed. Sir James Fergusson (Edinburgh: Edinburgh University Press, 1970), pp. 6–8.

39. *Foedera*, 1–4. 98, 100, 104, 109, 110, 141, 142.

40. C. Köhler and C. Langlois, "Lettres inédites concernant les croisades," *B.E.C.* 3 (1891): 61–62.

41. Simon Lloyd, "The Lord Edward's Crusade, 1270–1272: Its Setting and Significance," in *War and Government*, p. 120.

42. *Ordonnances des roys de France*, 1: 422.

43. *Documents relatifs aux états généraux et assemblées réunis sous Philippe le Bel*, ed. Georges Picot, Collection des documents inédits sur l'histoire de France, vol. 11 (Paris: Imprimerie nationale 1901), p. 490.

44. L. Carolus-Barré, "Les enquêtes pour la canonisation de Saint Louis," *R.H.E.F.* 57 (1971): 19–29.

45. See Pope Clement's enthusiastic descriptions of the Crusader zeal of Philip the Fair, *Registrum*, nos. 2986, 2988, 3218. 3219.

46. *Excerpta e memoriale historiarum auctore Johanne Parisiensi, (H.F.* 21. 635, 646).

47. *Ordonnances*, 1: 330, 333, 473.

48. *Histoire de St. Louis*, ed. M. Natalis de Wailly, p. 68.

49. *Ordonnances*, 1: 330, 472–473, 501.

50. *E memoriale historiarum*, pp. 656–57; *Geoffrey de Paris*, pp. 134–36.

51. *Chroniques de St. Denis* (*H.F.* 20. 689); *Anciennes chroniques de Flandre* (*H.F.* 22. 399).

52. Huizinga, *The Waning*, p. 239.

53. *Fragment d'une chronique anonyme finissant en 1328* (*H.F.* 21. 150).

54. "L'image du monde," ed. by Charles Langlois, *La vie en France*, 3: 159.

55. *Le roman de Fauvel par Gervais du Bus*, ed. Arthur Langfors, Société des anciens textes français (Paris 1914–1919), p. 116.

56. Gaston Paris, "Le roman de Fauvel," *Histoire littéraire de la France* 32 (1898): 138.

57. *Guillaume de Nangis*, p. 320.

58. *Paris et ses historiens aux XIVe et XVe siècles*, ed. Antoine Le Roux de Lincy

and L. M. Tisserand, Histoire générale de Paris (Paris: Imprimerie impériale, 1867), p. 25.

59. Henri de Ferriéres, "Livres du roi modus et de la reyne ratio," ed. Daniel Poiron, *Le moyen âge (1300–1480)*, 2 vols. (Paris: Arthaud, 1970–1971), 2: 249.

60. Tyerman, "*Sed nihil*," p. 172.

61. *La Chanson de Roland*, ed. Joseph Bédier, 3rd. ed. (1922, Paris: H. Piazza, 1964), p. 96.

62. Richier, *La vie de Saint-Remi poème du XIIIe siècle*, ed. W. N. Bolderston (London: H. Frowde, 1912), lines 46 ff. On the other hand, Pierre Jame wrote in the early-fourteenth century that the deaths of men perishing in unjust wars were imputed as homicide to the princes responsible for the wars; see Petrus Jacobus, *Aurea practica libellorum* (Cologne, 1574), p. 279.

63. Teófilo R. Ruiz, "Reaction to Anagni," *The Catholic Historical Review* 65 (1979): 385–401.

64. M. Langlois, "Une réunion publique à Paris sous Philippe le Bel," *Bulletin de la société d'histoire de Paris* 15 (1888): 134.

65. Pierre Dupuy, *Histoire du différend entre le pape Boniface VIII et Philippe le Bel* (Paris: S. et G. Cramoisy, 1655), pp. 324 ff.

66. Heinrich Finke, *Papsttum*, 2: 139, 145.

67. Ernest Renan, *Etudes sur la politique religieuse du règne de Philippe le Bel* (Paris: C. Levy, 1899), pp. 156 ff.

68. Jean Leclercq, "Un sermon prononcè pendant la guerre de Flandre sous Philippe le Bel," *Revue du moyen âge latin* 1 (1945): 170.

69. J. Schwalm, "Beitrage zur Reichsgeschichte," *Neues Archiv der Gesellschaft für altere deutsche Geschichtskunde* 25 (1899–1900): 564–65.

70. *Ordonnances*, 1: 643–44.

71. *Ibid.*, p. 643; Charles Taylor, "French Assemblies and Subsidy in 1321," *Speculum* 43 (1968): 228.

72. Philippe de Mézières, *Le Songe du vieil pélerin*, ed. George W. Coopland (London: Cambridge University Press, 1969), pp. 398–99.

73. *Foedera* 2–3. 130–31, 149.

74. Pierre Dubois, *De recuperatione Terrae Sanctae*, ed. Angelo Diotti (Florence: Leo S. Olschki, 1977), pp. 192 f.

75. *Ibid.*, pp. 133–34.

76. *Ibid.*, p. 159.

77. W. Brandt, "Pierre Dubois, Modern or Medieval?," *A.H.R.* 35 (1929–1930): 514 f.

78. Sylvia Schein, "The future *regnum Hierusalem*: A Chapter in Medieval State Planning," *Journal of Medieval History* 10 (1984): 103.

79. C. J. Tyerman, "Philip V of France, the Assemblies of 1319–1320 and the Crusade," *Bulletin of the Institute of Historical Research* 57 (1984): 15.

Chapter 9

1. Isidor Ginsburg, "National Symbolism," in *Modern Germany*, ed. Paul Kosok (Chicago: The Chicago University Press, 1933), p. 292.

2. Augustine, *On Christian Doctrine*, II.4, II.25, II.37, III.2, III.29.

3. Wilbur S. Howell, *The Rhetoric of Alcuin and Charlemagne*, Princeton Studies in English, vol. 23 (Princeton: Princeton University Press, 1941), pp. 134–35.

4. Harold D. Lasswell, "Nations and Classes: The Symbols of Identification," in *Public Opinion and Communication*, ed. Bernard Berelson and Morris Janowitz (New York: Free Press, 1966), p. 33.

5. Maurice H. Farbridge, *Studies in Biblical and Semitic Symbolism* (1923, reprint New York: Ktav, 1970), pp. 27–89.

6. Philippe de Thaon, *Bestiaire*, ed. Charles Langlois, *La vie en France*, 3: 14–17.

7. Jehan Bonnet, *Livre des Secrets aus Philosophes*, *Ibid.*, pp. 316–17.

8. Joseph Strayer, "The Promise of the Fourteenth Century," *Proceedings of the American Philosophical Society* 106–6 (1961): 609–11.

9. Enrico Fulchignoni, *La civilisation de l'image*, trans. G. Crescenzi et E. Darmouni (Paris: Payot, 1972), pp. 174–81.

10. W. Vinacke, "Stereotypes as Social Concepts," *Journal of Social Psychology* 46 (1957): 239–41.

11. W. Scott, "Psychological and Social Correlates of International Images," in *International Behaviour*, ed. Herbert Kelman (New York: Holt, Rinehart and Winston, 1965), p. 100.

12. Kenneth E. Boulding, *The Image: Knowledge in Life and Society* (Ann Arbor: University of Michigan Press, 1966), p. 112.

13. Walter Lippmann, *Public Opinion* (1921, New York: MacMillan Paperbacks, 1960), pp. 95–98.

14. Edouard Perroy, *The Hundred Years War*, trans. David C. Douglas (1945, New York: Capricorn Books, 1965), pp. xxvii–xxviii, 60–68.

15. Bernard Guillemain, "Les tentatives pontificales de médiation dans le litige franco-anglais de Guyenne au XIVe siècle," *Bulletin philologique et historique du comité des travaux historiques et scientifiques*, (1958), pp. 431–32.

16. Gerald R. Owst, *Preaching in Medieval England*, pp. 203–4.

17. Robertus de Avesbury, *De gestis mirabilibus regis Edwardi tertii*, ed. Edward M. Thompson (London: *R.S.*, 1889), p. 369; *Eulogium historiarum*, 3: 210.

18. Desmond Seward, *The Hundred Years War: The English in France 1337–1453* (New York: Atheneum, 1978), pp. 19–40.

19. J. Fishman, "An Examination of the Process and Function of Social Stereotyping," *Journal of Social Psychology* 43 (1956): 40.

20. Walter Lippmann, *Public Opinion*, p. 119.

21. On the origins of the name France and its territorial connotations in the Early Middle Ages, see Godefroid Kurth, "La France et les Francs dans la langue politique du moyen âge," *R.Q.H.* 57 (1895): 340–57.

22. A. Coville, "Poèmes historiques du début de la Guerre de Cent Ans," *Histoire littéraire de la France* 38 (1949): 264–66.

23. A. Piaget, "Un poème inédit de Guillaume de Digulleville: Le roman de la fleur de lis," *Romania* 62 (1936): 336 ff.

24. Gervais du Bus, *Le roman de Fauvel*, p. 116.

25. *Political Poems and Songs Relating to English History Composed During the Period from the Accession of Edward III to that of Richard III*, ed. Thomas Wright, 2 vols. (London: *R.S.*, 1859), 1: 35–40.

26. *Ibid.*, p. 41. In the poem "On Crécy and Neville's Cross," the leopard represented the collective symbol of the English warriors; see *Ibid.*, p. 52.

27. Thomas Wright, *Political Songs of England*, p. 93.

28. *Political Poems*, p. 126.

29. *Ibid.*, pp. 137–38.

30. *Ibid.*, pp. 205–6.

31. "Clauwaert, Clauwaert, hoet u van den Leliaert," Frans J. P. Van Kalken, *Histoire de la Belgique* (Brussels: J. Lebegue, 1944), p. 115; the same symbolism appears in contemporary chronicles; see *Annales Gandenses*, ed. Hilda Johnstone, Medieval Classics (London : T. Nelson, 1951), pp. 20, 21, 25, 26, 28.

32. R. Jervis, *Perception and Misperception in International Politics* (Princeton: Princeton University Press, 1976), p. 310.

33. "Edward the First's War With Scotland," in *Political Songs*, p. 273.

34. "The Battle of Bannockburn," *Ibid.*, p. 262.

35. *Ibid.*, p. 263, 267.

36. "Song on the Execution of Sir Simon Fraser," *Ibid.*, pp. 212–13, 223.

37. Thomas Wright, *Political Poems*, p. 26.

38. "On the Truce of 1347," *Ibid.*, p. 54.

39. *Ibid.*, p. 43.

40. *Ibid.*, p. 42.

41. On the first expressions of the French identification with the role of chosen people at the eleventh century, see the interesting study of Lèon Gautier, "L'idèe politique dans les chansons de geste," *R.Q.H.* 7 (1869): 79–86.

42. Thomas Wright, *Political Songs*, pp. 162–63.

43. "An Invective Against France," *Ibid.*, p. 30.

44. "The Battle of Bannockborn," *Ibid.*, p. 267.

45. *Gesta Edwardi Tertii auctore Bridlingtoniensi*, ed. William Stubbs, in *Chronicles of the Reigns of Edward I and II*, 2: 146.

46. "On the Battle of Neville's Cross," Thomas Wright, *Political Poems*, p. 49.

47. "On the Truce of 1347," *Ibid.*, p. 56.

48. *Ibid.*, p. 45, 56, 102.

49. Harold D. Lasswell, "Nations and Classes," p. 33.

50. *Chronique de Jean le Bel*, ed. Jules Viard and Eugène Deprez, 2 vols. (1904, reprint Paris: Champion, 1977), 1: 66–67.

51. *Chronique latine de Guillaume de Nangis*, ed. Hercule Géraud, 2 vols, Société de l'histoire de France (Paris: Renouard, 1843), pp. 100, 137; *Chronique de Richard Lescot*, ed. Jean Lemoine, Société de l'histoire de France (Paris: Renouard, 1896), p. 74.

52. It had to be noted, however, that the Trojan Myth already appeared in the writings of Bede and Frédégaire in the eighth century. On its development in the later Middle Ages, see Amnon Linder, "*Ex mala parentela bona sequi seu oriri non potest:* The Trojan Ancestry of the Kings of France and the *Opus Davidicum* of Johannes Angelus de Legonissa," *Bibliothèque d'humanisme et Renaissance* 40 (1978): 497–98.

53. Victor Werner, "Le pouvoir des mots et la guerre," in *La Communication sociale et la guerre*, ed. Victor Werner (Brussels: Emile, 1974), p. 60.

54. Thomas Wright, *Political Songs*, pp. 180–81.

55. *The Chronicles of Jean Froissart in Lord Berner's Translation,* ed. Gillian and William Anderson (Carbondale, Ill.: Southern Illinois University Press, 1963), pp. 3–4.

56. Ellen Seiter, "Stereotypes and the Media: A Re-evaluation," *Journal of Communication* 36–2 (1986): 15–16.

57. Froissart, *Chroniques: Dernière rédaction du premier livre: Edition du manuscrit de Rome, Reg. lat. 869*, ed. George Diller, Textes littéraires français, vol. 194 (Paris: Minard, 1972), p. 42.

58. *Chronique des quatre premiers Valois (1327–1393)*, ed. Siméon Luce, Société de l'histoire de France, vol. 109 (Paris: Renouard, 1862), p. 2.

59. Charles Langlois, "Les Anglais du moyen âge d'après les sources françaises," *R.H.* 52 (1893): 313.

60. Achille Jubinal, *Nouveau recueil de contes, dits, fabliaux et autres pièces inédites des XIIIe, XIVe et XVe siècles*, 2 vols. (Paris: E. Pannier, 1839), 1: 75. On the tendency to ascribe behavioral functions to human features, see Stuart A. Rice, "Stereotypes: A Source of Error in Judging Human Character," *Journal of Personnel Research* 5 (1926): 267–76.

61. Ernest Langlois, "Anciens proverbes français," *B.E.C.* 60 (1899): 569–601.

62. The term *furor Teutonicus* was used for the first time by Lucanus in *Pharsalia* (1:255–56) to suggest the temperamental mood characteristic of German tribes; it appears in medieval sources such as the *Chronicon universale* with negative connotations, see, (*M.G.H., Scriptores*, 6. 214).

63. Emil Du Boys-Reymond, *Über die Grenzen des Naturerkennens: Die sieben Weltrathsel, zwei Vorträge* (Leipzig: Veit, 1882), p. 51.

Chapter 10

1. Isidore of Seville, *Etymologiarum*, 8. 3.

2. *Decreti Secunda Pars*, 24. 3. 26–31, *C.I.C.*, 1: 997–98.

3. Gordon Leff, *Heresy in the Later Middle Ages*, 2 vols. (Manchester: Manchester University Press, 1967), 1: 1–4.

4. Gordon Leff, "Hérésies savantes et hérésies populaires au moyen âge," in *Hérésies et sociétés dans l'Europe pré-industrielle 11e–18e siècles*, Communications et débats du Colloque de Royaumont, ed. Jacques Le Goff (Paris-La Haye: Mouton and Co. 1968), p. 219; Jean Musy, "Mouvements populaires et hérésies au XIe siècle en France," *R.H.*, 253 (1975): 63–71.

5. Dom Jean Leclercq, *La royauté du Christ*, pp. 37–39.

6. *Id.*, "Les controverses sur la pauvreté du Christ, in *Etudes sur l'histoire de la pauvreté*, 2 vols. (Paris: Sorbonne, 1974), 1: 45–48; *Mansi*, 5, 18 ff.

7. Louis Duval-Arnould, "Une apologie de la pauvreté volontaire par un universitaire séculier de Paris (1256)," in *Etudes*, p. 428 f.

8. Cinzio Violante, "La pauvreté dans les hérésies du XIe siècle en Occident," in *Etudes*, pp. 350–68; T. Manteuffel, "Naissance d'une hérésie," in *Hérésies et sociétes*, pp. 100–101.

9. Georges Duby, "Les pauvres des campagnes dans l'Occident médiéval jusqu'au XIIIe siècle," *R.H.E.F.* 52 (1966): 32.

10. Norman Cohn, *Millenium*, pp. 22–32, 71–73.

11. Michel Cépède and Maurice Lengelle, *Economie alimentaire du globe: essai d'interprétation* (Paris: Librairie de Médicis, 1953), p. 14, 19.

12. Michel Mollat, "Le problème de la pauvreté au XIIe siècle," in *Vaudois Languedociens et pauvres Catholiques*, Cahiers de Fanjeaux (Toulouse: E. Privat, 1967), p. 26.

13. Christine Thouzellier, "Hérésie et pauvreté à la fin du XIIe et au début du XIIIe siècle," in *Etudes*, 1: 375–88.

14. Etienne Delaruelle, "Dévotion populaire et hérésie au moyen âge," in *La piété populaire*, p. 198; Gordon Leff, "Heresy and the Decline of the Medieval Church," *Past and Present* 20 (1961): 40–42.

15. In 1111, Pope Paschal II and Henry V reached an agreement according to which the Church will renounce all the regalian rights, in return to the king's readiness

to give up the rights of investiture. Yet the firm opposition of prelates forced the pope to retract; see Albert Hauck, *Kirchengeschichte Deutschlands*, 4 vols. (Leipzig: J. C. Hinrichs, 1887–1903), 3: 897–905.

16. *Sermones Vulgares of Jacques de Vitry*, ed. Thomas F. Crane, p. 46.

17. Owst, *Literature and Pulpit*, pp. 501–2.

18. Michel Mollat, "La notion de la pauvreté au moyen âge: position de problèmes," *R.H.E.F.* 52 (1966): 13–18.

19. In the Old Testament, the Hebrew term "evion" designed poor men who suffered from contempt and, consequently, God protected them most; Jean Leclercq, "Aux origines bibliques du vocabulaire de la pauvreté," in *Etudes*, 1: 36–42.

20. Brian Tierney, *Medieval Poor Law: A Sketch of Canonical Theory and Its Application in England* (Berkeley: University of California Press, 1959), pp. 11–12.

21. J. Yunk, "Economic Conservatism, Papal Finance and the Medieval Satires on Rome," *Medieval Studies* 23 (1961): 334–51; *Ecrivains anticonformistes*, ed. Renè Nelli, p. 295.

22. Hugues de Rouen, *Contra haereticos (P.L.* 192. 1256).

23. Bernard Gui, *Manuel*, p. 51.

24. *Ibid.*, p. 36.

25. *Ibid.*, pp. 85–97; Salimbene (*M.G.H., Scriptores*, 32. 257–58).

26. Guillaume de Nangis, *Chronique*, ed. Geraud, 2: 218.

27. *Chronicon rhythmicum Austriacum*, ed. W. Wattenbach (*M.G.H., Scriptores*, 25. 363); *Continuatio Praedicatorum Vindobonensium, Annales Austriae*, ed. D. Wilhelmus Wattenbach (*M.G.H., Scriptores,* 9. 728). On other heretical movements of the time, see the excellent survey of Gordon Leff (above, note 3) and the bibliography enclosed.

28. Reinerus, *Annales Sancti Jacobi Leodiensis (M.G.H., Scriptores,* 16. 654).

29. Robert E. Lerner, "An Angel of Philadelphia," in the Reign of Philip the Fair: The Case of Guiard of Cressonessart," in *Order and Innovation*, pp. 343–64.

30. Rainier Sacconi, in *Heresy and Authority in Medieval Europe*, ed. and trans. Edward Peters (Philadelphia: University of Pennsylvania Press, 1980), pp. 125–32.

31. Ignaz von Dollinger, *Beitrage zur Sektengeschichte des Mittelalters*, 2 vols. (Munchen: O. Beck, 1890), 2: 39.

32. *Le Registre d'inquisition de Jacques Fournier, évêque de Pamiers 1318–1325*, ed. Jean Duvernoy, 3 vols. (Paris: Mouton, 1978), 2: 305, 3: 871; see also Charles Molinier, "L'église et la société cathares," *R.H.* 95 (1907): 19–21.

33. Bernard Gui, *Manuel*, pp. 11–27.

34. The Passau Anonymous, *Heresy and Authority*, p. 152.

35. Etienne Delaruelle, "Le Catharisme en Languedoc vers 1200: une enquête," in *La piété populaire*, pp. 219–20.

36. *Le Registre de Jacques Fournier*, 2: 328.

37. *Ibid.*, 2: 367.

38. *Ibid.*, 2: 25–26, 54.

39. *Ibid.*, 3: 123.

40. Emmanuel Le Roy Ladurie, *Montaillou*, p. 575.

41. Innocent III, *Epistolae, (P.L.* 214. 71; 215. 915).

42. Agustin Fliche, Christine Thouzellier and Yvonne Azais, *La Chrétienté Romaine (1198–1274*, in *Histoire de l'église*, 13 vols. (Paris: Bloud & Gay, 1950), 10: 118.

43. *Mansi*, 21. 1177.

44. Eckbertus, *Sermones Contra Catharos*, 13 (*P.L.* 195. 13).

45. R. I. Moore, "Heresy as Disease," in *The Concept of Heresy in the Middle*

Ages, ed. W. Lourdaux and D. Verhelst (Louvain-Hague: Louvain University Press, 1976), p. 11.

46. Vincent of Lerins, *Commonitorium (P.L.* 1. 639).

47. Georges Duby, *Cathédrales*, pp. 44–45.

Chapter 11

1. Arno Borst, "La transmission de l'hérésie au moyen âge," in *Hérésies et sociétés, pp. 273–80.*

2. Cinzio Violante, "Hérésies urbaines et hérésies rurales en Italie du XIe au XIIIe siècle," in *Hérésies et sociétés*, pp. 176–78.

3. René Nelli, *La vie quotidienne des Cathares du Languedoc au XIIIe siècle*, 3rd ed. (Paris: Hachette, 1969), pp. 27–28.

4. As early as in the sixth century, Benedict of Nursia emphasized his contempt for those gyrovagui who "move about all their lives through various countries, staying as guests for three or four days at different monasteries. They are always in the move and never settle down, and are slaves to their own wills and to the enticements of gluttony," *The Rule of Saint Benedict*, trans. Cardinal Gasquet (New York: Cooper Square Publishers, 1966), p. 8.

5. Herimanni, *Monumenta historiae Tornacensis, Liber de restauratione monasterii sanctii Martini Tornacensis (M.G.H., Scriptores,* 14. 274–75).

6. William of Malmesbury, *De Gestis rerum Anglorum*, f. 284 "De Berengario," (*P.L.* 179. 1257).

7. René Nelli, *La Vie quotidienne*, pp. 260–61.

8. *Ibid.*, pp. 264–66.

9. Bernard Gui, *Manuel*, p. 58.

10. The Passau Anonymous, *Heresy and Authority*, p. 151.

11. *Le Registre de Jacques Fournier*, 1: 151.

12. *Ibid.*, 1: 191, 3: 871.

13. *Ibid.*, 1: 172.

14. *Ibid.*, 1: 199.

15. *Ibid.*, 1: 235.

16. M. D. Lambert, *Medieval Heresy: Popular Movements from Bogomil to Hus* (London: E. Arnold, 1977), p. 115.

17. *Le Registre de Jacques Fournier*, 2: 525; 3: 871, 1190, 1200.

18. *Ibid.*, 1: 141.

19. Elie Griffe, *Le Languedoc cathare de 1190 à 1210* (Paris: Letouzey et Ane, 1971), pp. 125–26.

20. *Le Registre de Jacques Fournier*, 3: 755.

21. Bernard Gui, *Manuel*, p. 115.

22. R. I. Moore, "The Origins of Medieval Heresy," *History* 55 (1970): 27.

23. Charles Molinier, "L'église et la société cathares," p. 7.

24. Bernard Gui, *Manuel*, pp. 51–53.

25. Vaso Leodiensis episcopus, *Vita Vasonis (P.L.* 142. 751).

26. *The Passau Anonymous*, pp. 150–51.

27. *Innocentii III, Epistolae (P.L.* 214. 695–99).

28. Bernard Gui, *Manuel*, pp. 35–37.

29. Cinzio Violante, "*Hérésies urbaines*," p. 186.

30. *The Passau Anonymous, loc. cit.*

31. Bernard Gui, *Manuel*, p. 37.

32. Christine Thouzellier, "Hérésie et croisade au XIIe siècle," *Revue d'histoire ecclésiastique* 49 (1954): 855–72.

33. Bernard Gui, *Manuel*, p. 113.

34. Emmanuel Le Roy Ladurie, *Montaillou*, p. 562.

Chapter 12

1. "*Speculum Perfectionis*," ed. Paul Sabatier, *Collection d'études et de documents pour l'histoire religieuse et littéraire du moyen âge*, 4 vols. (Paris 1898–1902), 1: 72.

2. Thomas de Celano, *Vita Prima S. Francisci Analecta Franciscana* (Quaracchi: Collegium S. Bonaventurae, 1926–) 10. 1. 99.

3. Brother Thomas d'Eccleston, *De adventu fratrum Minorum in Angliam*, ed. Marie T. Laureilhe, *Sur les routes d'Europe au XIIIe siècle* (Paris: ed. Franciscaines, 1959), pp. 98–99.

4. Jourdain de Giano, *Chronicles*, c. V. *Ibid.*, p. 28.

5. Sylvia Schein, "Latin Hospices in Jerusalem in the Late Middle Ages," *Zeitschrift des Deutschen Palästina* 101 (1985): 91–92.

6. John Moorman, *A History of the Franciscan Order From Its Origin to the Years 1517* (Oxford: Clarendon Press, 1968), p. 30.

7. The term *Spirituals* was first applied in the early-fourteenth century to the rigorist section of the Franciscans in southern France. The *Conventuals* were their opponents, and the majority of the order, who observed the rule according to papal interpretation, see Franz Ehrle, "Spiritualen und Fraticellen," *Archiv für Litteratur und Kirchengeschichte* 4 (1888): 138–80.

8. *Bullarium Franciscanum*, 1: 68–70.

9. *Ibid.*, 3: 404–16.

10. Berard Marthaler, O.F.M., "Forerunners of the Franciscans: The Waldenses," *Franciscan Studies* 18 (1958): 133–42.

11. M. D. Lambert, *Franciscan Poverty: The Doctrine of the Absolute Poverty of Christ and the Apostles in the Franciscan Order (1210–1323)* (London: S.P.C.K., 1961), pp. 209–11; Id. "The Franciscan Crisis Under John XXII," *Franciscan Studies* 32 (1972): 125–28.

12. *Bullarium Franciscanum*, 5: 128–30.

13. *Continuateurs de Guillaume de Nangis*, p. 620; *Richard Lescot*, pp. 49–50.

14. *Jean de Paris*, p. 664.

15. Walsingham, *Historia Anglicana*, pp. 112–13; *Flores historiarum*, ad a. 1322.

16. Moorman, *Franciscan Order*, p. 314.

17. *Contra Beguinos et fratres minores qui dicuntur Spirituales*, ed. Stephanus Baluze and Joannis D. Mansi, *Miscellanea novo ordine digesta*, 4 vols. (Lucca: V. Junctinium, 1761–64), 2: 272–74.

18. Lester K. Little, "L'utilité sociale de la pauvreté volontaire," in *Etudes*, pp. 447–59; see also Jacques Le Goff, "Apostolat mendiant et fait urbain dans la France médiévale: l'implantation géographique et sociologique des ordres mendiants (XIIIe–XVe siécles)," *R.H.E.F.* 54 (1968): 69–76; M. de Fontette, "Villes médiévales et ordres mendiants," *Revue historique de droit français et étranger* 48 (1970): 390–407.

19. Franz Ehrle, "Die Spiritualen, ihr Verhaltniss zum Franciscanerorden und zu den Fraticellen," *Archiv für Literatur und Kirchengeschichte des Mittelalters* 2 (1886): 144–46.

20. *Lettres Communes*, nos. 64272, 6216, 8710.

21. C. Partee, "Peter John Olivi: An Historical and Doctrinal Study," *Franciscan Studies* 20 (1960): 215–60.

22. Bernard Gui, *Manuel*, pp. 132–35.

23. *Nicolaus Minorita*, in Baluze, *Miscellanea*, 3: 76.

24. *Lettres Communes*, nos. 16134, 16215; *Bullarium Franciscanum*, 5: pp. 224–25.

25. *Ibid.*, pp. 234–35; Baluze, *Miscellanea*, 3: 208–11.

26. Bryan Tierney, *Papal Infallibility*, pp. 93 f.

27. *Extravagantis Johannis XXII, C.I.C.*, 2: 1224.

28. This was argued by Zenzellinus de Cassanis; see Walter Ullmann, *Medieval Papalism*, pp. 38–51; this approach, however, contravened thirteenth-century patterns; see Bryan Tierney, "Grosseteste and the Theory of Papal Sovereignty," *Journal of Ecclesiastical History* 6 (1955): 1–17.

29. The essence of John XXII's approach is indicated by his former support of the Franciscans against Jean de la Palu who denied the pope's right to allow them to hear confession, see, *Lettres Communes*, no. 42480, Raynaldus, *Annales Ecclesiastici*, ad a. 1327, XXVII–XXXVI.

30. *Bullarium Franciscanum*, 5: 238–46; *Lettres Communes*, nos. 18088, 18140.

31. *Bullarium Franciscanum*, 5: 256–259; *Lettres Communes*, nos. 20406, 20343, 23228.

32. *Continuateurs de Guillaume de Nangis*, pp. 622–23; *Richard Lescot*, p. 51.

33. John of Winterthur, ed. F. Baethgen and C. Brun (*M.G.H., Scriptores*, n.s. vol. 3 (Berlin, 1955), p. 95.

34. William Ockham, *Epistola ad fratres minores*, p. 6; all Ockham's writings, the *Dialogus* excepted, are cited from the edition of Hilary S. Offler, *Guillelmi de Ockham: Opera politica*, 3 vols. (Manchester: Manchester University Press, 1940).

35. Ockham, *Opus nonaginta dierum*, c. 94, pp. 708–9.

36. *Ibid.*, p. 469.

37. *De Imperatore et pontificia potestate*, p. 33; *Opus nonaginta dierum*, p. 707.

38. *Ibid.*, p. 772, p. 477.

39. *Ibid.*, p. 669; Felice Tocco, *La Questione della povertá nel secolo XIV secondo nuovi documenti* (Naples: F. Perrella, 1910), pp. 51–57, 249.

40. *Opus nonaginta dierum*, p. 310.

41. Ockham, *Opus nonaginta dierum*, pp. 469, 477, 669, 707–9, 772; *Id., Dialogus de imperio et pontificia potestate*, p. 33; *Id., Epistola ad fratres minores*, p. 15.

42. Ockham, *Opus nonaginta dierum*, p. 412.

43. *Ibid.*, p. 620.

44. *Ibid.*, p. 785.

45. *Ibid.*, p. 606.

46. Ockham, *Opus nonaginta dierum*, p. 310, 412, 620, 785; in terms of comparison, see the more extremist approach of Marsilius of Padua, *Defensor Pacis*, ed. P. Scholz, *Fontes Iuris Germanici (M.G.H.*, Hanovre 1932), 2. 11. 3; 2. 13 . 6. 20; 2. 14. 1. 13; 2. 25. 1. 8; 3. 2. 28; Georges Lagarde, *Le defensor pacis*, in *La naissance de l'esprit laïque au déclin du moyen âge*, 5 vols. (Louvain—Paris: E. Nauwelaerts, 1956–1970), 3: 340 f.

47. Aegidius Romanus, *De ecclesiastica potestate*, pp. 37–38, 48–50, 59–60, 79, 96.

48. Robert Lerner, "Refreshment of the Saints: The Time After Antichrist as a Station for Earthly Progress in Medieval Thought," *Traditio* 32 (1976): 141–44.

49. *Acta Aragonensia*, 2: 676.

50. A. Pelzer, "Les 51 articles de Guillaume d'Occam censurés en Avignon en

1326," *Revue d'histoire de l'église* 18 (1922): 240–70; C. Brampton, "Personalities at the Process Against Ockham at Avignon," *Franciscan Studies* 26 (1966): 4–25; D. Burr, "Ockham, Scotus and the Censure at Avignon," *Church History* 37 (1968): 144–59.

51. *Lettres Communes*, nos. 30189, 42491–93, 42418–26, 42428, 42436–76, 54817.

52. *Bullarium Franciscanum*, 6: 50, 67.

53. Ockham, *Breviloquium de potestate papae*, p. 60.

54. Bryan Tierney, "Ockham, the Conciliar Theory and the Canonists," *Journal of the History of Ideas* 15 (1954): 48.

55. Ockham, *An Princeps*, pp. 244–46.

56. Ockham, *Octo Quaestiones*, p. 32.

57. Ockham, *Contra Benedictum*, p. 262; see, also, Georges Lagarde, "Marsile de Padoue et Guillaume d'Ockham," in *Etudes d'histoire du droit canonique dédiées à G. Le Bras*, ed. G. Vadel et al., 2 vols. (Paris: Sirey, 1965), pp. 593–605.

58. Ockham, *Dialogus inter clericum et militem*, Ia, VI, c. 66 (Paris: Guidone Mercatore, 1498); *Id., Octo Quaestiones*, p. 82.

59. Ockham, *Dialogus*, Ia, V, c. 15.

60. *Ibid.*, IIIae, IV, c. 6; see, also, Marsilius von Padua, *Defensor Pacis*, p. 541.

61. Ockham, *Dialogus*, Ia, VI, c. 3.

62. *Ibid.*, Ia, II, c. 25; *Tractatus contra Johannes*, p. 49, 72.

63. Ockham, Dialogus, Ia, VI, c. 90; *Octo Quaestiones*, p. 82; Georges Lagarde, *Guillaume d'Ockham: Critique des structures ecclésiastiques, Id., La naissance*, 5: 86. It is rather doubtful, however, if Ockham could be regarded as the promotor of the conciliar movement as claimed by C. Bayley; see "Pivotal Concepts in the Political Philosophy of William of Ockham," *Journal of the History of Ideas* 10 (1949): 215–17.

64. M. D. Lambert, *Franciscan Poverty*, p. 137.

65. C. Brampton, "Ockham at Avignon," p. 18.

66. Raoul Manselli, *Spirituali e Beghini in Provenza* (Rome: Instituto Storico Italiano per il Medio Evo, 1959), p. 137.

Chapter 13

1. H. Busse, "Vom Felsendom zum Templum Domini," in *Das Heilige Land im Mittelalters*, ed. Wolfdietrich Fisher and Jürgen Schneider (Neustadt: Degener, 1982), pp. 19–32.

2. *Mansi*, 21. 357 ff.

3. *De Laude Novae Militiae*, III, 4, (*P.L.* 182. 923); Malcolm Barber, "The Social Context of the Templars," *Transactions of the Royal Historical Society*, 5th series, 34 (1984): 32–36.

4. Norman Housley, "Saladin Triumph Over the Crusader States: The Battle of Hattin, 1187," *History Today* 37 (1987): 17–23.

5. Sophia Menache, *The Papal Policy With Regard to the Templar Order Between 1240–1312*, [Hebrew] (M.A. thesis: University of Haifa, 1973), pp. 55–120.

6. Jonathan Riley-Smith, *The Knights of St. John*, p. 377.

7. E. Ashtor, "Investments in Levant Trade in the Period of the Crusades," *The Journal of European Economic History* 14 (1985): 427–41.

8. Hans G. Prutz, *Die geistlichen Ritterorden* (Berlin: E.S. Mittler und Sohn, 1908), p. 207.

9. H. F. Delaborde, *Recueil des Actes de Philippe Auguste*, 1: 418–20.

10. Laura Di Fazio, *Lombardi e templari nella realta socioeconomica durante il regno di Filippo il Bello (1285–1314)* (Milan: Coop. Libraria IULM, 1986), pp. 57 f.

11. The interest aroused by the affair of the Templars gave rise to rich research on the issue. Already at the end of the nineteenth century, Charles Langlois called attention to this fact on his review on "Livres sur l'histoire des Templiers," *R.H.* 40 (1889): 169. From 1980 onwards appeared the following books, A. Fleige and B. Lafille, *Les Templiers et leur mystère*, 2 vols. (La Seyne-sur-Mer 1980); Peter Partner, *The Murdered Magicians: The Templars and their Myth* (Oxford: Oxford University Press, 1982); Michel Picar, *Les Templiers* (Paris: M.A., 1986); Roger Sève and Anne M. Chagny-Sève, *Le Procès des Templiers d'Auvergne, 1309–1311* (Paris: Comité des travaux historiques et scientifiques, 1986).

12. *Jean de Paris*, p. 649.

13. Georges Lizerand, *Le dossier de l'affaire des Templiers* (Paris: E. Champion, 1923), p. 21.

14. Edgard P. Boutaric, "Clément V, Philippe le Bel et les Templiers," *R.Q.H.* 10 (1871): 331–32. The consternation caused by the royal policy on the papal curia was confirmed by contemporary chroniclers; see *Jean de Paris*, p. 650; Bernard Gui, *Vitae*, col. 66.

15. *Le procès des Templiers*, ed. M. Jules Michelet, 2 vols., Collection des documents inédits sur l'histoire de France (Paris: Imp. nationale, 1841–1851), 1: 89–96; see also A. Gilmour-Bryson, "L'eresia e i Templari: 'Oportet et haereses esse'," *Richerche di Storia sociale e religiosa*, n.s., 24 (1983): 101–14.

16. Ernest Renan, *Philippe le Bel*, pp. 361–62.

17. Heinrich Finke, *Papsttum*, 2: 139, 145.

18. *Chronicon Guillelmi Scoti (H.F.* 21. 205); Richard Lescot, *Chronique*, p. 29; *Fragment d'une chronique anonyme finissant en 1328*, p. 150; *Chronique anonyme finissant en 1356 (H.F.* 21. 139); *Continuateurs de Guillaume de Nangis*, pp. 595–96; Sexta Vita, *Vitae*, cols. 99–102.

19. Bernard Gui, *Vitae*, col. 65; *Fragment d'une chronique . . . 1328,* p. 649; Sexta Vita, *Vitae*, cols. 100–101.

20. *Anciennes chroniques de Flandre*, p. 398; *Chroniques de St. Denis*, p. 682; *Chronique Normande du XIVe siècle*, ed. Auguste et E. Molinier, Société de l'histoire de France (Paris: Renouard, 1882), p. 29; *Annales Gandenses*, p. 88; *Chronographia Regum Francorum*, ed. Henri Moranville, 3 vols., Société de l'histoire de France (Paris: H. Laurens 1891–1897), 1: 179–82; *Extraits d'une chronique . . . 1383*, p. 142.

21. Amalricus Augerius, *Vitae*, col. 95; *Chroniques de St. Denis*, p. 686.

22. *Chronicon Angliae Petriburgense*, ed. John A. Giles, (London: Caxton Society, 1845), p. 159; *Annales Londoniensis*, ed. William Stubbs, *Edward I and II*, 1: 152, 158; *Flores historiarum*, p. 331; *Annales Hibernie*, ed. John T. Gilbert, *Chartularies of St. Mary's Abbey, Dublin*, 2 vols. (London: *R.S.*, 1884), 2: 336.

23. *La cronica di Dino Compagni delle cose occorrenti ne' tempi suoi*, ed. Isidoro del Lungo, 1907–1916 (*R.I.S.* 9–2. 219); Heinricus Rebdorfensis, *Annales Imperatorum et Paparum (F.R.G.* 4. 552); *Marco Battagli da Rimini*, p. 65; Iohannis de Bazano, *Chronicon Mutinense (1188–1363)*, ed. Tommaso Casini, 1917–1919 (*R.I.S.*, 15–4. 71).

24. *Johannes Victoriensis (F.R.G.* 1. 352); Agnolo di Tura, *Cronaca Senese*, ed. Alessandro Lisini and Fabio Iacometti, *Cronache Senesi,* 1939 (*R.I.S.* 15–6. 299); *Cronaca Senese di Paolo di Tommaso Montauri, Ibid.*, p. 233; Matthias Nuewenburgensis, *Cronica (F.R.G.* 4. 237).

25. Galfridi le Baker de Swynebroke *Chronicon*, ed. John A. Giles (London: Caxton Society, 1847), p. 53; Heinrich Finke, *Papsttum*, 2: 51.

26. *Litterae ad regem Franciae de causa Templariorum*, ed. J. Schwalm (*M.G.H.*, *Constitutiones*, vol. 4, Hanover, 1906, p. 196).

27. *Foedera*, 1–4. 94–5, 100.

28. Malcolm Barber, "Propaganda in the Middle Ages: The Charges Against the Templars," *Nottingham Medieval Studies* 17 (1973): 57.

29. *Chartularium Universitatis Parisiensis*, ed. Heinrich Denifle and Aemilio Chatelain, 4 vols. (1898–1899, reprint Bruxelles: Culture et civilisation, 1964), 2: 125–30. This approach was shared by Augustinus Triumphus; see Richard Scholz, *Die Publizistik zur Zeit Philipps des Schönen und Bonifaz' VIII* (Stuttgart: F. Enke, 1903), pp. 508–16.

30. *Acta Imperii Angliae et Franciae ab anno 1267 ad annum 1313*, ed. Fritz Kern (1911, reprint Hildesheim: G. Olms, 1973), pp. 120–22; *Acta Henrici VII Romanorum Imperatoris*, ed. Francesco Bonaini (Florence: M. Cellinii, 1877), pp. 6–8.

31. *Regestum Clementis Papae V*, nos. 2352, 2371, 2387, 2614, 3381, 3400–3401, 5544, 7508.

32. Georges Picot, *Etats Généraux et Assemblées*, pp. 487–91.

33. Georges Lizerand, *Clément V et Philippe IV, le Bel* (Paris: Hachette, 1910), pp. 119–20.

34. Georges Picot, *Etats généraux et assemblées*, p. LVI.

35. *Jean de Paris*, p. 650; Finke, *Papsttum*, 2: 135.

36. *Richard Lescot*, p. 30; *Continuateurs de Guillaume de Nangis*, p. 597.

37. Lizerand, *Le dossier*, pp. 110–37.

38. Ptolomeo de Lucca, *Vitae*, cols. 29–30; *Jean de Paris*, p. 651; *Chroniques de St. Denis*, p. 683.

39. *Continuateurs de Guillaume de Nangis*, p. 600.

40. *Foedera*, 1–4. 138–39; Clarence Perkins, "The Wealth of the Knights Templars in England and the Disposition of It After Their Dissolution," *A.H.R.* 15 (1910): 254.

41. *Annales Paulini*, p. 265; Henry Knighton, *Chronicon*, ed. Joseph R. Lumby, 2 vols. (London: *R.S.*, 1889–1895), 1: 407; Thomas Walsingham, *Historia Anglicana*, p. 120; *Chronica Pontificum Ecclesiae Eboracensis partes 3*, ed. James Raine, in *The Historians of the Church of York and Its Archbishops*, 3 vols. (London: *R.S.*, 1879–1894), 2: 414; *Flores historiarum*, p. 143, 331–32; *Annales Londoniensis*, pp. 180–98; *Gesta Edwardi de Carnavan auctore canonico Bridlingtoniensi*, ed. William Stubbs, in *Chronicles*, 2: 28–31.

42. *Councils and Synods*, pp. 1277–1278; Walter of Hemingburgh, *Chronicon*, pp. 391–95; *Chronica Pontificum Ecclesiae Eboracensis*, p. 413.

43. *Councils and Synods*, p. 1338.

44. *Walter of Hemingburgh*, p. 395.

45. *Councils and Synods*, p. 1339.

46. Langland, *The Vision of William Concerning Piers the Plowman: Text B*, ed. Walter W. Skeat, 3 vols., English Text Society (Oxford: Oxford University Press, 1869), 2: 282.

47. *Annales Paulini*, p. 270.

48. Finke, *Papsttum*, 2: 251.

49. *Walter of Hemingburgh*, p. 396.

50. Finke, *Papsttum*, 2: 258; Ptolomeo de Lucca, *Vitae*, col. 43.

51. Ptolomeo de Lucca, *Vitae*, col. 43.

52. *Conciliorum Oecumenicorum Decreta*, ed. Josepho Alberigo et al., 2nd ed. (Basel: Herder, 1962), pp. 312–19.

53. *Ibid.*, pp. 319–23.

54. *Le Roman de Fauvel*, p. 38.

55. *Chronographia Regum Francorum*, p. 209.

56. *Flores historiarum*, p. 147; *Gesta Edwardi de Carnavan*, p. 37.

57. *Annales Paulini*, p. 271; Walsingham, *Historia Anglicana*, p. 127; Adam Murimuth, *Continuatio chronicarum*, ed. Eduard M. Thompson (London: *R.S.*, 1889), p. 16; Thomas de la More, *Vita et mors Edwardi regis Angliae*, ed. William Stubbs, *Chronicles*, 2: 298.

58. *Agnolo di Tura*, p. 315; Finke, *Papsttum*, 2: 245; *Vitae*, col. 590; Trithenius, *Chronico Hirsaugiensi, Mansi*, 25. 408.

59. *Iohannes Victoriensis*, pp. 369–70.

60. Dante Alighieri, *Divina Comedia*, Purgatorio, c. XX, v. 91–93.

61. *Storie Pistoresi: 1300–1348*, ed. Silvio A. Barbi, 1907–1927 (R.I.S. 9–5. 224).

62. Bernard Gui, *E Floribus Chronicorum seu catalogo Romanorum pontificum necnon e chronico Regum Francorum (H.F.* 21. 723); *Richard Lescot*, p. 40; *Jean de Paris*, p. 658; *Continuateurs de Guillaume de Nangis*, p. 609; *Geffroi de Paris*, p. 145; *Annales Hibernie*, p. 341; *Agnolo di Tura*, p. 300; *Chronica S. Petri Erfordensis Moderna*, ed. O. Holder-Egger (*M.G.H., Scriptores*, 30–1. 446); *Chronica Reinhardsbrunnensis, Ibid.*, p. 651; see, also, J. Fried, "Wille, Freiwilligkeit und Gestandnis um 1330.: Zur Beurteilung des letzen Templer-Grossmeisters Jacques de Molay," *Historisches Jahrbuch* 105 (1985): 388–425.

BIBLIOGRAPHY

PRIMARY SOURCES

Acta Aragonensia. Edited by Heinrich Finke. 3 vols. Berlin: W. Rothschild, 1908–1922.

Acta Henrici VII Romanorum Imperatoris. Edited by Francesco Bonaini. Florence: M. Cellinii, 1877.

Acta Imperii Angliae et Franciae ab anno 1267 ad annum 1313. Edited by Fritz Kern. 1911, reprint Hildesheim: G. Olms, 1973.

Acta Imperii inedita saeculi XIII et XIV. Edited by Eduard Winkelmann. Innsbruck: Wagner'sche Universitats-Buchhandlung, 1885.

Acta Imperii Selecta. Edited by Johann F. Böhmer. Innsbruck: Wagner, 1870.

Actes des notaires génois de Péra et de Caffa de la fin du treizième siècle (1281–1290). Edited by Georg I. Bratianu. Bucarest: Cultura nationala, 1927.

Adam Murimuth. *Continuatio Chronicarum*. Edited by E. Thompson. London: *R.S.*, 1889.

Aegidius Romanus. *De Ecclesiastica Potestate*. Edited by Richard Scholz. Stuttgart: Druck der Union deutsche Verlagsgesellschaft, 1902.

Agnolo di Tura. *Cronaca Senese*. In *Cronache Senesi*, edited by Alessandro Lisini and Fabio Iacometti, pp. 255–564. *R.I.S.* 15–6. 1939.

Alani de Insulis. *Summa de arte praedicatoria*. *P.L.* 210. 109–97.

Alcuini. *Epistolae*. in *Epistolae karolini aevi*. Edited by Ernst Dümmler et al. 5 vols. 4: 1–481. Berlin: *M.G.H.*, 1895.

Alexandri III. *Epistolae et privilegia*. *P.L.* 200.

Ambrosius. *Opera Omnia*. *P.L.* 16. 875–1288.

Anciennes chroniques de Flandre. *H.F.* 22. 329–429.

Annales de Dunstaplia. In *Annales Monastici*, edited by H. Luard. 5 vols. 3: 3–420. London: *R.S.*, 1864–1869.

Annales Gandenses. Edited by Hilda Johnstone. Medieval Classics. London : T. Nelson, 1951.

Annales Herbipolenses. Edited by Georgius H. Pertz. *M.G.H. Scriptores*. 16. 1–12.

Annales Hibernie. In *Chartularies of St. Mary's Abbey, Dublin*, edited by John T. Gilbert. 2 vols., 2: 303–98. London: *R.S.*, 1884.

Annales Londoniensis. In *Chronicles of the Reigns of Edward I and II*, edited by William Stubbs. 2 vols. 1: 1–251. London: *R.S.*, 1882.

Annales Paulini. In *Chronicles of the Reigns of Edward I and II*. 1: 253–370.

Arnold de Liége. *Alphabetum Narracionum*. Edited by Colette Ribaucourt. Ph. D. diss.: Ecole des Hautes Etudes, 1985.

Augustine. *On Christian Doctrine*. Edited and translated by Durant W. Robertson. New York: Liberal Art Press, 1958.

———. *De civitate Dei contra paganos*. In *Oeuvres de St. Augustin*, edited by Gustave Bardy. 37 vols, vols. 33–37. Paris: Desclée de Brouwer, 1959.

———. *De Resurrectione mortuorum*. *P.L.* 39. 1599–1611.

———. *The Confessions*. Translated by Edward B. Pusey. In *Medieval Classics*. E dited by Thomas Symons, London-New York: Nelson, 1953.

Baldric of Bourgueil. *Historia Jerosolimitana*. *RHC Occ*. 4. 9–111.

Bede. *Historia ecclesiastica*. Edited by Robert Hussey. Oxford: Oxford University Press, 1846.

Benedict. *The Rule of Saint Benedict*. Translated by Cardinal Gasquet. New York: Cooper Square Publishers, 1966.

———. *Benedicti regula*. Edited by R. Hanslik. Vienna: Corpus Scriptorum Latinorum editum consilio et impensis Academicae Scientiarum Austricae 75, 1960.

Benedict of Peterborough. *Gesta regis Henrici secundi*. Edited by William Stubbs. 2 vols. London: *R.S.*, 1867.

Bernard de Clairvaux. *Epistolae*. *P.L.* 182. 67–721.

———. *De Consideratione*. *P.L.* 182. 727–808.

———. *De Laude Novae Militiae ad milites Templi*. *P.L.* 182. 922–40.

Bernard Gui. *Manuel de l'Inquisiteur*. Edited by Guillaume Mollat. 2 vols. 1926. Paris: Les Belles Lettres, 1964.

———. *E Floribus Chronicorum seu catalogo Romanorum pontificum necnon e chronico Regum Francorum*. *H.F.* 21. 690–734

Bonaventura. *De perfectione evangelica*. In *St. Bonaventurae Opera Omnia*. 11 vols. 5: 117–98. Quaracchi: Collegii S. Bonaventurae, 1891.

Boniface VIII. *Les registres de*. Edited by G. Digard et al. 4 vols. Paris: Bibliothèque des écoles françaises d'Athènes et de Rome, 1884–1891.

Bullarium Franciscanum Romanorum Pontificum. Edited by Joannes H. Sbaralea. 5 vols. Rome: Typ. Sacrae congregationis de propaganda fidei, 1759–1921.

Bullarium Ordinis Militiae de Calatrava. Edited by D. Ignatius J. de Ortega et Cotes et al. 1747–1761, reprint Barcelona: El Albir, 1981.

Caesarius von Heisterbach. *Caesarii Heisterbacensis monachi ordinis Cisterciensis Dialogus Miraculorum*. Edited by Josephus Strange. 3 vols. Cologne: J. M. Heberle, 1851–1857.

———. *Die Fragmente des Libri VIII Miraculorum*. Rome: Römische Quartalschrift für Christliche Altertumskunde und für Kirchengeschichte, 1901.

Charlemagne. *Capitulare missorum generale*. *M.G.H.. Legum*. 1: 183–84.

Chartularium Universitatis Parisiensis. Edited by Heinrich Denifle and Aemilio Chatelain. 4 vols. 1891–1899, reprint Bruxelles: Culture et civilisation, 1964.

Chronica Pontificum Ecclesiae Eboracensis partes 3. In *The Historians of the Church of York and Its Archbishops*, edited by James Raine. 3 vols. London: *R.S.*, 1879–1894.

Chronica Reinhardsbrunnensis. *M.G.H. Scriptores*. 30–1. 490–656.

Chronica S. Petri Erfordensis Moderna. Edited by O. Holder-Egger. *M.G.H. Scriptores*. 30–1. 335–457.

Chronica Sigeberti Gemblacensis. Edited by D. L. C. Bethmann. *M.G.H. Scriptores*. 6. 300–74.

Chronicon Angliae Petriburgense. Edited by John A. Giles. London: Caxton Society, 1845.

Chronicon Estense. Edited by Giulo Bertoni and Emilio P. Vicini. *R.I.S.* 15–3. s.d.

Chronicon Girardi ab Arvernia. *H.F.* 21. 213–19.

Chronicon Guillelmi Scoti. *H.F.* 21. 202–11.

Chronicon rhythmicum Austriacum. Edited by W. Wattenbach. *M.G.H. Scriptores*. 25. 349–68.

Chronique anonyme finissant en 1356. *H.F.* 21. 137–40.

Chronique de Jean le Bel. Edited by Jules Viard and Eugène Déprez. 2 vols. 1904, reprint Paris: Champion, 1977.

Chronique de Richard Lescot. Edited by Jean Lemoine. Socièté de l'histoire de France. Paris: Renouard, 1896.

Chronique des quatre premiers Valois (1327–1393). Edited by Simeon Luce. Socièté de l'histoire de France, vol. 109. Paris: Renouard, 1862.

Chronique latine de Guillaume de Nangis. Edited by Hercule Geraud. 2 vols. Socièté de l'histoire de France. Paris: Renouard, 1843.

Chronique Normande du XIVe siècle. Edited by Auguste et E. Molinier. Société de l'histoire de France. Paris: Renouard, 1882.

Chronique rimée attribuée à Geffroi de Paris. *H.F.* 22. 87–166.

Chroniques de St. Denis. *H.F.* 20. 654–729.

Chronographia Regum Francorum. Edited by Henri Moranville. 3 vols. Socièté de l'histoire de France. Paris: H. Laurens 1891–1897.

Clausula de unctione Pippini. Edited by Arndt. *M.G.H. Scriptores Rerum Merovingiarum*. 1: 465–66.

Clément VI: Lettres closes, patentes et curiales. Edited by Eugène Déprez et al. Bibliothèque des écoles françaises d'Athènes et de Rome. vol. 3. Paris 1901.

Collection d'études et de documents pour l'histoire religieuse et littéraire du moyen âge. Edited by Paul Sabatier. 4 vols. Paris 1898–1902.

Concilia Magnae Britanniae et Hiberniae. Edited by David Wilkins. 4 vols. London: R. Gosling, 1737.

Conciliorum Oecumenicorum Decreta. Edited by Josepho Alberigo et al. 2nd ed. Basel: Herder, 1962.

Continuateurs de Guillaume de Nangis (*Les*). *H.F.* 20. 583–646.

Continuatio chronici Odoranni monachi. *H.F.*. 10. 165–69.

Continuatio Praedicatorum Vindobonensium, Annales Austriae. Edited by D. Wilhelmus Wattenbach. *M.G.H. Scriptores*. 9. 724–32.

Contra Beguinos et fratres minores qui dicuntur Spirituales. In *Miscellanea novo ordine digesta*, edited by Stephanus Baluze and Joannis D. Mansi. 4 vols. Lucca: V. Junctinium, 1761–1764.

Coronatio Philippi I seu ordo qualiter is in regem coronatus est. *H.F.* 11. 32–33.

Councils and Synods with Other Documents Relating to the English Church. Edited by F. Maurice Powicke and Christopher Cheney. 2 vols. Oxford: Clarendon Press, 1964.

Cronaca Senese di autore anonimo. In *Cronache Senesi*, edited by A. Lisini and F. Iacometti, pp. 41–172. *R.I.S.* 15–6. s.d.

Dante Alighieri. *La Divina Commedia*. Edited by L. Magugliani. 3 vols. Milan: Biblioteca universale Rizzoli, 1949.

De Expugnatione Lyxbonensi. Edited and translated by Charles Wendell David. 1936, New York: Octagon Books, 1976.

De necessariis observantiis scaccarii dialogus. Edited by A. Hughes, C G. Crump and Charles Johnson. London 1902.

Die Kreuzzugsbriefe aus den Jahren 1088–1100. Edited by H. Hagenmeyer. Innsbruck: Wargner'sche Universitats-Buchhandlung 1901.

Dino Compagni. *La cronica di Dino Compagni delle cose occorrenti ne' tempi suoi*. Edited by Isidoro del Lungo. *R.I.S.* 9–2. 1907–1916.

Documents relatifs aux états généraux et assemblées réunis sous Philippe le Bel. Edited by Georges Picot. Collection des documents inédits sur l'histoire de France, vol. 11. Paris: Imprimerie nationale, 1901.

Eckbertus. *Sermones Contra Catharos. P.L.* 195. 11–102.

Einhard. *The Life of Charlemagne.* Edited and translated by Sidney Painter. Michigan: University of Michigan Press, 1960.

Ekkehardi Chronicon Universale. Edited by D.G. Waitz and P. Kilon. *M.G.H. Scriptores.* 6. 33–231.

English Constitutional Documents: 1307–1485. Edited by Eleanor Lodge and Gladys Thornton. New York: Octagon Books, 1972.

English Historical Documents: 1189–1327. General editor David C. Douglas. 11 vols. vol. 3, edited by Harry Rothwell. London: Eyre & Spottiswoode, 1975.

Epistolae Henrici IV. Edited by Carl Erdmann. Deutches Mittelalter. Kritische Studien Texte. Weimar: *M.G.H.*, 1950.

Etienne de Bourbon. *Anecdotes historiques . . . d'.* Edited by A. Lecoy de la Marche. Société d'histoire de la France. Paris: Librairie Renouard, 1877.

Eulogium historiarum sive temporis . . . a monacho quodam Malmesburiensi exeratum. Edited by F. Haydon. London: *R.S.*, 1858–1863.

Ex actibus pontificum Cenomannensium. H.F. 10. 384–86.

Excerpta e memoriale historiarum auctore Johanne Parisiensi. H.F. 21. 630–89.

Ex delatione Corporis S. Juliani. H.F. 10. 360–61.

Flores historiarum. Edited by H. Luard. 3 vols. London: *R.S.*, 1890.

Fragment d'une chronique anonyme finissant en 1328. H.F. 21. 146–58.

Froissart. *Chroniques.* Edited by H. Kervyn de Lettenhove. 25 vols. Brussels, 1867–1877.

———. *Chroniques.* Edited by Lèon and Albert Mirot. 17 vols. Paris: C. Klincksieck, 1952–.

———. *The Chronicles of Jean Froissart in Lord Berner's Translation.* Edited by Gillian and William Anderson. Carbondale, Ill.: Southern Illinois University Press, 1963.

———. *Chroniques: Dernière rédaction du premier livre: Edition du manuscrit de Rome, Reg. lat. 869.* Edited by George Diller. Textes littéraires français, vol. 194. Paris: Minard, 1972.

Fulcher of Chartres. *Historia Iherosolymitana: Gesta Francorum Iherusalem peregrinantium. RHC Occ.* 3. 315–485.

Galfridi le Baker de Swynebroke. *Chronicon.* Edited by John A. Giles. London: Caxton Society, 1847.

Geoffrey Chaucer. *The Canterbury Tales.* Edited Paul G. Ruggiers. Oklahoma: Oklahoma University Press, 1979.

Gervais du Bus. *Le roman de Fauvel par.* Edited by Arthur Langfors. Société des anciens textes français. Paris 1914–1919.

Gesta Edwardi de Carnavan auctore canonico Bridlingtoniensi. In *Chronicles of the Reigns of Edward I and II*, 2: 25–92.

Gesta Edwardi Tertii auctore Bridlingtoniensi. In *Chronicles of the Reigns of Edward I and II*, 2: 93–152.

Gesta Francorum expugnantium Iherusalem. RHC Occ. 3. 489–585.

Gesta Romanorum. Edited and translated by Charles Swan and W. Hooper. New York: Dover Publications, 1959.

Giovanni Boccaccio. *The Decameron.* Translated by Richard Aldington. New York: Garden City, 1930.

Giovanni Villani. *Cronica*. Edited by Franco G. Dragomanni. 4 vols. Firenze: S. Coen, 1844–1845.

Giraldus Cambrensis. *Speculum duorum*. Edited by Yves Lefevre and R.B. Huygen. Cardiff: University of Wales Press, 1974.

______. *De rebus a se gestis*. In *Opera*, edited by J.S. Breuer. 8 vols. 1: 1–123. London: *R.S.*, 1861.

______. *De principis institutione*. In *Opera*. Edited by George F. Warner. 8: 5–329.

Glanville. *The Treatise on the Laws and Customs of the Realm of England Commonly Called Glanvill*. Edited by G. D. Hall. London: Nelson, 1965.

Gregorii Magni Registrum Epistolarum. Edited by D. Norberg. *Corpus Christianorum Series Latina 140A*. Turnhout, 1982.

Gregorius Turonensis. *Historiae Francorum*. Edited by Henri Omont. Paris: Berthaud, 1905.

Gregory VII. *Das Register Gregors VII*. Edited by E. Caspar. 1929, reprint Berlin: *M.G.H. Epistolae Selectae*, 1967.

Grosseteste. *Epistolae*. Edited by H. R. Luard. London: *R.S.*, 1861.

Guibert of Nogent. *Historia quae dicitur Gesta Dei per Francos*. *RHC Occ*. 4. 119–263.

Guillaume de Nangis. *Gestae Sanctae Memoriae Ludovici Regis Franciae*. *H.F.* 20. 309–465.

Guillaume Durand le Jeune. *Tractatus de modo generalis Concilii celebrandi*. Paris: F. Clousier, 1671.

Guillaume le Maire. *Le Livre de*. Edited by Celestin Port. Paris 1874.

Heinricus Rebdorfensis. *Annales Imperatorum et Paparum*. *F.R.G.* 4. 507–68.

Henry Knighton. *Chronicon*. Edited by Joseph R. Lumby. 2 vols. London: *R.S.*, 1889–1895.

Herimanni. *Monumenta historiae Tornacensis: Liber de restauratione monasterii sanctii Martini Tornacensis*. *M.G.H. Scriptores*. 14. 274–318.

Histoire du différend entre le pape Boniface VIII et Philippe le Bel. Edited by Pierre Dupuy. Paris: S. and G. Cramoisy, 1655.

Historia de Hierosolymitano Itinere. Edited by J. H. and L. Hill. Paris: Paul Geuthner, 1977.

Hugh de Fleury. *Tractatus de regia potestate et sacerdotali dignitate*. Edited by Ernest Sackur. *M.G.H. Libelli de Lite*. 2: 465–94.

Hugh de St. Victor. *De institutione novitiorum*. *P.L.* 176. 925–52.

______. *De Sacramentis Christianae Fidei*. *P.L.* 176. 173–618.

Hugues de Rouen. *Contra haereticos*. *P.L.* 192. 1255–98.

Humbert de Romans. *De modo prompte audendi sermones*. Edited by M. de La Bigne. Lyon: Maxima Bibliotheca Veterum Patrum, vol. 25, 1677.

______. *De eruditione religiosorum praedicatorum libri duo*. Rome: A. de Rubeis, 1739.

______. *Opus Tripartitum*. In *Appendix ad fasciculum rerum expetendarum*, edited by C. Brown. London 1690.

Hyeronimus. *Epistulae*. Edited by Isidorus Hilberg, 3 vols. *Corpus Scriptorum Ecclesiasticarum Latinorum*, vols. 54–56.

Innocent III. *Epistolae*. *P.L.* 216.

Innocent IV. *Les registres d'*. Edited by Elie Berger. 4 vols. Paris: Bibliothèque des écoles françaises d'Athènes et de Rome, 1884.

Iohannes de Bazano. *Chronicon Mutinense (1188–1363)*. Edited by Tommaso Casini. *R.I.S.* 15–4. 1917–1919 .

Iohannes de Tulbia. *De Iohanne Rege Ierusalem*. In *Quinti Belli Sacri Scriptores*

Minores, edited by R. Röhricht, pp. 117–40. Geneve: Sociètè de l'Orient latin, 1879.

Isidore of Seville. *Etymologiarum*. Edited by Wallace M. Lindsay. 2 vols. Oxford: Clarendon Press, 1911.

Ivo of Chartres. *Decretum*. *P.L.* 161. 9–1037.

———. *Epistolae*. *P.L.* 162. 11–286.

Jacques de Viterbe. *De Regimine Christiano*. Edited by H. X. Arquilliére. *Le plus ancien traitè de l'èglise*. Paris, 1926.

Jacques de Vitry. *The Exempla or Illustratives Stories from the Sermones Vulgares of*. Edited by Thomas F. Crane. 1890. reprint New York: B. Franklin, 1971.

———. *Lettres de*. Edited by R. Huygens. Leyden: E. J. Brill, 1960.

Jacques de Voragine. *Legenda aurea*. Translated by J. B. M. Roze. Paris: Garnier, 1967.

Jacques Fournier. *Le Registre d'inquisition de Jacques Fournier, évêque de Pamiers 1318–1325*. Edited by Jean Duvernoy. 3 vols. Paris: Mouton, 1978.

Jean de Joinville. *Histoire de St. Louis*. Edited by J. Natalis de Wailly. Paris: Renouard, 1872.

———. *The Life of St. Louis*. In *Chronicles of the Crusades*, edited and translated by M. Shaw. Baltimore: Penguin Books, 1963.

Jean Gobi. *Scala Coeli*. Edited by Marie Anne Mercoyrol. Ph. D. diss.: Ecole des hautes études, 1984–1985.

Jerónimo Zurita. *Anales de la Corona de Aragón*. Edited by Angel Canellas Lopez. 8 vols. Zaragoza: Fernando el Católico, 1977.

Joannis Chrysostomi. *Commentarius in Sanctum Matthaeum Evangelistam*. *P.G.* 57.

———. *Commentarius in Epistolam ad Romanos*. *P.G.* 60. 391–682.

Johannes Quidort von Paris. *De Regia potestate et papali*. Edited by Fritz Bleienstein. Stuttgart: E. Klett, 1969.

Johannes Victoriensis. *F.R.G.* 1. 271–450 .

John VIII. *Epistolae*. *M.G.H. Epistolae*. 7. 30–333.

John of Salisbury. *Policraticus*. Edited by Clement Webb. 2 vols. 1909, reprint Frankfurt: Minerva, G.M.B.H. 1965.

John of Trokelowe. *Annales*. In *Chronica Monasterii S. Albani*, edited by H. Riley. 12 vols, 3: 61–127. London: *R.S.*, 1866.

John of Winterthur. Edited by F. Baethgen and C. Brun. *M.G.H. Scriptores*. n.s. vol. 3. Berlin 1955.

John Pontissara. *Registrum*. Edited by C. Deedes. Surrey Record Society vols. I and VI. London: Canterbury & York Society, 1915–1924.

Jordanes. *Getica*. Edited by Theodorus Mommsen. *M.G.H. Auctores antiquissimorum*. vol. 5. Berlin: Weidmann, 1882.

Julien of Vézelay. *Sermons*. Edited by Damien Vorreux. Collections sources chrétiennes, vol. 13. Paris: Cerf, 1972.

Langland. *The Vision of William Concerning Piers the Plowman: Text B*. Edited by Walter W. Skeat. 3 vols. English Text Society. Oxford: Oxford University Press, 1869.

Layettes du Trèsor des chartes. Edited by Alexandre Teulet et al.. 2 vols. Paris: H. Plon, 1863–1909.

Leo IV. *Epistolae*. *M.G.H. Epistolae*. 5. 585–615.

Lettere di mercanti a Pignol Zucchello (1336–1350). Edited by R. Morozzo della Rocca. Venice: Comitato per la pubblicazione delle fonti relative alla storia di Venezia, 1957.

Lettres communes de Jean XXII (1316–1334). Edited by Guillaume Mollat. 16 vols. Paris: Bibliothèque des écoles françaises d'Athènes et de Rome, 1921–1947.

Lettres secrètes et curiales du pape Jean XXII relatives à la France. Edited by Auguste Coulon and S. Clemencet. Bibliothèque des écoles françaises d'Athènes et de Rome. 9 vols. in 4. Paris: E. de Boccard, 1900–1960.

Liber duelii Christiani in obsedione Damiate. In *Quinti Belli*, pp. 141–66.

Liber Exemplorum ad usum praedicantium saeculo XIII compositus. Edited by Andrew G. Little. Aberdeen: British Society of Franciscan Studies, 1918.

Litterae ad regem Franciae de causa Templariorum. Edited by J. Schwalm. *M.G.H. Constitutiones*. vol. 4. Hanover 1906.

Majus Chronicon Lemovicense a Petro Coral et aliis conscriptum. H.F. 21. 761–88.

Manegold of Lautenback. *Ad Gebehardum Liber*. Edited by K. Francke. *M.G.H. Libelli de Lite*. 1: 300–430.

Manuel des péchés: Etude de littérature religieuse anglo-normande (XIIIe siècle). Edited by Emile J. Arnould. Paris 1940.

Marcha di Marco Battagli da Rimini. Edited by A. Massera. *R.I.S.* 16–3. 1913.

Marsilius of Padua. *Defensor Pacis*. Edited by P. Scholz. *Fontes Iuris Germanici. M.G.H.* Hanovre 1932.

Matthaeus Parisiensis. *Chronica Majora*. Edited by H. Luard. 7 vols. London: *R.S.*, 1872–1883.

Matthias Nuewenburgensis. *Cronica. F.R.G.* 4. 149–276.

Miscellanea novo ordine digesta. Edited by Stephanus Baluze and Joannis D. Mansi. 4 vols. Lucca: V. Junctinium, 1761–1764.

Olim, ou registres des arrêts rendus par la cour du roi (Les). Edited by Arthur Beugnot. 4 vols. Paris: Collection de documents inédits sur l'histoire de France, 1839–1848.

Orderic Vitalis. *Historia ecclesiastica*. Edited by Marjorie Chibnall. 7 vols. London 1969–1979.

Ordinacio de predicatione S. Crucis in Anglia. Edited by R. Röhricht. In *Quinti belli*, pp. 1–26.

Ordonnances des roys de France de la troisième race. Edited by E. Laurière. 21 vols. 1727–1734, reprint Farnborouth: Gregg Press, 1967–1968.

Paolo di Tommaso Montauri. *Cronaca Senese*. In *Cronache Senesi*, pp. 179–252.

Paris et ses historiens aux XIVe et XVe siècles. Edited by Antoine Le Roux de Lincy and L. M. Tisserand. Histoire générale de Paris. Paris: Imprimerie impériale, 1867.

Peter of Blois. *Epistolae. P.L.* 207. 1–559.

______. *Opera omnia*. Edited by John A. Giles. 4 vols. Oxford: Patres Ecclesiae Anglicanae, 1843–1848.

Petri Cantinelli. *Chronicon*. Edited by F. Torraca. *R.I.S.* 28–2. 1902.

Petri Cantoris. *Verbum abbreviatum. P.L.* 205. 22–370.

Petri Damiani. *Epistolarum Libri Octo. P.L.* 144. 206–498.

Petrus Jacobus. *Aurea practica libellorum*. Cologne, 1574.

Petrus Lombardi. *Liber de Ecclesiasticis Sacramentis. P.L.* 192. 1089–1112.

Philippe de Comynes. *Mémoires*. Edited by Joseph Calmette. 3 vols. Paris: Les Classiques de l'histoire de France, 1924–1925.

Philippe de Mezières. *Le Songe du vieil pélerin*. Edited by George W. Coopland. London: Cambridge University Press, 1969.

Philippe de Novare. *Les quatre âges de l'homme*. Edited by M. de Freville. Paris: F. Didot, 1888.

Pierre Dubois. *De recuperatione Terrae Sanctae*. Edited by Angelo Diotti. Florence: Leo S. Olschki, 1977.

Pilgrimage of S. Silvia of Aquitania to the Holy Land (The). Edited by John H. Bernard. New York: AMS Press, 1971.

Political Poems and Songs Relating to English History Composed During the Period from the Accession of Edward III to that of Richard III. Edited by Thomas Wright. 2 vols. London: *R.S.*, 1859.

Political Songs of England from the Reign of John to that of Edward II. Edited by Thomas Wright. Camden Society, vol. 6. London, 1839.

Possidonius. *Vita Augustini*. Edited by Mary M. Muller and Roy J. Deferrari. The Fathers of the Church, vol. 15. New York, 1953.

Procés des Templiers (Le). Edited by M. Jules Michelet. 2 vols. Collection des documents inèdits sur l'histoire de France. Paris: Imp. nationale, 1841–1851.

Quellen zur Geschichte des Papsttums und des romischen Katholizismus. Edited by C. Mirbt and K. Aland. Tubingen: Siebeck, 1967.

Rainier Sacconi. In *Heresy and Authority in Medieval Europe*, translated by Edward Peters, pp. 125–32. Philadelphia: University of Pennsylvania Press, 1980.

Raymond of Aguilers. *Historia francorum qui ceperunt Iherusalem. RHC Occ.* 3. 231–309.

Recueil des actes de Philippe Auguste. Edited by Henri F. Delaborde et al.. 2 vols. Paris: Académie des Inscriptions et Belles Lettres, 1916.

Regesta Imperii. Edited by Johann F. Böhmer. 11 vols. Innsbruck: Wagner, 1893.

Regestum Clementis Papae V. Edited by A. Tosti et al. 8 vols. Rome: Typographia Vaticana: 1885–1892.

Registrum John Peckham. Edited by C. T. Martin. 3 vols. London: *R.S.*, 1882.

Registrum Palatinum Dunelmense: The Register of Richard de Kellawe, Bishop of Durham (1311–1316). Edited by T. Hardy. 4 vols. London: *R.S.*, 1873–1878.

Registrum Ralph Baldock, G. Segrave, R. Newport et S. Gravesend. Edited by R. C. Fowler. Canterbury and York Society, 1911.

Reinerus. *Annales Sancti Jacobi Leodiensis. M.G.H. Scriptores*. 16. 651–80.

Richeri Historiarum libri III. Edited by Georgius Waitz. Hanovre: Hahnianus, 1839.

Richier. *La vie de Saint-Rémi: poème du XIIIe siècle*. Edited by W. N. Bolderston. London: H. Frowde, 1912.

Robert of Reims. *Historia Iherosolimitana. RHC Occ.* 3. 719–882.

Roberti regis diplomata. H.F. 10. 573–626.

Robertus de Avesbury. *De gestis mirabilibus regis Edwardi tertii*. Edited by Edward M. Thompson. London: *R.S.*, 1889.

Roger Bacon. *Opus Majus*. Edited by John H. Bridges. 1897, London: Williams & Norgate, 1914.

Roger of Howden. *Gesta Regis Henrici Secundi*. Edited by William Stubbs. 2 vols. London: *R.I.S.*, 1867.

Roland. *La Chanson de Roland*. Edited by Joseph Bédier. 3rd. ed. 1922, Paris: H. Piazza, 1964.

Salimbene de Adam. *Cronica*. Edited by G. de Scalia. Bari: G. Laterza, 1966.

———. *Cronica fratris Salimbene de Adam ordinis Minorum. M.G.H. Scriptores*. 32. 1–652.

Select Charters and Other Illustrations of English Constitutional History . . . to the Reign of Edward the First. Edited by William Stubbs. 9th ed. 1870, Oxford, Clarendon Press, 1929.

Selected Letters of Pope Innocent III Concerning England (1198–1216). Edited by Christopher R. Cheney and W. H. Semple. London: Th. Nelson, 1953.

Statutes of Lincoln Cathedral. Edited by Henry Bradshaw and C. Wordsworth. Cambridge: Cambridge University Press, 1892.

Storie Pistoresi: 1300–1348. Edited by Silvio A. Barbi. *R.I.S.* 11-5. 1907–1927.

Tacitus. *Annales*. Edited by Henry Furneaux. Oxford: Clarendon Press, 1956.

______. *De Germania*. Edited by Henry Furneaux. Oxford: Clarendon Press, 1894.

______. *De Germania*. Translated by Alfred J. Church and William Jackson Brodribb. London: Macmillan, 1882.

Textbook of Roman Law from Augustus to Justinian. Edited by W. Buckland. Cambridge: Cambridge University Press, 1921.

The Declaration of Arbroath. Edited by Sir James Fergusson. Edinburgh: Edinburgh University Press, 1970.

Thomas Aquinas. *Brevem principium in Sacram Scripturam Opusculum*. Edited by Pierre Mandonnet. 7 vols. Paris: Desclée, de Brouwer et Cie., 1938.

______. *Summa Theologica*. Edited by the Fathers of the Dominican Province. 3 vols. 1920 reprint: New York: Benziger Brothers Inc., 1948.

Thomas Becket. *Materials for the History of*. Edited by James Craigie Robertson. 7 vols. London: *R.S.*, 1875.

Thomas Cobham. *Summa Confessorum*. Edited by F. Bloomfield. Louvain: Analecta medievalia Namurcensia, 1968.

Thomas d'Eccleston. *De adventu fratrum Minorum in Angliam*. In *Sur les routes d'Europe au XIIIe siècle*, edited by Marie T. Laureilhe. Paris: ed. Franciscaines, 1959.

Thomas de la More. *Vita et mors Edwardi regis Angliae*. In *Chronicles of the Reigns of Edward I and II*, 2: 297–319.

Thomas de Celano. *Vita Prima S. Francisci*. In *Analecta Franciscana*, vol. 10. Quaracchi: Collegium S. Bonaventurae, 1926–.

Thomas Walsingham. *Historia Anglicana*. In *Chronica Monasterii S. Albani*. 2 vols.

Trithenius. *Chronico Hirsaugiensi*. *Mansi*. 25. 408–12.

Urbain IV. *Les registres d'*. Edited by Joseph Guiraud. 5 vols. Paris: Bibliothèque des écoles françaises d'Athènes et de Rome, 1901–1906.

Vaso Leodiensis episcopus. *Vita Vasonis*. *P.L.* 142. 726–63.

Vincent de Beauvais. *Speculum Quadruplex, sive Speculum Maius Naturale, Doctrinale, Morale, Historiale*. 4 vols, reprint Graz 1964–1965.

Vincent of Lerins. *Commonitorium*. *P.L.* 1. 637–86.

Vita Prima S. Francisci. In *Analecta Franciscana*, vol. 10. Quaracchi: Collegium S. Bonaventurae, 1926.

Walter Map. *De Nugis Curialum*. Edited by Montagne R. James. London: Honourable Society of Cymmrodion, 1923.

Walter of Hemingburgh. *Chronicon*. Edited by H. Hamilton. 2 vols. English Historical Society. London, 1848–1849.

William Ockham. *Guillelmi de Ockham: Opera politica*. Edited by Hilary S. Offler. 3 vols. Manchester: Manchester University Press, 1940.

______. *Dialogus inter clericum et militem*. Paris: Guidone Mercatore, 1498.

William of Malmesbury. *De gestis regum Anglorum*. *P.L.* 179. 959–1390.

William Rishanger. *Chronica et annales regnantibus Henrico Tertio et Edwardo Primo*. Edited by H. T. Riley. London: *R.S.*, 1865.

SECONDARY SOURCES

Adam, Paul. *La vie paroissiale en France au XIVe siècle*. Paris: Sirey, 1964.

Alef, Gustave. "The Origin and Early Development of the Muscovite Postal Service." *Jahrbucher für Geschichte Osteuropas* 15 (1967): 1–15.

Allard, Guy H. "Dante et la mort." In *Le sentiment de la mort au moyen âge*, edited by Claude Sutto. Etudes présentées au cinquième colloque de l'Institut d'études médiévales de l'Université de Montréal, pp. 211–27. Montréal: L'Aurore, 1979.

Allport, Gordon W. and Postman, Leo J. "The Basic Psychology of Rumor." In *Public Opinion, Propaganda and Politics*, edited by N. Kies, pp. 170–75. Jerusalem: Hebrew University, 1971.

Alphandery, Paul. "Les citations bibliques chez les historiens de la première croisade." *Revue de l'histoire des religions* 99 (1929): 139–57.

Alphandery, Paul and Dupront, Alphonse. *La Chrétienté et l'idée de croisade*. 2 vols. Paris: Albin Michel, 1954–1959.

Amstrong, C. A. J. "Some Examples of the Distribution and Spreed of News in England at the Time of the Wars of the Roses." In *Studies in Medieval History Presented to F. M. Powicke*, edited by R. W. Hunt, et al. Oxford: Clarendon Press, 1948.

Arasse, Daniel. "Entre dévotion et culture: Fonctions de l'image religieuse au XVe siécle." In *Faire Croire: Modalités de la réception et de la diffusion des messages religieux du XIIe au XVe siècles*, pp. 131–46. Rome: Ecole Française de Rome, 1981.

Aries, Philippe. "Huizinga et les thèmes macabres." In *Bijdragen en mededelingen Betreffende de geschiendenis der Nederlanden*. Colloque Huizinga. Gravenhage, 88–2 (1973): 246–57.

———. *Essais sur l'histoire de la mort en Occident du moyen âge à nos jours*. Paris: Seuil, 1975.

———. *L'homme devant la mort*. Paris: Seuil, 1977.

Ashtor, E. "Investments in Levant Trade in the Period of the Crusades." *The Journal of European Economic History* 14 (1985): 427–41.

Aubert, Felix. "Les huissiers du parlement de Paris, 1300–1420." *B.E.C.* 47 (1886): 370–93.

Aydelotte, William O. "Quantification in History." In *Quantitative History*, edited by D. Karl Rowney and James Q. Graham Jr., pp. 3–22. Homewood, Ill.: The Dorsey Press, 1969.

Baer, Yitzhak F. "Criticism of the Disputations of Rabbi Yehiel of Paris and Nahmanides." In *Studies in the History of the Jewish People*. 2 vols, 2: 128–43. 1931, reprint Jerusalem: The Historical Society of Israel, 1985.

Baillie, Donald M. *The Theology of the Sacraments and Other Papers*. New York: Scribner, 1957.

Baras, Zvi. "Jewish-Christian Disputes and Conversions in Jerusalem." In *Jerusalem in the Middle Ages*, edited by Benjamin Z. Kedar, pp. 27–38. Jerusalem: Yad Ben Zvi, 1979.

Barber, Malcolm. "Propaganda in the Middle Ages: The Charges Against the Templars." *Nottingham Medieval Studies* 17 (1973): 42–57.

———. "The Social Context of the Templars." *Transactions of the Royal Historical Society*. 5th series. 34 (1984): 27–46.

———. "The Crusade of the Shepherds in 1251." *Proceedings of the 10th Annual Meeting of the Western Society for French History*. Lawrence, 1984: 1–23.

Bayley, C. "Pivotal Concepts in the Political Philosophy of William of Ockham." *Journal of the History of Ideas* 10 (1949): 199–218.

Beckingham, C. F. *Between Islam and Christendom: Travellers, Facts and Legends in the Middle Ages and the Renaissance*. London: Variorum Reprints, 1987.

Bennet, M. "The First Crusaders' Image of Muslims: The Influence of Vernacular Poetry." *Forum for Modern Language Studies* 22 (1986): 101–22.

Benson, Lee. "An Approach to the Scientific Study of Past Public Opinion." In *Quantitative History*, pp. 23–63.

Benzinger, Joseph. "Zum Wesen und zu den Formen von Kommunikationen und Publizistik im Mittelalter." *Publizistik* 15 (1970): 295–318.

Berger, E. "Les préparatifs d'une invasion anglaise et la descente de Henry III en Bretagne (1229–1230)." *B.E.C.* 54 (1893): 5–44.

Bériou, Nicole. "L'art de convaincre dans la prédication de Ranulphe d'Homblières." In *Faire croire*, pp. 39–65.

Bischoff, Bernard. "The Study of Foreign Languages in the Middle Ages." *Speculum* 36 (1961): 209–24.

Bisson, Thomas N. "Consultative Functions in the King's Parliament (1250–1314)." *Speculum* 44 (1969): 353–73.

______. "The General Assemblies of Philip the Fair, Their Character Reconsidered." *Studia Gratiana* 15 (1972): 537–64.

Blake, E. O. "The Formation of the Crusade Idea." *Journal of Ecclesiastical History* 21–1 (1970): 11–31.

Blake, E. O. and Morris, C. "A Hermit goes to War: Peter and the Origins of the First Crusade." *Studies in Church History* 22 (1985): 79–108.

Bloch, Marc. *The Royal Touch*. Translated by J. E. Anderson. 1923, London : Routledge & K. Paul, 1973.

Blumenkranz, B. "Les auteurs chrétiens latins du moyen âge sur les Juifs et le Judaisme." *Revue des études Juives* 109 (1948–1949): 3–67; 111 (1951–1952): 5–61; 113 (1954): 5–36; 114 (1955): 37–90; 117 (1958): 5–58.

Boase, T. S. R. "Fontevrault and the Plantagenêts." *Journal of the British Archeological Association* 24 (1971): 1–10.

Borst, Arno. *Der Turmbau von Babel: Geschichte der Meinungen über Ursprung und Viefalt der Sprachen und Volker*. Stuttgart: A. Hiersemann, 1957.

______. "La transmission de l'hérésie au moyen âge." In *Hérésies et sociétés dans l'Europe pré-industrielle, 11e–18e siècles*, edited by Jacques Le Goff. Communications et débats du Colloque de Royaumont, pp. 273–80. Paris-La Haye: Mouton and Co, 1968.

Boulding, Kenneth E. *The Image: Knowledge in Life and Society*. Ann Arbor: University of Michigan Press, 1966.

Bourgain, Louis. *La chaire française au XIIe siècle*. 1879, reprint Genève: Slatkine, 1973.

Boutaric, Edgard P. "Clément V, Philippe le Bel et les Templiers." *R.Q.H.* 10 (1871): 301–42.

Bowsky, W. *Henry VII in Italy: The Conflict of Empire and City-State*. Lincoln: University of Nebraska Press, 1960.

Brampton, C. "Personalities at the Process Against Ockham at Avignon." *Franciscan Studies* 26 (1966): 4–25.

Brandt, W. "Pierre Dubois, Modern or Medieval?" *A.H.R.* 35 (1929–1930): 507–21.

Brown, Elizabeth. " 'Cessante Causa' and the Taxes of the Last Capetians: The Political Applications of a Philosophical Maxim." *Studia Gratiana* 15 (1972): 565–87.

———. "Taxation and Morality in the XIIIth and XIVth Centuries: Conscience and Political Power and the Kings of France." *French Historical Studies* 8–1 (1973): 1–28.

Brundage, James A. "Holy War and the Medieval Lawyers." In *The Holy War*, edited by Thomas P. Murphy, pp. 99–140. Ohio: Ohio State University Press, 1976.

Bryant, W. N. "Matthew Paris, Chronicler of St. Albans." *History Today* 19 (1969): 772–82.

Burns, Robert I. "A Medieval Income Tax: The Tithe in the Thirteenth Century Kingdom of Valencia." *Speculum* 41 (1966): 438–52.

———. "Christian Islamic Confrontation in the West: The XIIIth Century Dream of Conversion." *A.H.R.* 76 (1971): 1386–1434.

Burr, D. "Ockham, Scotus and the Censure at Avignon." *Church History* 37 (1968): 144–59.

Busse, H. "Vom Felsendom zum Templum Domini." In *Das Heilige Land im Mittelalters*, edited by Wolfdietrich Fisher and Jürgen Schneider, pp. 19–32. Neustadt: Degener, 1982.

Caillet, Louis. *La papauté d'Avignon et l'église de France*. Paris: Presses Universitaires de France, 1975.

Cam, Helen M. "Medieval Representation in Theory and Practice." *Speculum* 29 (1954): 347–55.

———. "Shire Officials: Coroners, Constables and Bailiffs." In *The English Government at Work 1327–1336*, edited by James F. Willard and William A. Morris. 3 vols. 3: 143–83. Cambridge, Mass.: The Medieval Academy of America, 1940.

Caplan, Harry. "The Four Senses of Scriptural Interpretation and the Mediaeval Theory of Preaching." *Speculum* 4 (1929): 282–90.

Carolus-Barré, L. "Les enquêtes pour la canonisation de Saint Louis." *R.H.E.F.* 57 (1971): 19–29.

Carozzi, Claude and Traviani-Carozzi, Huguette. *La fin des temps: Terreurs et prophéties au moyen âge*. Paris: Stock, 1982.

Caspar, E. "Die Kreuzzugsbullen Eugens III." *Neues Archiv . . . für altere deut. Geschichtskunde* 45 (1924): 285–305.

Cépède, Michel and Lengelle, Maurice. *Economie alimentaire du globe: essai d'interprétation*. Paris: Librairie de Médicis, 1953.

Chadwick, Henry. *The Early Church*. London: Penguin Books, 1967.

Chailley, Jacques. *L'école musicale de Saint-Martial de Limoges jusqu'à la fin du XIe siècle*. Paris : Les livres essentiels, 1960.

Chartier, Roger s.v. "Populaire." In *La nouvelle histoire*, edited by Jacques le Goff et al., pp. 458–60. Paris: Retz, 1978.

Chazan, Robert. "A Jewish Plaint to Saint Louis." *Hebrew Union College Annual* 45 (1974): 287–305.

———. "The Barcelona 'Disputation' of 1263: Christian Missionizing and Jewish Response." *Speculum* 52 (1977): 824–42.

———. "From Friar Paul to Friar Raymond: The Development of Innovative Missionizing Argumentation." *Harvard Theological Review* 76:3 (1983): 289–306.

Cheney, Christopher R. *English Synodalia of the Thirteenth Century*. London: Oxford University Press, 1940.

———. "Cardinal John of Ferentino, Papal Legate in England in 1206." *E.H.R.* 76 (1961): 654–60.

Chenu, Marie D. *Introduction à l'étude de Saint Thomas d'Aquin*. Montréal: Institut d'études médiévales, 1950.

Cherry, C. *On Human Communication: A Review, a Survey, and a Criticism*. Cambridge, Mass.: M.I.T. Press, 1978.

Chiffoleau, Jacques. "Sur l'usage obsessionnel de la messe pour les morts à la fin du moyen âge." In *Faire croire*, pp. 235–56.

Clark, James M. *The Dance of Death in the Middle Ages and the Renaissance*. Glasgow: Jackson, Son & Company, 1950.

Cobban, Alan B. *The Medieval Universities, their development and Organization*. London: Methuen, 1975.

Cohn, Norman. *The Pursuit of the Millennium*. 2nd ed. New York: Harper Torchbooks, 1961.

Constable, Giles. "The Second Crusade as seen by Contemporaries." *Traditio* 9 (1953): 213–79.

______. *Letters and Letter-Collections*. In *Typologie des sources du moyen âge occidental*. Edited by L. Genicot, vol. 17. Turnhout: Brepols, 1976.

______. "Monachisme et pélerinage au moyen âge." *R.H.* 258 (1977): 3–27.

Cooper, Eunice and Jahoda, Marie. "The Evasion of Propaganda: How Prejudiced People Respond to Anti-Prejudice Propaganda." In *Public Opinion*, pp. 176–79.

Coulton, George G. *Medieval Village, Manor and Monastery*. New York: Harper & Brothers, Torchbooks, 1960.

Coville, A. "Poèmes historiques du début de la Guerre de Cent Ans." *Histoire littéraire de la France* 38 (1949): 259–333.

Cowdrey, H. E. J. "Pope Urban II's Preaching of the First Crusade." *History* 5 (1970): 177–88.

______. "The Genesis of the Crusades: The Spring of Western Ideas of Holy War." In *The Holy War*, pp. 9–32.

______. "Pope Gregory VII's 'Crusading' Plans of 1074." In *Outremer: Studies in the History of the Crusading Kingdom of Jerusalem Presented to Joshua Prawer*, edited by Benjamin Z. Kedar et al., pp. 27–40. Jerusalem: Yad Ben Zvi, 1982.

Crosby, Ruth. "Oral Delivery in the Middle Ages." *Speculum* 11 (1936): 88–110.

Curschmann, Michael. "Oral Poetry in Medieval English, French, and German Literature: Some Notes on Recent Research." *Speculum* 42 (1967): 36–52.

David, Henry. "Opinion Research in the Service of the Historian." In *Common Frontiers of the Social Sciences*, edited by Mina Komarovsky, pp. 269–78. Glencoe, Ill.: Free Press, 1957.

D'Avenel, Georges. *L'évolution des moyens de transport*. Paris 1919.

Deighton, H. "Clerical Taxation by Consent, 1279–1301." *E.H.R.* 68 (1953): 161–92.

Delaruelle, Etienne. "La culture religieuse des laïcs en France aux XIe et XIIe siècles." In *I Laici nella Societas Christiana dei secoli XI e XII*. Miscellanea del Centro di Studi Medievali, 5: 548–81. Milan, 1968.

______. "La pietá populare nel secolo XI." In *La piété populaire au moyen âge*. Turin: Bottega d'Erasmo, 1975.

Delisle, Léopold. "Preuves de la Préface." *H. F.* 24. 271–368.

Delumeau, Jean. *La peur en Occident (XIVe—XVIIIe siècles)*. Paris: Fayard, 1978.

______. *Le péché et la peur: La culpabilité en Occident*. Paris: Fayard, 1983.

De Pange, J. *Le roi trés chrétien*. Paris: Sireil, 1949.

Déprez, Eugène. *Les préliminaires de la guerre de cent ans*. Paris: A. Fontemoing, 1902.

Desimoni, C. "Actes passés à Famagouste de 1299 à 1301 par devant le notaire Lamberto di Sambuceto." *Revue de l'Orient Latin* 1 (1893): 58–139, 275–312, 321–53.

Dickinson, Jocelyne G. "Blanks and Blank Charters in the Fourteenth and Fifteenth Century." *E.H.R.* 66 (1951): 375–87.

Di Fazio, Laura. *Lombardi e templari nella realta socioeconomica durante il regno di Filippo il Bello (1285–1314)*. Milan: Coop. Libraria IULM, 1986.

Digard, Georges. *Philippe le Bel et le Saint Siège de 1285 à 1304*. Paris: Librairie du Recueil Sirey, 1936.

Dollinger, Ignaz von. *Beitrage zur Sektengeschichte des Mittelalters*. 2 vols. Munchen: O. Beck, 1890.

Dreiling, Raymundus. *Der Konzeptualismus in der universalienlehre des franziskanerzbischofs Petrus Aureoli*. Münster: W. Aschendorff, 1913.

Du Boys-Reymond, Emil. *Über die Grenzen des Naturerkennens: Die sieben Weltrathsel, zwei Vorträge*. Leipzig: Veit, 1882.

Duby, Georges. *Fondements d'un nouvel humanisme, 1280–1440*. Genève: Skira, 1966.

———. "Les pauvres des campagnes dans l'Occident médiéval jusqu'au XIIIe siècle." *R.H.E.F.* 52 (1966): 25–32.

———. "The Diffusion of Cultural Patterns in Feudal Society." *Past and Present* 39 (1968): 3–10.

———. *Le temps des cathédrales: L'art et la société 980–1420*. Paris: Gallimard, 1976.

Dudden, Frederick H. *Gregory the Great: His Place in History and Thought*. London-New York: Longmans, Green & Co., 1905.

Dufourcq, Charles E. "Les communications entre les royaumes chrétiens ibériques et les pays de l'occident musulman dans les derniers siècles du moyen âge." In *Les communications dans la péninsule Ibérique au moyen âge*. Actes du colloque tenu à Pau les 28–29 Mars 1980 sous la direction de P. Tucoo-Chala, pp. 29–44. Paris: Centre national de la recherche scientifique, 1981.

Dunham, Barrows. *Heroes and Heretics: A Political History of Western Thought*. New York: Alfred A. Knopf, 1964.

Dunham, William Huse Jr. and Wood, Charles. "The Right to Rule in England: Depositions and the Kingdom's Authority, 1327–1485." *A.H.R.* 81 (1976): 738–61.

Duval-Arnould, Louis. "Une apologie de la pauvreté volontaire par un universitaire séculier de Paris (1256)." In *Etudes sur l'histoire de la pauvreté*, edited by Michel Mollat. 2 vols. 1: 412–45. Paris: Sorbonne, 1974.

Edwards. L. "The Political Importance of the English Bishops During the Reign of Edward II." *E.H.R.* 59 (1944): 311–47.

Ehrismann. G. *Bibliothek des Literarischen Vereins*. vol. 247. Tübingen, 1908.

Ehrle, Franz. "Die Spiritualen, ihr Verhältniss zum Franciscaneorden und zu den Fraticellen." *Archiv für Literatur und Kirchengeschichte des Mittelalters* 1 (1885) 509–69; 2 (1886) 106–64, 249–336; 3 (1887) 553–623; 4 (1888): 1–190.

Ellul, Jacques. *Propaganda: The Formation of Men's Attitudes*. Translated by Konrad Kellen and Jean Lerner. New York: A.A. Knopf, 1965.

Elyada, Uzi. *Manipulation et Théâtralité: Le père Duchesne, 1788–1791*. Ph.D. Diss.: Ecole des hautes études, 1984–1985.

Erdmann, Carl. *The Origins of the Idea of Crusade*. Translated by Marshall W. Baldwin and Walter Goffart. 1935, Princeton: Princeton University Press, 1977.

Eschmann, T. "St. Thomas on the Two Powers." *Medieval Studies* 20 (1958): 177–205.

Farbridge, Maurice H. *Studies in Biblical and Semitic Symbolism*. 1923, reprint New York: Ktav, 1970.

Farrer, W. "An Outline Itinerary of Henry I." *E.H.R.* 34 (1919): 303–82, 505–79.

Ferreira-Priegue, Elisa M. "Circulación y red viaria en la Galicia medieval." In *Les communications*, pp. 65–71.

Finke, Heinrich. *Papsttum und Untergang des Templerordens*. 2 vols. Münster: Aschendorffsche Buchhandlung, 1907.

Finn, Rex Weldon. *The Domesday Inquest and the Making of Domesday Book*. London: Longmans, 1981.

Finucane, Ronald C. *Miracles and Pilgrims: Popular Beliefs in Medieval England*. London: J. M. Dent & Sons, 1977.

Fishman, J. "An Examination of the Process and Function of Social Stereotyping." *Journal of Social Psychology* 43 (1956): 27–64.

Fleige, A. and Lafille, B. *Les Templiers et leur mystère*. 2 vols. La Seyne-sur-Mer, 1980.

Fliche, Agustin, Thouzellier, Christine and Azais, Yvonne. *La Chrétienté Romaine (1198–1274)*. In *Histoire de l'église*, vol. 10. Paris: Bloud & Gay, 1950.

Folz, Robert. "La sainteté de Louis IX d'après les textes liturgiques de sa fête." *R.H.E.F.* 57 (1971): 31–45.

Fontette, M. de. "Villes médiévales et ordres mendiants." *Revue historique de droit français et étranger* 48 (1970): 390–407.

Fried, J. "Wille, Freiwilligkeit und Gestandnis um 1330.: Zur Beurteilung des letzen Templer-Grossmeisters Jacques de Molay." *Historisches Jahrbuch* 105 (1985): 388–425.

Fritz, Ernst. "Über Gesandtschaftswesen und Diplomatie an der Wende vom Mittelalter zur Neuzeit." *Archiv für Kulturgeschichte* 33 (1950): 64–95.

Fulchignoni, Enrico. *La civilisation de l'image*. Translated by G. Crescenzi et E. Darmouni. Paris: Payot, 1972.

Furnivall, Frederick. *King Edward II's Household and Wardrobe Ordinances, 1323*. London: Chaucer Society, 1876.

Fustel de Coulanges, Numa Denis. *Saint Louis et le prestige de la royauté*. Colombes: Martin Morin, 1970.

Galbraith, Vivian H. *The Making of Domesday Book*. Oxford: Clarendon Press, 1961.

Ganshof, Francois L. *Le Moyen Age, Histoire des relations internationales*, vol. 1. Paris: Hachette, 1953.

______. "Merowingisches Gesandtschaftswesen." In *Aus Geschichte und Landeskunde*. Forschungen und Darstellungen. Franz Steinbach zum 65. Geburstag gewidmet von seinen Freunden und Schulern, pp. 166–83. Bonn 1960.

García y García, A. and Kuttner, S. "A New Eyewitness Account of the Fourth Lateran Council." *Traditio* 20 (1964): 115–78.

Gaudemet, Jean. "La paroisse au moyen âge: état des questions." *R.H.E.F.* 59 (1973): 5–22.

Gautier, Leon. "L'idée politique dans les chansons de geste." *R.Q.H.* 7 (1869): 79–114.

Gilchrist, John. "The Erdmann Thesis and the Canon Law, 1083–1141." In *Crusade and Settlement*, edited by P. W. Edbury, pp. 37–45. Cardiff: University College Cardiff Press, 1985.

Gilmour-Bryson, A. "L'eresia e i Templari: 'Oportet et haereses esse'." *Richerche di Storia sociale e religiosa* 24 (1983): 101–14.

Ginsburg, Isidor. "National Symbolism." In *Modern Germany*, edited by Paul Kosok, pp. 292–324. Chicago: The Chicago University Press, 1933.

Glorieux, P. "L'enseignement au moyen âge: Techniques et méthodes en usage à la faculté de théologie de Paris au XIIIe siècle." *Archives d'histoire doctrinale et littéraire du moyen âge* 43 (1968): 65–186.

Gougaud, L. "La danse dans les églises." *Revue d'histoire de l'église* 15 (1914): 5–22.

Grabois, Aryeh. "De la trêve de Dieu à la paix du roi: Etude sur les transformations du mouvement de la paix au XIIe siècle." In *Mélanges offerts à René Crozet*. 2 vols. Edited by Pierre Gallais and Yves-Jean Rion, 1:585–96. Poitiers: Société d'Etudes Médiévales, 1966.

———. "The Crusade of King Louis VII: A Reconsideration." In *Crusade and Settlement*. Edited by P.W. Edbury. pp. 94–104. Cardiff: University College Cardiff Press, 1985.

———. "Medieval Pilgrims. The Holy Land and Its Image in European Civilization." In *Pillars of Smoke and Fire*, pp. 65–79. Edited by Moshè Sharon. Johannesburg: Southern Books Pub., 1987.

———. "La royauté sacrée au XIIe siècle: manifestation de propagande royale." In *Idéologie et Propagande in France*. Edited by Myriam Yardeni, pp.31–41. Paris: Picard, 1987.

Gravier, H. "Essai sur les prévôts royaux du XIe au XIVe siècle." *Nouvelle revue historique du droit français et étranger* 27 (1903): 648–64.

Griffe, Elie. *Le Languedoc cathare de 1190 à 1210*. Paris: Letouzey et Ane, 1971.

Guenée, Bernard. "Etat et nation en France au moyen âge." *R. H.* 237 (1967): 17–30.

———. *Histoire et culture historique dans l'Occident médiéval*. Paris: Aubier Montaigne, 1980.

Guillemain, Bernard. "Chiffres et statistiques pour l'histoire ecclésiastique du moyen âge." *Le moyen âge* 8 (1953): 341–65.

———. "Les tentatives pontificales de médiation dans le litige franco-anglais de Guyenne au XIVe siècle." *Bulletin philologique et historique du comité des travaux historiques et scientifiques* (1958): 423–32.

Guterbock, F. *Die Gelnhauser Urkunde*. Berlin, 1920.

Habermas, Jurgen. *Communication and the Evolution of Society*. Translated by Thomas McCarthy. Boston: Beacon Press, 1979.

Hallam, Elizabeth M. "Royal Burial and the Cult of Kingship in France and England, 1060–1330." *Journal of Medieval History* 8 (1982): 359–80.

Halphen, Louis. "La place de la royauté dans le système féodal." *R. H.* 172 (1933): 249–56.

Harding, Carol. "Acting with Intention: A Framework for Examining the Development of the Intention to Communicate." In *The Origins and Growth of Communication*, edited by Lynne Feagans, et al., pp. 123–35. New Jersey: Ablex Publishing Corporation, 1984.

Hardt, Hanno. "Communication and History: The Dimensions of Man's Reality." In *Approaches to Human Communication*, edited by Richard W. Budd and Brent D. Ruben, pp. 145–55. Rochelle Park: Hayden Book Co., 1972.

Harper, Nancy. *Human Communication Theory: The History of a Paradigm*. Rochelle Park: Hayden Book Co., 1979.

Haskins, Charles H. "The Spread of Ideas in the Middle Ages." *Speculum* 1 (1926): 19–30.

Hauck, Albert. *Kirchengeschichte Deutschlands*. 4 vols. Leipzig: J. C. Hinrichs, 1887–1903.

Heath, Sidney H. *Pilgrim Life in the Middle Ages*. New York: Kennikat Press, 1971.

Hébert, Michel. "La mort: impact réel et choc psychologique." In *Le sentiment de la mort*, pp. 17–30.

Héfèle, Charles J. and Leclercq, H. *Histoire des conciles*. 16 vols. in 8. Paris: Letouzey et Ane, 1913.

Hendrix, S. "In Quest of the *vera ecclesia*: The Crises of Late Medieval Ecclesiology." *Viator* 7 (1976): 347–78.

Henneman, John B. "Enquêteurs-réformateurs and Fiscal Officers in Fourteenth Century France." *Traditio* 24 (1968): 309–49.

Heutsch, A. A. *La littérature didactique au moyen âge*. Halle, 1903.

Hill, Mary C. "King's Messengers and Administrative Developments in the Thirteenth and Fourteenth Centuries." *E.H.R.* 61 (1946): 315–28.

Horowitz, Jeannine. *La peur en milieu ecclésiastique: XIIIe—première moitié du XIVe siècle*. Ph. D. diss.: Ecole des hautes études, 1985.

Housley, Norman. *The Italian Crusades: The Papal Angevin Alliance and the Crusades Against Christian Powers, 1254–1343*. Oxford: Clarendon Press, 1982.

______. "Pope Clément V and the Crusades of 1309–1310." *Journal of Medieval History* 8 (1982): 29–43.

______. "Saladin Triumph Over the Crusader States: The Battle of Hattin, 1187." *History Today* 37 (1987): 17–23.

Howell, Wilbur S. *The Rhetoric of Alcuin and Charlemagne*. Princeton Studies in English, vol. 23. Princeton: Princeton University Press, 1941.

Hrabar, V. E. *De Legatis et Legationibus Tractatus Varii*. Dorpat: Typographeo Mattieseniano, 1905.

Huard, Georges. "Considérations sur l'histoire de la paroisse rurale des origines à la fin du moyen âge." *R.H.E.F.* 24 (1938): 5–22.

Huber, Alfons and Dopsch, Alfons. *Österreichische Reichsgeschichte*. Vienna: Neudruck der Ausgabe, 1901.

Huizinga, Johan. *The Waning of the Middle Ages*. Translated by F. Hopman. 1924, reprint, Middlesex: Penguin Books, 1982.

Jacoby, Zehava. "Le chapiteau allégorique et le sermon: deux voies parallèles dans le processus créateur de l'imagerie romane." *Atti del Convegno Romanico mediopadano-romanico europeo*. Parma (1982): 382–90.

Jegou, F. "Confrérie de Saint Nicolas de Guérande, statuts de 1350." *Revue de Bretagne et de Vendée* 6–2 (1874): 1–35.

Jenkinson, Hilary C. and Mills, Mabel H. "Rolls from a Sheriff's Office of the Fourteenth Century." *E.H.R.* 43 (1928): 21–32.

Jervis, R. *Perception and Misperception in International Politics*. Princeton: Princeton University Press, 1976.

Jordan, William C. "The Lamb Triumphant and the Municipal Seals of Western Languedoc in the Early Thirteenth Century." *Revue Belge de Numismatique* 123 (1977): 213–19.

______. *Louis IX and the Challenge of the Crusade*. Princeton: Princeton University Press, 1979.

Jowett, Garth S. "Propaganda and Communication: The Re-emergence of a Research Tradition." *Journal of Communication* 37–1 (1987): 97–114.

Jubinal, Achille. *Nouveau recueil de contes, dits, fabliaux et autres pièces inedites des XIIIe, XIVe et XVe siècles*. 2 vols. Paris: E. Pannier, 1839.

Jusserand, J. J. *English Wayfaring Life in the Middle Ages*. Translated by Lucy Toulmin Smith. 1889, London: Ernest Benn, 1950.

Kahrl, Stanley J. "Introduction." In *The Holy War*, pp. 1–8.

Kann, Robert A. "Public Opinion Research: A Contribution to Historical Method." In *Quantitative History*, pp. 64–80.

Kantorowicz, Ernst H. *Pro patria mori* in Medieval Thought." *A.H.R.* 56 (1951): 472–92.

———. *The King's Two Bodies: A Study in Mediaeval Political Theology*. Princeton: Princeton University Press, 1957.

———. *Laudes Regiae: A Study in Liturgical Acclamations and Mediaeval Ruler Worship*. Berkeley: University of California Press, 1958.

Katzir, Yael. "The Vicissitudes of the True Cross of the Crusaders." In *The Crusaders in their Kingdom: 1099–1291*, edited by Benjamin Z. Kedar, pp. 243–53. Jerusalem: Yad Ben Zvi, 1987.

Kecskemeti, Paul. "Propaganda." In *Handbook of Communication*, edited by Ithiel de Sola Pool et al., pp. 844–70. Chicago: Rand McNally, 1973.

Kedar, Benjamin Z. "The Passenger List of a Crusader Ship, 1250: Towards the History of the Popular Element in the Seventh Crusade." *Studi Medievali* 13 (1972): 267–79.

Keen, M. H. *England in the Later Middle Ages*. London: Meuthen, 1973.

Kenaan, Nurith. "The Sculptural Lintels of the Crusader Church of the Holy Sepulchre in Jerusalem." In *Jerusalem in the Middle Ages*, pp. 316–26.

Köhler, C. and Langlois, C. "Lettres inédites concernant les croisades." *B.E.C.* 3 (1891): 46–63.

Krause, V. "Geschichte der Missi dominici." *Mitteilungen des Instituts für osterreichische Geschichtsforschung* 11 (1890): 193–214.

Krey, A. C. "Urban's Crusade: Success or Failure?" *A.H.R.* 53 (1948): 235–50.

Kurth, Godefroid. "La France et les Francs dans la langue politique du moyen âge." *R.Q.H.* 57 (1895): 337–99.

Lagarde, Georges. *La naissance de l'esprit laïque au déclin du moyen âge*. 5 vols. Louvain—Paris: E. Nauwelaerts, 1956–1970.

———. "Marsile de Padoue et Guillaume d'Ockham." In *Etudes d'histoire du droit canonique dédiées à G. Le Bras*, ed. G. Vadel et al. pp. 593–605. 2 vols. Paris: Sirey, 1965.

Lambert, M. D. *Franciscan Poverty: The Doctrine of the Absolute Poverty of Christ and the Apostles in the Franciscan Order (1210–1323)*. London: S.P.C.K., 1961.

———. "The Franciscan Crisis Under John XXII." *Franciscan Studies* 32 (1972): 123–43.

———. *Medieval Heresy: Popular Movements from Bogomil to Hus*. London: E. Arnold, 1977.

Langlois, Charles V. "Livres sur l'histoire des Templiers." *R.H.* 40 (1889): 168–79.

———. "Les Anglais du moyen âge d'après les sources françaises." *R.H.* 52 (1893): 298–315.

———. "Doléances recueillies par les enquêteurs de Saint Louis et des derniers Capétiens directs." *R.H.* 100 (1909): 63–95.

———. *La vie en France au moyen âge*. 4 vols. Paris: Hachette, 1926.

Langlois, Ernest. "Anciens proverbes français." *B.E.C.* 60 (1899): 569–601.

Langlois, M. "Une réunion publique à Paris sous Philippe le Bel." *Bulletin de la société d'histoire de Paris* 15 (1888): 135–68.

La nouvelle histoire. Edited by Jacques Le Goff, et al. Paris: CEPL, 1978.

Lapparent, P. *L'oeuvre politique de François de Meyronnes*. Paris, 1940–1942.

Lasswell, Harold D. and Blumenstock, Dorothy. *World Revolutionary Propaganda*. New York: A. A. Knopf, 1939.

———. "Nations and Classes: The Symbols of Identification." In *Public Opinion and*

Communication, edited by Bernard Berelson and Morris Janowitz, pp. 27–42. New York: Free Press, 1966.

Laureilhe, Marie T. *Sur les routes d'Europe au XIIIe siècle: Jourdain de Giano, Thomas d'Eccleston et Salimbene d'Adam*. Paris: ed. Franciscaines, 1959.

Le Bras, Gabriel. "Les confréries chrétiennes: problèmes et propositions." In *Etudes de sociologie religieuse*. 2 vols, 2: 423–62. Paris: Presses universitaires de France, 1956.

______. *Institutions ecclésiastiques de la Chrétienté médiévale*. In *Histoire de l'église*, edited by Augustin Fliche and V. Martin, vol. 12. Tournai 1959–1964.

Leclercq, H. s.v. *Ancilla Dei*." *Dictionnaire d'archéologie chrétienne et de liturgie*, 1–1: 1973–1993.

______. s.v. *Servus servorum Dei*." *ibid.*, 15–1: 1360–63.

Leclercq, Jean. "Un sermon prononcé pendant la guerre de Flandre sous Philippe le Bel." *Revue du moyen âge latin* 1 (1945): 166–73.

______. *L'idée de la royauté du Christ au moyen âge*. Paris: Cerf, 1959.

______. "Deux témoins de la vie des cloîtres au moyen âge." *Studi Medievali* 12 (1971): 987–95.

______. "Lettres de S. Bernard: Histoire ou littérature?." *Studi Medievali* 12 (1971): 1–74.

______. "Modern Psychology and the Interpretation of Medieval Texts." *Speculum* 48 (1973): 476–90.

______. "Aux origines bibliques du vocabulaire de la pauvreté." In *Etudes sur l'histoire de la pauvreté*, 1: 35–43.

______. "Les controverses sur la pauvreté du Christ." In *Etudes sur l'histoire de la pauvreté*, 1: 45–55.

Lecoy de la Marche, A. "La prédication de la croisade au treizième siècle." *R.Q.H.* 48 (1890): 5–28.

Leff, Gordon. "Heresy and the Decline of the Medieval Church." *Past and Present* 20 (1961): 36–51.

______. *Heresy in the Later Middle Ages*. 2 vols. Manchester: Manchester University Press, 1967.

______. "Hérésies savantes et hérésies populaires au moyen âge." In *Hérésies et sociétés,* pp. 219–25.

Le Goff, Jacques. "Au moyen âge: Temps de l'église et temps du marchand." *Annales E.S.C.* 15–3 (1960): 417–33.

______. "Apostolat mendiant et fait urbain dans la France médiévale: l'implantation géographique et sociologique des ordres mendiants (XIIIe–XVe siècles)." *R.H.E.F.* 54 (1968): 69–76.

______. *The Birth of Purgatory*. Translated by Arthur Goldhammer. 1981, Chicago : Chicago University Press, 1984.

______. and Schmitt, Jean C. "Au XIIIe siècle: Une parole nouvelle." In *Histoire vécue du peuple chrétien*, edited by Jean Delumeau. 2 vols. 1: 257–80. Paris: Privat, 1979.

Lerner, Robert E. " 'An Angel of Philadelphia' in the Reign of Philip the Fair: The Case of Guiard of Cressonessart." In *Order and Innovation in the Middle Ages*. Essays in Honor of Joseph R. Strayer, edited by William C. Jordan, Bruce McNab and Teofilo Ruiz, pp. 343–64. Princeton: Princeton University Press, 1976

______. "Refreshment of the Saints: The Time After Antichrist as a Station for Earthly Progress in Medieval Thought." *Traditio* 32 (1976): 97–144.

Le Roy Ladurie, Emmanuel. *Montaillou, village occitan de 1294 à 1324.* Paris: Gallimard, 1975.

Levett, Ada E. "The Summons to a Great Council, 1213." *E.H.R.* 31 (1916): 85–90.

Lewis, Ewart. *Medieval Political Ideas.* 2 vols. London: Routlege & Paul, 1954.

Lietzmann, Hans. *The Era of the Church Fathers.* Translated by Bertram Lee Wolf. New York: Meridian Books, 1950.

Linder, Amnon. *Ex mala parentela bona sequi seu oriri non potest*: The Trojan Ancestry of the Kings of France and the Opus Davidicum of Johannes Angelus de Legonissa." *Bibliothèque d'humanisme et Renaissance* 40 (1978): 497–512.

Linehan, Peter. "The Church, the Economy and the *reconquista* in Early XIVth Century Castile." *Revista española de teología* 43 (1983): 275–303.

Lippmann, Walter. *Public Opinion.* 1921, New York: MacMillan Paperbacks, 1960.

Little, Lester K. "Saint Louis Involvement with the Friars." *Church History* 33 (1964): 125–48.

———. "L'utilité sociale de la pauvreté volontaire." In *Etudes sur l'histoire de la pauvreté*, pp. 447–59.

Lizerand, Georges. *Clément V et Philippe IV, le Bel.* Paris: Hachette, 1910.

———. *Le dossier de l'affaire des Templiers.* Paris: E. Champion, 1923.

Lloyd, Simon. "The Lord Edward's Crusade, 1270–1272: Its Setting and Significance." In *War and Government in the Middle Ages: Essays in Honour of J. O. Prestwich*, edited by J. Gillingham and J. C. Holt, pp. 120–33. Woodbridge: Boydell Press, 1984.

———. "Political Crusades in England c. 1215–1217 and c. 1263–1265." In *Crusade and Settlement*, pp. 113–20.

Loeb, Isidore. "La controverse de 1240 sur le Talmud." *Revue des études Juives* 1 (1880): 247–61; 2 (1881) 248–70; 3 (1881) 39–57.

Lucas, Henry S. "The Great European Famine of 1315, 1316 and 1317." *Speculum* 5 (1930): 343–77.

———. "Diplomatic Relations Between England and Flanders, 1329–1336." *Speculum* 11 (1936): 59–87.

———. "The Machinery of Diplomatic Intercourse." In The *English Government at Work, 1327–1336*, 1: 300–31.

Lumley, Frederick E. *The Propaganda Menace.* New York-London: The Century Co., 1933.

Lunt, William E. *Financial Relations of the Papacy With England to 1327.* Cambridge, Mass.: Medieval Academy of America, 1939.

Luttrell, Anthony. "The Crusade in the Fourteenth Century." In *Europe in the Late Middle Ages*, edited by J. R. Hale et al., pp. 122–54. London: Faber & Faber, 1965.

Mackinney, Loren C. "The People and Public Opinion in the Eleventh Century Peace Movement." *Speculum* 5 (1930): 181–206.

Mâle, Emile. *L'art religieux de la fin du moyen âge en France.* Paris: A. Colin, 1908.

———. *The Gothic Image: Religious Art in France of the Thirteenth Century.* Translated by Dora Nussey. 1913, New York: Harper & Row, 1958.

Manselli, Raoul. *Spirituali e Bighini in Provenza.* Rome: Instituto Storico Italiano per il Medio Evo, 1959.

Manteuffel, T. "Naissance d'une hérésie." In *Hérésies et sociétés*, pp. 97–101.

Markowski, M. *Crucesignatus*: Its Origins and Early Usage." *Journal of Medieval History* 10 (1984): 157–65.

Marongiu, Antonio. "Il principio della democrazia e del consenso: *quod omnes tangit ab omnibus approbari debet* nel secolo XIV." *Studia Gratiana* 8 (1962): 555–75.

______. *Medieval Parlements*. Translated by S. J. Woolf. London: Eyre & Spottiswoode, 1968.

Marthaler, Berard. "Forerunners of the Franciscans: The Waldenses." *Franciscan Studies* 18 (1958): 133–42.

Mattingly, Garret. "The First Resident Embassies: Mediaeval Italian Origins of Modern Diplomacy." *Speculum* 12 (1937): 423–39.

McCready, W. "Papalists and Antipapalists: Aspects of the Church/State Controversy in the Late Middle Ages." *Viator* 6 (1975): 241–273.

McGinn, Bernard. "*Iter Sancti Sepulchri:* The Piety of the First Crusaders." In *Essays in Medieval Civilization*, edited by R. E. Sullivan, et al., pp. 33–71. Austin-London: University of Texas Press, 1978.

McKenzie, John L. *The Power and the Wisdom*. Milwaukee: Bruce Publishing Co., 1965.

Meersseman, G. "La prédication dominicaine dans les congrégations mariales en Italie au XIIIe siècle." *Archivum Fratrum Praedicatorum* 18 (1948): 131–61.

______. "Etudes sur les anciennes confréries Dominicaines." *Archivum Fratrum Praedicatorum* 20 (1950): 5–113; 21 (1951): 51–196; 22 (1952): 5–176; 23 (1953): 275–308.

Menache, Sophia. *The Papal Policy With Regard to the Templar Order Between 1240–1312*. M.A. thesis: University of Haifa, 1973.

______. *The Status of the Papacy and the Image of the Popes at the Beginning of the Avignonese Period (1335–1334)*. Ph. D. diss.: Hebrew University, 1980.

______. "La naissance d'une nouvelle source d'autorité: L'université de Paris." *R.H.* 268-2 (1982): 305–27.

______. "Contemporary Attitudes Concerning the Templars' Affair: Propaganda Fiasco?." *Journal of Medieval History* 8 (1982): 135–47.

______. "Isabelle of France, Queen of England: A Reconsideration." *Journal of Medieval History* 10 (1984): 107–24.

______. "Philippe le Bel: Genèse d'une image." *Revue Belge de philologie et d'histoire* 72-4 (1984): 689–702.

______. "Un peuple qui a sa demeure à part: Boniface VIII et le sentiment national français." *Francia* 12 (1985): 193–208.

______. "Un ideal caballerezco en el medioevo tardío: Don Alonso de Aragón." *Revista de la Universidad de Alicante* 6 (1987): 9–29.

Miccoli, G. "La crociata del fanciulli del 1212." *Studi medievali* 2 (1961): 407–43.

Molinier, Charles. "L'église et la société cathares." *R.H.* 94 (1907): 225–48; 95 (1907) 1–22.

Mollat, Michel. "La notion de la pauvreté au moyen âge: position de problèmes." *R.H.E.F.* 52 (1966): 5–23.

______. "Le problème de la pauvreté au XIIe siècle." In *Vaudois Languedociens et pauvres Catholiques*, pp. 23–47. Cahiers de Fanjeaux. Toulouse: E. Privat, 1967.

Moore, R. I. "The Origins of Medieval Heresy." *History* 55 (1970): 21–36.

______. "Heresy as Disease." In *The Concept of Heresy in the Middle Ages*, edited by W. Lourdaux and D. Verhelst, pp. 1–11. Louvain-Hague: Louvain University Press, 1976.

Moorman, John. *A History of the Franciscan Order From Its Origins to the Year 1517*. Oxford: Clarendon Press, 1968.

Morris, Colin. *Medieval Media: Mass Communication in the Making of Europe*. Southampton: University of Southampton, 1972.

———. "Propaganda for War and the Dissemination of the Crusading Ideal in the Twelfth Century." *Studies in Church History* 20 (1983): 79–101.

Munro, Dana C. "The Speech of Pope Urban II at Clermont, 1095." *A.H.R.* 11 (1906): 231–42.

Mury, P. "La bulle Unam sanctam." *R.Q.H.* 26 (1879): 41–130, 46 (1889) 253–57.

Musy, Jean. "Mouvements populaires et hérésies au XIe siècle en France." *R.H.* 253 (1975): 33–76.

Neale, J. E. "The Diplomatic Envoy." *History* 13 (1928): 204–18.

Nelli, René. *La vie quotidienne des Cathares du Lànguedoc au XIIIe siècle*. 3rd ed. Paris: Hachette, 1969.

———. *Ecrivains anti-conformistes du moyen âge occitan*. Paris: Phoebus, 1977.

———. *La philosophie du catharisme: Le dualisme radical au XIIIe siècle*. Paris: Payot, 1978.

Offler, H. "Empire and Papacy: The Last Struggle." *Transactions of the Royal Historical Society*. 5 ser. 6 (1956): 21–47.

Olson, Clair C. "The Minstrels at the Court of Edward III." *Publications of the Modern Language Association* 56 (1941): 601–11.

Ortega, Fernando. "La paz y la guerra en el pensamiento augustiniano." *Revista española de derecho canónico* 20 (1965): 5–35.

Owst, Gerald R. *Preaching in Medieval England: An Introduction to Sermon Manuscript (1350–1450)*. Cambridge: Cambridge University Press, 1926.

———. "The People's Sunday Amusements in the Preaching of Medieval England." *Holborn Review* 68 (1926): 32–45.

———. *Literature and Pulpit in Medieval England*. Oxford: Basil Blackwell, 1966.

Pantin, W. A. *The English Church in the Fourteenth Century*. 1955, reprint Notre Dame-Indiana: Notre-Dame University Press, 1962.

Paris, Gaston. "Le roman de Fauvel." *Histoire littéraire de la France* 32 (1898): 108–53.

Partee, C. "Peter John Olivi: An Historical and Doctrinal Study." *Franciscan Studies* 20 (1960): 215–60.

Partner, Peter. *The Lands of St. Peter*. Berkeley: University of California Press, 1972.

———. *The Murdered Magicians: The Templars and their Myth*. Oxford: Oxford University Press, 1982.

Payen, J. C. "Le *Dies irae* dans la prédication de la mort et des fins dernières au moyen âge." *Romania* 86 (1965): 48–76.

Pelzer, A. "Les 51 articles de Guillaume d'Occam censurés en Avignon en 1326." *Revue d'histoire de l'église* 18 (1922): 240–70.

Perkins, Clarence. "The Wealth of the Knights Templars in England and the Disposition of It After Their Dissolution." *A.H.R.* 15 (1910): 252–63.

Perroy, Edouard. *The Hundred Years War*. Translated by David C. Douglas. 1945, New York: Capricorn Books, 1965.

Peters, Edward. *Heresy and Authority in Medieval Europe*. Philadelphia: University of Pennsylvania Press, 1980.

Petit Dutaillis, Charles. *The Feudal Monarchy in France and England*. Translated by E. D. Hunt. 1936, London: Routledge & Kegan, 1966.

Piaget, A. "Un poème inédit de Guillaume de Digulleville: Le roman de la fleur de lis." *Romania* 62 (1936): 317–58.

Picar, Michel. *Les Templiers*. Paris: M.A., 1986.

Picard, Barbara Leonie. *French Legends, Tales and Fairy Stories*. London: Oxford University Press, 1960.

Poiron, Daniel. *Le moyen âge (1300–1480)*. 2 vols. Paris: Arthaud, 1970–1971.

Pollard, A. F. "History, English and Statistics." *History* 11 (1926–1927): 15–24.

Post, Gaines. "A Roman-Canonical Maxim: *Quod omnes tangit* in Bracton and Early Parliaments." *Traditio* 4 (1946): 197–251.

______. "Two Notes on Nationalism in the Middle Ages." *Traditio* 9 (1953): 296–320.

______. "Law and Politics in the Middle Ages: The Medieval State as a Work of Art." In *Perspectives in Medieval History*, edited by Katherine Fischer Drew and Floyd Seyward Lear, pp. 59–76. Chicago: University of Chicago Press, 1963.

Powicke, Maurice. *The Christian Life in the Middle Ages*. Oxford: Clarendon Press, 1935.

Prawer, Joshua. *A History of the Latin Kingdom of Jerusalem*. 2 vols. Jerusalem: Mosad Bialik, 1963.

______. "Jerusalem in the Christian and Jewish Perspectives of the Early Middle Ages." in *Settimane di Study sull' Alto Medio Evo*, pp.1–57. Spoleto, 1980.

Prutz, Hans G. *Die geistlichen Ritterorden*. Berlin: E.S. Mittler und Sohn, 1908.

Purcell, Maureen. "Changing Views of Crusade in the Thirteenth Century." *The Journal of Religious History* 7 (1973): 3–19.

Pye, Lucian. "Communication and Political Development." In *Public Opinion*, pp. 368–77.

Queller, Donald E. *The Office of Ambassador in the Middle Ages*. Princeton: Princeton University Press, 1967.

Renan, Ernest. *Etudes sur la politique religieuse du règne de Philippe le Bel*. Paris: C. Levy, 1899.

Reviron, J. *Les idées politico-religieuses d'un évêque au IXe siècle et le "De Institutione Regia."* Paris: J. Vrin, 1930.

Rice, Stuart A. "Stereotypes: A Source of Error in Judging Human Character." *Journal of Personnel Research* 5 (1926): 267–76.

Richard, Jean. "La papauté et la direction de la première croisade." *Journal des savants* (1960): 49–59.

Richardson, H. G. "The Coronation Oath in Medieval England: The Evolution of the Office and the Oath." *Traditio* 16 (1960): 111–202.

Richter, Michael."Kommunikationsprobleme im lateinischen Mittelalter." *Historische Zeitschrift* 222 (1976): 43–80.

Riley-Smith, Jonathan. *The Knights of St. John in Jerusalem and Cyprus, c. 1050–1310*. London: Macmillan, 1967.

______. *What were the Crusades?* London: MacMillan Press, 1977.

______. *The First Crusade and the Idea of Crusading*. London: Athlone Press, 1986.

Riley-Smith, Louise and Jonathan. *The Crusades: Idea and Reality 1095–1274*. London: Edward Arnold, 1981.

Rivière, Jean. *Le problème de l'église et de l'état au temps de Philippe le Bel*. Paris-Louvain: E. Champion, 1926.

Röhricht, R. "Die Pilgerfahrten nach dem Heiligen Lander vor den Kreuzzugen." *Historisches Taschenbuch*, fifth series, vol. 5 (Leipzig, 1875).

Root, Robert K. "Publication Before Printing." *Publications of the Modern Language Association* 28 (1913): 417–31.

Rosenau, James N. "The Opinion Policy Relationship." In *Public Opinion*, pp. 196–200.

Roskell, J. S. "The Problem of the Attendance of the Lords in Medieval Parliaments." *Bulletin of the Institute of Historical Research* 29 (1956): 153–204.

Ruiz, Teofilo R. "Reaction to Anagni."*The Catholic Historical Review* 65 (1979): 385–401.

Runciman, Steven. *The First Crusade*. Cambridge: Cambridge University Press, 1980.

Sachs, Arieh. "Religious Despair in Mediaeval Literature and Art." *Mediaeval Studies* 26 (1964): 231–56.

Saint-Pierre, Bernard. "Mourir au XVe siècle: le testament de Jeanne d'Entrecasteaux." In *Le sentiment de la mort*, pp. 79–96.

Salusbury-Jones, G. T. *Street Life in Medieval England*. Sussex: Harvester Press, 1975.

Samaran, Charles and Mollat, Guillaume. *La fiscalité pontificale en France au XIVe siècle*. 1905, reprint Paris: E. de Boccard, 1968.

Schein, Silvia. "Gesta Dei per Mongolos 1300: The Genesis of a Non Event." *E.H.R.* 94 (1979): 805–19.

———. "The Crusades as a Messianic Movement." In *Messianism and Eschatology*, edited by Zvi Baras, pp. 177–89. Jerusalem: Israel Historical Society, 1984.

———. "The future *regnum Hierusalem*: A Chapter in Medieval State Planning." *Journal of Medieval History* 10 (1984): 95–105.

———. "Latin Hospices in Jerusalem in the Late Middle Ages." *Zeitschrift des Deutschen Palästina* 101 (1985): 82–92.

Scherer, Wilhelm. *Geschichte der deutschen Litteratur*. 3 vols. Berlin: Weidmann, 1905.

Schmitt, Jean C. "Religion populaire et culture folklorique." *Annales E.S.C.* 31 (1976): 941–53.

———. "Le suicide au moyen âge."*Annales E.S.C.* 31 (1976): 3–28.

———. "La geste, la cathédrale et le roi." *L'arc* (numéro consacré à G. Duby, 1978): 9–12.

———. "Du bon usage du 'Credo'." In *Faire croire*, pp. 337–61.

Scholz, Richard. *Die Publizistik zur Zeit Philipps des Schönen und Bonifaz' VIII*. Stuttgart: F. Enke, 1903.

Schramm, Percy Ernst. *Herrschaftszeichen und Staatssymbolik. Schriften der M.G.H.* Stuttgart, 1956.

Schramm, Wilbur. "Two Concepts of Mass Communication." In *Public Opinion*, pp. 262–68.

Schwalm, J. "Beitrage zür Reichsgeschichte." *Neues Archiv der Gesellschaft für altere deutsche Geschichtskunde* 25 (1899–1900): 520–85.

Scott, W. "Psychological and Social Correlates of International Images." In *International Behaviour*, edited by Herbert Kelman, pp. 70–103. New York: Holt, Rinehart and Winston, 1965.

Seiter, Ellen. "Stereotypes and the Media: A Re-evaluation." *Journal of Communication* 36–2 (1986): 14–26.

Sève, Roger and Chagny-Sève, Anne M. *Le Procès des Templiers d'Auvergne, 1309–1311*. Paris: Comité des travaux historiques et scientifiques, 1986.

Seward, Desmond. *The Hundred Years War: The English in France 1337–1453*. New York: Atheneum, 1978.

Shahar, Shulamit. *The Fourth Estate: A History of Women in the Middle Ages*. Translated by Chaya Galai. London: Methuen, 1983.

Siberry, Elizabeth. "Missionaries and Crusaders, 1095–1274: Opponents or Allies?" *Studies in Church History* 20 (1983): 103–110.

Siedl, L. "Images of the Crusades in Western Art: Models as Metaphors." In *The Meeting of Two Worlds*. Kalamazoo (1986): 377–92.

Smoje, Dujka. "La mort et l'au-delá dans la musique médiévale." In *Le sentiment de la mort*, pp. 249–67.

Southern, Richard W. *Western Society and the Church in the Middle Ages*. 1970, Middlesex: Penguin Books, 1972.

Spiegel, Gabrielle M. "The Cult of St. Denis and Capetian Kingship." *Journal of Medieval History* 1 (1975): 43–69.

Stapleton, Thomas. "A Brief Summary of the Wardrobe Accounts of 10, 11, 14 Edward II." *Archaelogia* 26 (1836): 318–45.

Stenton, Frank. "The Road Systems of Medieval England." *Economic History Review* 7 (1936–1937): 1–21.

Stenton, L. "Communications." In *Medieval England*, edited by Austin Lane Poole. 2 vols., 1: 196–208. Oxford: Clarendon Press, 1958.

Stock, Brian. *The Implications of Literacy: Written Language and Models of Interpretation in the Eleventh and Twelfth Centuries*. Princeton: Princeton University Press, 1983.

Stoianovich, T. *French Historical Method: The Annales Paradigm*. Ithaca-London: Cornell University Press, 1976.

Strayer, Joseph R. "The Promise of the Fourteenth Century." *Proceedings of the American Philosophical Society* 106–6 (1961): 609–11.

______. *Medieval Statecraft and the Perspectives of History*. Princeton: Princeton University Press, 1969.

______. "The Historian's Concept of Public Opinion." In *Common Frontiers*, pp. 263–68.

______. *The Reign of Philip the Fair*. Princeton: Princeton University Press, 1980.

Strayer, Joseph and Taylor, C. H. *Studies in Early French Taxation*. Cambridge, Mass.: Harvard University Press, 1939.

Stretton, Grace. "The Travelling Household in the Middle Ages." *Journal of the British Archaeological Association* 40 (1940): 75–103.

Sumption, Jonathan. *Pilgrimage: An Image of Medieval Religion*. London: Faber and Faber, 1975.

Taylor, Charles. "French Assemblies and Subsidy in 1321." *Speculum* 43 (1968): 217–44.

Taylor, J. "The Manuscripts of the *Modus Tenendi Parliamentum*." *E.H.R.* 83 (1968): 673–88.

Thorndike, Lynn. "Public Readings of New Works in Medieval Universities." *Speculum* 1 (1926): 101–3.

Thouzellier, Christine. "Hérésie et croisade au XIIe siècle." *Revue d'histoire ecclésiastique* 49 (1954): 855–72.

______. "Hérésie et pauvreté à la fin du XIIe et au début du XIIIe siècle." In *Etudes sur l'histoire de la pauvreté*, 1: 371–87.

Throop, Palmer A. *Criticism of the Crusade: A Study of Public Opinion and Crusade Propaganda*. 1940, Philadelphia: Porcupine Press, 1975.

Thurston, H. "The Bells of the Mass." *The Month* 123 (1914): 389–401.

Tierney, Bryan. "Ockham, the Conciliar Theory and the Canonists." *Journal of the History of Ideas* 15 (1954): 40–70.

______. "Grosseteste and the Theory of Papal Sovereignty." *Journal of Ecclesiastical History* 6 (1955): 1–17.

———. *Medieval Poor Law: A Sketch of Canonical Theory and Its Application in England*. Berkeley: University of California Press, 1959.

———. *Origins of Papal Infallibility (1150–1350)*. Leiden: E. J. Brill, 1972.

Tocco, Felice. *La Quistione della povertá nel secolo XIV secondo nuovi documenti*. Naples: F. Perrella, 1910.

Toussaert, Jacques. *Le sentiment religieux, la vie et la pratique religieuse des laïcs en Flandre maritime et au "West Hoeck" de langue Flamande aux XIVe, XVe et début du XVIe siècle*. Paris: Librairie Plon, 1963.

Tyerman, C. J. *Sed nihil fecit?*: The Last Capetians and the Recovery of the Holy Land." In *War and Government in the Middle Ages*, pp. 170–81.

———. "Philip V of France, the Assemblies of 1319–1320 and the Crusade." *Bulletin of the Institute of Historical Research* 57 (1984): 15–34.

———. "Philip VI of France and the Recovery of the Holy Land." *E.H.R.* 100 (1985): 25–51.

Ullmann, Walter. "A Medieval Document on Papal Theories of Government." *E.H.R.* 61 (1946): 180–201.

———. *Medieval Papalism: The Political Theories of the Medieval Canonists*. London: Methuen, 1949.

———. *A Short History of the Papacy in the Middle Ages*. 2nd ed. London: Methuen & Co., 1974.

Underhill, Evelyn. *The Life of the Spirit and the Life of To-day*. New York: E. P. Dutton, 1922.

Urbach, E.E. "Etudes sur la litterature polémique au moyen âge." *Revue des études Juives* 100 (1935): 49–77.

Urbieto Arteta, Antonio. "Los caminos que unían a Aragón con Francia durante la Edad Media." *Les communications*, pp. 21–27.

Van Kalken, Frans J. P. *Histoire de la Belgique*. Brussels: J. Lebégue, 1944.

Vauchez, André. "Présentation." In *Faire croire*, pp. 7–16.

Verger, Jacques. *Les universités au moyen âge*. Paris: Presses universitaires de France, 1973.

Viard, Paul. *Histoire de la dîme ecclésiastique principalement en France jusqu'au décret de Gratien*. Dijon: Jobard, 1909.

Vinacke, W. "Stereotypes as Social Concepts." *Journal of Social Psychology* 46 (1957): 229–43.

Vinay, Gustavo. *Il dolore e la morte nella spiritualita dei secoli XII e XIII*. Todi, 1967.

Violante, Cinzio. "Hérésies urbaines et hérésies rurales en Italie du XIe au XIIIe siècle." In *Hérésies et sociétés*, pp. 171–202.

———. "La pauvreté dans les hérésies du XIe siècle en Occident." In *Etudes sur l'histoire de la pauvreté*, pp. 347–69.

Watt, J. "The Theory of Papal Monarchy in the XIIIth Century: The Contribution of the Canonists." *Traditio* 20 (1964): 179–317.

Werner, Victor. "Le pouvoir des mots et la guerre." In *La Communication sociale et la guerre*, edited by Victor Werner, pp. 17–61. Brussels: Emile, 1974.

White, Albert B. "Was There a "Common Council" Before Parliament?." *A.H.R.* 25 (1919): 1–17.

Willard, James F. "Inland Transportation in England During the Fourteenth Century." *Speculum* 1 (1926): 361–74.

Windlesham, David. *Communication and Political Power*. London: J. Cape, 1966.

Wood, Charles. "The Mise of Amiens and Saint Louis' Theory of Kingship." *French Historical Studies* 6 (1970): 300–310.

______. "Queens, Queans and Kingship: An Inquiry into Theories of Royal Legitimacy in Late Medieval England and France." In *Order and Innovation in the Middle Ages*, pp. 385–400.

Wright, J. Robert. *The Church and the English Crown, 1305–1334*. Toronto: Pontifical Institute of Mediaeval Studies, 1980.

Wyse, Alexander. "The enquêteurs of Louis IX." *Franciscan Studies* 25–1 (1944): 34–62.

Yunk, J. "Economic Conservatism, Papal Finance and the Medieval Satires on Rome." *Medieval Studies* 23 (1961): 334–51.

Zink, Michel. *La prédication en langue romane avant 1300*. Paris: Honoré Champion, 1976.

______. "Détachement du monde et soumission au monde dans la prédication et la littérature édifiante en français du XIIe et du XIIIe siècle." In *Idéologie et propagande*, pp. 43–54.

INDEX